D0915425

TECHNOLOGY AND SOCIETY

TECHNOLOGY AND SOCIETY

T.

HISTORY OF AMERICAN INDUSTRIAL SCIENCE

By

COURTNEY ROBERT HALL

ARNO PRESS
A NEW YORK TIMES COMPANY
New York • 1972

Reprint Edition 1972 by Arno Press Inc.

Reprinted from a copy in The Newark
Public Library

Technology and Society
ISBN for complete set: 0-405-04680-4
See last pages of this volume for titles.

Manufactured in the United States of America

Library of Congress Cataloging in Publication Data

Hall, Courtney Robert, 1894-
History of American industrial science.

(Technology and society)
Reprint of the 1954 ed.
Bibliography: p.
1. United States--Industries--History.
2. Industrial arts--History. 3. Technology--
History. I. Title. II. Series.
[HC103.H175 1972] 338'.0973 72-5052
ISBN 0-405-04704-5

HISTORY OF
AMERICAN INDUSTRIAL SCIENCE

HISTORY OF AMERICAN INDUSTRIAL SCIENCE

By

COURTNEY ROBERT HALL

History Department, Queens College

LIBRARY PUBLISHERS
NEW YORK

8 West 40th Street, New York 18, N Y

Printed in the United States of America

To Jane, Nana and Pounce.

Keen observers of the wonders of industrial science.

HISTORY OF AMERICAN INDUSTRIAL SCIENCE

ACKNOWLEDGMENTS

The author desires to express his gratitude to the following business and industrial organizations for aid in the gathering of materials included in the volume:

Air Reduction Company
Allied Chemical and Dye Corporation
Aluminum Company of America
American Bemberg
American Can Company
American Car and Foundry Company
American Cyanamid Company
American Enka Corporation
American Locomotive Company
American Paper and Pulp Association
American Radiator and Standard Sanitary Corporation
American Viscose Corporation
Anaconda Copper Mining Company
Armour and Company
Association of American Railroads
Bausch & Lomb Optical Company
Bethlehem Steel Corporation

Celanese Corporation of America
Champion Paper and Fibre Company
Colgate-Palmolive-Peet Company
Container Corporation of America
Continental Baking Company
Corning Glass Works
Curtis-Wright Corporation
Douglas Aircraft Company Incorporated
Dow Chemical Company
E. I. du Pont de Nemours and Company
Firestone Tire and Rubber Company
General Aniline and Film Corporation
General Dynamics Corporation
General Electric Company
General Motors Corporation
B. F. Goodrich Company
Goodyear Tire and Rubber Company
Industrial Rayon Corporation
International Business Machines Corporation
International Harvester Company
International Paper Company
Kalamazoo Vegetable Parchment Company
Kennecott Copper Corporation
Libbey-Owens-Ford Glass Company
Masonite Corporation
Monsanto Chemical Company
National Association of Manufacturers
National Biscuit Company
Newport News Shipbuilding and Dry Dock Company
Pillsbury Mills Incorporated
Proctor and Gamble Company
Radio Corporation of America
Revere Copper and Brass Incorporated
Sinclair Oil Corporation
Standard Oil Company of New Jersey

Swift and Company
The Texas Company
Time Incorporated
Todd Shipyards Corporation
Union Carbide and Carbon Corporation
United States Gypsum Company
United States Plywood Corporation
United States Rubber Company
United States Smelting and Refining Company
United States Steel Corporation
Western Union Telegraph Company
Westinghouse Air Brake Company
Westinghouse Electric Company

Thanks are also due the American Association for the Advancement of Science and Dr. James Creese, President of Drexel Institute of Technology, for permission to use the latter's able statement concerning industrial science.

Special acknowledgment is due Jane S. H. Hall, wife of the author, who made an extensive study of food and clothing and wrote the chapter covering those matters. Her counsel and judgment regarding many other portions of the book were at all times most helpful.

Industrial science is a happy and convenient phrase. It describes an area of knowledge and discipline which is now peculiarly promising, as broad as science and as productive as modern industry. . . .

Today the words "science" and "technology" are on every page of current history. We speak the two words almost in one breath. Their alliance is so generally and mutually profitable that in everyday speech we often fail to make a distinction between the two words.

Perhaps the new phrase "industrial science" will relieve us of an ordeal in semantics and permit us to recognize frankly the fact that every enterprise of science is of practical significance in our economy.

Science has permeated our entire industrial structure. No company and no community can dare to neglect the advancement in science of its own industrial interests.

PRESIDENT JAMES CREESE
Drexel Institute of Technology

(from an address at the 1951 meeting of the American Association for the Advancement of Science, and printed in *Industrial Science—Present and Future,* Washington, 1952.)

HISTORY OF AMERICAN INDUSTRIAL SCIENCE

PREFACE

It should be a matter of considerable interest to all Americans that in this country has been developed the most extensive and intricate industrial organization that the world has ever known. The fact that this is true is taken for granted by many people and is unnoticed by others. Students of American history should be aware that this massive industrial growth has a meaning for them, for it points out what is perhaps the chief characteristic of the American nation: its industrial "know-how."

Whether many of us fully appreciate the significance of American industrial science or not, it is certainly important that we begin to do so, for it has done more to make the United States a strong nation than any other field of activity that we can name. Furthermore, as our defense authorities are well aware, industrial science has contributed more to our national security than has any other asset which we have developed. Adequate preparedness, in a military sense, we seldom have had at the occurrence of an emergency. Since 1860, however, we have always possessed the industrial and technical resources to build a defense which has met the demands of the time.

PREFACE

It is quite difficult for many Americans to grasp easily just how this prime element in national life has come to be. In the first place, the very mass of information which is available is beyond the sight of the average person; likewise much of the material which would inform us about industrial development is garbed in a form which is clear only to the technically-trained specialist. An air of mystery seems to pervade the great factories and laboratories in which the miracles of modern industrial science are performed. The process of specialization has gone so far as to leave the ordinary person far behind in his effort to understand something of what has taken place. Yet there should be some way in which such a person can readily comprehend a few of the basic developments which have been made, or which are in progress, to keep the United States in the forefront of the technological and industrial effort.

A single purpose has actuated publisher and author in preparing this volume: to help make the general public aware of the need for the continued improvement of our industrial system, along the lines of greater productivity and efficiency and into new lines of helpful development, and to tell the story in terms which most people can understand.

To say that this is an ambitious undertaking is to understate the case. The author, while he had an extensive exposure to many of the details of our industrial mobilization during the last war, claims no more than some experience as a teacher and writer in the historical field, much of which has been gained in the exploration of American scientific progress. For such analysis of the industrial field as he has been able to make, he has had to rely heavily upon much of the printed material which bears on the subject from the large area of formal industrial history, from government publications, and from the extensive publications of the industrial companies themselves.

It is possible, therefore, that an occasional error of interpretation has been made, and in such event, the author alone is responsible. Materials of great value, of real significance, have been freely furnished him by some sixty of the leading industrial companies of the nation. In no instance has any effort whatever been made to induce the writer to beam his story in any particular direction. Their aid has been heartily extended, with the attitude: "Here are the facts about us; if you should wish further aid, we will be most happy to give it." It might be added here that industrial companies have gone a very long way, in their public information and historical departments, to produce understandable descriptions of their particular industries, for the information of the general public. These are, of course, limited to individual industries and do not present a general picture of industrial science, yet the quality of "company histories" is far higher than the usual person would suspect.

The most difficult task of any author who seeks to write a book of such comprehensive sweep is the selection of what must be said and what may be omitted; of what is basic and what is merely "more of the same." Obviously, no pretense can be made of coverage of all industry in this single volume; many kinds of activity can be merely mentioned; some must be omitted entirely. This is not so inaccurate a procedure as one might suppose, for it is the judgment of the author, after much pondering, that the story of American industry, especially in the twentieth century (which concerns us most) is mainly one of "large" industry, since the integration and combination of smaller industry into larger groups is one of the chief characteristics of our age. Perhaps this is regrettable in some particulars, but it is a fact.

The grouping of smaller into larger organizations, while occurring to a great degree, has never, in any case, proceeded to the point of absolute monopoly, though there may have been times when some danger of such a result existed. Indeed, it is more than likely that there is more small industry under what may be called "oligopoly," or control by several groups rather than a single one, than would be the case had we continued to live under the primitive economy described in the first chapter of this volume. Great competitive companies, by their very nature, stimulate rather than stifle the continuance of numberless small, specialized firms; these smaller companies will probably continue to serve industry in general in a very vital way. Both large and small industrial activities thrive under what we call the American system of free enterprise, and would not thrive under any other system which we know.

The plan of the book has been, therefore, to develop rather broadly the basic and inventive phases of our larger categories of industry, such as transportive, chemical, electrical, and so on, into their present state, with speculation as to their futures. As this plan has unfolded, it has been the aim of the author to include as many of the ramifications of these great industrial categories as space will allow. Such is the scope, and also the limitation, of this brief volume.

There may be some who will ask: where are the many social implications which are inherent in the industrial picture? What of the labor movement, of the employee, of industrial strife? The answer is simple. This is not a book on labor problems, or on general social or economic history. There are many and ample studies of that kind, to which a reader should refer for such information. If some feel that no book of this kind can be written without going extensively into what they may regard as a massive struggle for the control of the American economy, a simple contradiction should suffice. The American system (as the author sees it) is one of relatively free enterprise, and despite all efforts to change it into something else, whether into monopoly on the one hand, or into a socialistic economy on the other, it remains to this day basically free. This freedom has its limitations, of course; those limitations which great size and our huge stake in its future put upon it. The real question is: how many of our leading industrialists, labor leaders, or government officials, knowing the facts of the past and present, would willingly change our system into another? Imperfections can be corrected, and they should be. To change the American economic system into a different kind would be a disaster of great national and international consequence.

COURTNEY ROBERT HALL

Queens College, New York City
1954

TABLE OF CONTENTS

CHAPTER		PAGE
	Preface	
I	From the Beginning	1
	Colonial Industry	2
	The Revolution and Early Industry	5
	The Industrial Revolution	8
	Development of the Industrial System	15
II	Enter the Industrial State	24
	The Civil War and American Industry	26
	Conversion to Peacetime	36
	The Age of Steel Begins	38
	Tying Together the Loose Ends of a Continent	42
	Sources of Power, New and Old	47
	Science Comes to the Aid of Industry	49
III	Transportation in the Twentieth Century: Land and Water	55
	The Railroad Makes its Major Contributions	56
	Improvements in Railroad Equipment and Operations	61
	Water Transportation	68
	The Motor Car Enters the Picture	78
IV	Transportation in the Air	90
	Men Attempt to Fly	91
	Heavier-than-Air Flight	97
	The Wright Brothers Turn the Trick	101
	The Development of Commercial Aviation	108
	The Impact of the Airplane upon American Industry	119
V	The New World of Chemicals	125
	Some Early Beginnings	126
	The Nature of Modern Chemical Industry	131
	The Place of Research in Chemical Industry	136
	Some Recent Chemical Industrial Developments	140
	The Future of Chemical Industry	143

TABLE OF CONTENTS

CHAPTER PAGE

VI The Electrical and Communications Industries 150
Telephonic Communication 151
Electric Power and Lighting 157
Radio and Electronics 165
Recent Progress in Electricity and Electronics 175

VII Modern Mining and Metallurgical Industry 185
The Technology of Mining 187
Recent Developments in Mining and Metallurgy 198

VIII The Non-Metallic Minerals 217
Petroleum and the Other Fuels 218
Non-Metallic Minerals, other than Fuels 226

IX Rubber and Rubber Products 235
The Early History of Rubber 236
Modern Rubber Technology and Production 239
The Search for Chemical Rubber 243
Big Rubber 250
The Industry in Mid-century 267

X Pulp, Paper and Print 271
The Manufacturing Process 274
Some Characteristics of the Paper Industry 279
Printing and Publishing 285
Paper and Print: Present and Future 296

XI Feeding, Cleaning and Clothing the Millions 302
Agriculture and the Sciences 303
The Farms are Mechanized 310
Meat From Hoof to Market 317
The Perfect Food 322
The Staff of Life 329
Shortenings and Spreads 332
Preserving and Packaging 333
Soaps and Cleaners 337
Clothing the Masses 339

XII Precision in American Industrial Science 342
Business Machines 344
Meters, Gauges, and Communication Instruments 349
Precision in Glass 352

TABLE OF CONTENTS

CHAPTER PAGE

XIII Industrial Science and National Defense 362
The Wartime Capabilities of the United States 364
The Lessons of World War I 367
Industrial Science Faces a Supreme Test 379

XIV A Year of Industrial Science: Conclusion 400
The General Industrial Picture 402
Recent Milestones 409
A Glance into the Future 414

Bibliographical Note 419

Index 425

CHAPTER ONE

FROM THE BEGINNING

The first Americans were simple people. Their emigration to the North American continent may be readily accounted for by even a superficial understanding of the basic economic conditions of seventeenth century Europe. It is fair to assume that among the many motives which drove them to forsake their native lands, whether England, France, Germany or Holland, the economic urge was a powerful one. To landless peasants and underpaid workmen, even to the thrifty commercial classes, America loomed as a land of opportunity.

Recent scholarship in American colonial history has altered considerably the classic picture of this movement which established sizable settlements along the seaboard of the continent. While it is no doubt true that many setttlers were actuated by a desire for political and religious liberty, it is more accurate to regard them as primarily in search of better economic conditions than the class-conscious Old World afforded them. It is also very likely that without the strong feeling that plantations overseas would afford the ruler and his more prosperous subjects some tangible profits, the commercial companies which planted the first colonies would have refused to risk large sums in the projects. The fact that the ventures did not pay the original backers anything like the profits which they had expected had nothing to do with their original attitude.

Colonial Industry

The seventeenth century was an age of a simple, largely rural economy, whether one speaks of Europe or of colonial America. Most persons of that period could obtain the basic necessities of life by labor in the fields, by spinning, weaving and other household arts, or by following some one or other of the rude though ingenious crafts acquired through the ancient apprentice system. Even in western Europe, and more particularly in England, whence most of the early settlers had come, scientific industry in a modern sense was unknown until well into the eighteenth century. While England, first among modern nations, underwent a revolution in manufacturing about the middle of the century, the changes which took place there had no immediate repercussions in the colonies. So far as England was concerned, her improved methods of mining and of the manufacture of textiles and iron goods made her the chief supplier of such products to other countries and added materially to her importance as a commercial and shipping nation.

The American colonies, which stretched along the Atlantic seaboard from Maine to Georgia, had little or no part in this important development, except to become suppliers of modest amounts of raw materials to the mother country and to serve as markets for the disposal of cargoes of manufactured goods or of those luxuries which the wealthy planters or the prosperous merchants could afford. Early efforts to develop colonial manufactures were usually frowned upon by the British government because of the possibility of dangerous competition with its own industries. In the case of scarce raw materials, such as naval stores or ship timber, it was another story. Encouragement to produce items needed for British commercial expansion was standard procedure.

Colonial manufactures were of several kinds; there were the products made in the household for home use, as for instance. textiles of cotton, wool, linen and hemp; or leather, home-dressed and tanned, for breeches, vests, shoes and saddles, made from hides of both wild and domestic animals. A wide variety of other home manufactures of the colonial period is familiar to many a visitor to the numerous museums of early Americana or to the more elaborate restorations of colonial communities, such as Sturbridge Village or Colonial Williamsburg. More and more Americans are thus made aware of the extent and variety of household products, including not only the many items of leather and cloth, but also domestic utensils of iron, soap from wod ashes, and candles of beef tallow. The extensive processing and preserving of food was characteristic of every colonial household. Meat was slaughtered, cured, smoked and salted, frequently in the latter process, with the salt evaporated on the spot. Many of the farm tools, the baskets, brooms and buckets, even the wooden dishes, were made by the family for its own use.

Such self-sufficient industry had no economic importance beyond the household, and was the result rather of necessity than of deliberate choice. In default of a supply of manufactured goods, and lacking ready cash with which to purchase the costly imported products, colonial people were driven to depend upon local raw materials and native ingenuity. As the colonies grew, and as trained workmen settled in them, a second stage in manufacturing developed: the actual making of goods for the home market.

The large supply of valuable raw materials of forest, field and sea suggested to colonial workmen the processing of these, usually one step or two from their original form, for sale or trade with their immediate neighbors. Timber existed in large quantities, especially in New England and the uplands of the South, and it furnished the basis for thriving local industries. The great forests of Maine and New Hampshire afforded excellent ship timbers for the builders who began to operate yards along the navigable rivers. Coopers opened shops to make barrels in which to pack the flour milled nearby, or casks for the rum which the New Englanders were beginning to distill for shipment. In the colonies further south, the barrels were made to hold the tobacco of Maryland and Virginia. At camps in the great forest belt of the South the colonists prepared the valuable naval stores: the pitch, tar, turpentine and resin for which government bounties were paid.

Iron was found in many of the colonies; bog ore was dug from the swamps and ponds of the seacoast; this product, after smelting, would yield a small percentage of metal which served well enough for pots and kettles, but was inferior to the ore which was dug from the hills of western New England and eastern New York. Some of this better product was made into wrought iron for a wide variety of utensils.

The capital gathered for enterprises such as these was small and the size of the manufacturing operations correspondingly so. The only available power, other than hand or animal power, was in the fall of the streams as they proceeded to the sea, or along the seacoast, in the tidal rise and fall of captured water. One or two workmen, in small establishments, ground grain into meal or flour, sawed logs into lengths, fulled cloth, or operated a small iron forge. Such early factories seldom developed into large affairs because of the limited money economy of the time, the small demand for goods, and the failure of the raw product or of the power supply. The general equality of the period restricted the rise of an investing class. Labor was scarce; even such aid as the miller or iron master could obtain sometimes served him only on a part-time basis, the balance of the workman's time being required on his own farm or to care for his subsistence garden.

It seems clear that local colonial authorities were anxious

to extend these modest efforts in manufacturing. At an early date Massachusetts passed legislation requiring spinning by each family in the Commonwealth. Virginia mechanics were forbidden to plant corn or tobacco and various Virginia counties were required by law to maintain public tanneries. In the northern colonies gifts of land were offered to persons who would operate gristmills or who would establish iron forges; tax exemptions were granted and loans made from public funds for improved manufacturing processes, as for methods of preparing salt. The American patent system was initiated, it is said, by the grant of exclusive rights to produce an improved scythe to a certain Joseph Jenks of Massachusetts, in 1646 Early recognition was given the public importance of water power sources by regulating their use, and some colonial governments even inspected the product of gristmills and fixed the charges made for grinding.

The Revolution and Early Industry

The American Revolution did not stimulate manufacturing as much as would seem logical. The United Colonies, endeavoring to resist an enemy far superior to themselves in military, naval and economic power, resorted at first to non-importation agreements to bring Great Britain to terms. The inadequacy of American industrial production, however, soon forced Congress to authorize the importation of arms and munitions from abroad in exchange for American produce. Commercial firms like the Browns of Providence were specially licensed to carry on voyages to foreign ports to secure the necessary supplies and specie for carrying on the war. Such activity, along with much illegal trade across the lines and the relatively successful privateering by Americans, added to the family fortunes of many and may have benefitted the American cause. But the losses and delays of such a system left the Continental Army bereft of clothing, arms and powder at critical stages, and thus contributed to the length of the conflict. Speculation and wartime inflation increased the difficulties of the military authorities.

Industrially the war encouraged the continuance of household manufactures, as a substitute for the imported goods which had been coming to America in greater quantities in the years just before the war began. But with the British naval blockade and American non-importation rules in force, imports had been largely cut off from the home market; those products which did arrive, through capture or by privateering, were usually diverted to war purposes. Congress and the new state governments sought to encourage manufactures. Virginia offered premiums on textiles, textile machinery, nails and gunpowder. Many states discouraged the slaughter of sheep, in order to conserve the wool supply, and following the lead of Virginia, offered bounties for the production of war material. Social pressure was used to induce Americans to wear homespun clothing. Litttle could be done to increase the labor supply, however, as thousands of men were withdrawn from civilian labor into military service, or entered the more profitable occupation of privateering.

Shipbuilding suffered as the American fishing industry felt the effects of the naval blockade, but such a loss was partly offset by the heavy demand for vessels to serve as armed privateers; many ships, even naval vessels up to seventy-four guns, were built in yards temporarily free from British naval attack. The iron industry boomed; forges owned by patriots soon had backlogs of orders from Congress and the various states; forges owned by British sympathizers were frequently confiscated and turned over to persons loyal to the American cause. Virginia had a shop at Fredericksburg where small arms were made, and maintained at Westham a forge for the casting of cannon and cannon balls. In Springfield, Massachusetts there were several private shops for the making of weapons, and these were later acquired and operated by the Continental

Army. Important iron masters of Pennsylvania and New Jersey, having placed their forges at the service of the Army, helped Washington maintain resistance during the campaigns in those areas.

With the close of the war and the attainment of independence, much of the artificial stimulation which had livened American commerce and manufacturing was withdrawn, and the inevitable postwar lag set in. The government was weak, the currency disordered, and the brief commercial prosperity of the coast cities, where large armies had spent their payrolls, was a thing of the past after the British and French forces departed for Europe and the American Army disintegrated. Deficiencies in transportation and lack of capital deferred for a time the introduction in this country of the industrial revolution.

There were indications, however, that the depression would be a short one. In Georgia and South Carolina the cultivation of short staple cotton cut the price of the fibre in half, and the introduction of Whitney's gin, after 1794, increased cotton production to such an extent that American fibre soon became an important factor in supplying the spindles of England, and to a lesser extent, the new cotton mills of New England. This situation, in turn, fastened the slave system upon the South, and in some ways, by encouraging the large-scale production of cotton as the staple of many plantations, discouraged the introduction of industry in the South. The application of steam power to manufacturing and to water transportation, developments which were underway at the turn of the century, stimulated in New England the movement which first made the United States a factor in machine industry.

Before these changes in American agriculture and industry could affect the average citizen, the people of the United States lived just about as their ancestors had, isolated from one another by the barriers of distance and the slow, dangerous transportation afforded by stage and the coastal sailing vessel.

America was still in a state of rural economy, and the inhabitants lived mostly on farms, in small market villages, or near crossroads where the small town or county governments were centered. For clothing they wore homespun, as in the colonial days. They bought little which was not produced within a few miles of their homes. Only the well-to-do could afford clothes from London, wines from the Spanish islands, coffee from the Indies. Those who had surplus funds to invest seldom ventured to place them in industry at home, but chose rather to speculate in western lands, purchase tickets in the many lotteries, or share in commercial ventures overseas.

The Industrial Revolution

The factory system was introduced into the United States partly because of the interference with American trade before and during the War of 1812. The great Napoleonic struggle, of longer duration than the Revolution, made the position of the United States, first as a neutral and then as a belligerent, even more difficult than before. Both Great Britain and France, by ship seizures and condemnations in prize courts, did serious damage to American trade and shipping. As the long struggle continued, complicated by the entry of the United States near its end, important economic changes began in America. Manufacturing expanded as the demand grew for goods which became extremely expensive or whose importation was completely cut off. A great market for manufactured goods was in the making.

There were even more important reasons for the extension of industrialism to the United States. The development in cotton production, already noted, taken in connection with Whitney's important device, not only stimulated foreign trade but inspired a new type of American, the industrial entrepreneur, to take quick advantage of the situation. The time was ripe, after 1815, to introduce mechanical methods into manufacturing. Population had increased to more than seven millions by 1810, and a shift of Americans to large towns was beginning. Despite the losses of the war just ended, commerce had grown enormously. Capital earned in shipping and internal trade was now available, if practical means for its profitable investment could be demonstrated.

Some phases of the factory system had been started even before the War of 1812. Samuel Slater, a young English apprentice of Jedediah Strutt, partner of Sir Richard Arkwright, the industrialist, had arrived in the United States in disguise, thus evading the English laws against emigration of artisans. He soon became engaged with Moses Brown of Rhode Island in plans for the manufacture of textiles by power machinery. Slater's amazing memory was an important factor in the plans, for he was able, in 1790, the year after his arrival here, to reproduce the complicated Arkwright power-driven frame and other machines in Pawtucket Falls, which soon became a center for cotton cloth manufacture. Three years later the Scholfield brothers, from Yorkshire, matched Slater's feat and introduced into Massachusettts similar machinery for the manufacture of woolens.

Although Eli Whitney never profited from his invention of the cotton gin, his influence on the course of American manufacturing was very great. When the United States, in 1798, expected to raise a sizable army for service in the impending war with France, Whitney obtained a contract to manufacture ten thousand muskets for the United States government, the total to be completed in a two year period. After assembling workmen, and undertaking the mission, Whitney was able to produce only five hundred of the guns at the end of the first year, and thus faced cancellation of the contract.

The first guns of Whitney had been manufactured by individual skilled gunsmiths, working under strong pressure in order to accomplish an impossible task. Whitney, however, had been preparing for such a crisis; for two years he had labored to produce patterns or jigs which when fitted to his cutting and finishing machines would make each barrel or lock exactly like every other. Having secured the cooperation of the War Department, he was able to set up in his Whitneyville plant a production system which turned out all the parts for the necessary muskets at record speed, after which the parts could be assembled from the identical components into finished products. Thus was born, at the turn of the century, the vital principle of building large-scale production by the manufacture of standard, interchangeable parts.

An equally impressive step in industrialization was taken in the eastern Massachusetts area by 1814, when the merchant Francis Lowell, assisted by a clever mechanic named Paul Moody, perfected the power loom. When this was installed in their mill at Waltham with other machinery already available, Lowell and Moody were able to operate the first cotton factory in America in which every operation, from the treatment of the raw fibre to the making of finished cloth, was accomplished in one place. Rapid growth of the textile business proceeded thereafter. By 1830 one-half of the cottons, woolens and linens used in eastern communities of the United States came from domestic factories. Household production of the more common varieties of textiles declined sharply, except in the thinly-settled frontier communities.

The woolen industry had long been impeded by the inferiority of the breeds of sheep raised in this country. It has been stated that scarcely any improvement in the breeding of "wool" sheep had taken place for a century. The striking improvement in this direction began shortly after 1800, and owes much to the pioneer activity of several prominent Americans. The best breed of sheep for wool was the Spanish merino. The trouble was that the Spanish government, well aware of the quality of merino wool, had placed an embargo upon this animal, and it was next to impossible to obtain breeding animals from this source. But in 1802, Robert R. Livingston,

American minister to France and also a scientific farmer, obtained two pairs from French flocks and sent them to his large Hudson Valley estate. His diplomatic colleague in Spain, Colonel David Humphreys, did still bettter. When he returned from his diplomatic post (also in 1802), he was permitted by the Spanish government to send directly to America a flock of seventy-five ewes and twenty-five rams, which formed the nucleus of a very large herd on his Connecticut estate. Du Pont de Nemours, the Delaware powder maker, also imported examples of this choice breed.

The introduction of merino sheep might have had little importance beyond the improvement of the private stock owned by a few gentlemen farmers, but for the fact that international complications arising after 1805 practically cut off the supply of English woolens and made the manufacture of woolen cloth profitable in this country. The demand for the raw material far exceeded the available supply; prices for pure bred merinos rose to $1000 and $1500 and even higher, while the price of their wool reached the then fantastic figure of two dollars per pound. As a result of the Peninsular War in Spain, the embargo on merinos was raised, and owners made haste to sell their flocks before they should be scattered or destroyed by the invading armies. William Jarvis, American consul at Lisbon, quickly acquired large numbers for the American market, and in sixteen months nearly 20,000 animals were shipped to the United States. Despite the breaking of the price which accompanied such wholesale shipments, sheep raising and wool production became profitable and a main enterprise in eastern United States. Livingston and Humphreys both started woolen factories. The latter opened several woolen and cotton hosiery mills in the Naugatuck Valley in Connecticut. Interesting features of Humphreys' mills were his introduction of labor-saving devices, his evening school for his employees (many of whom were children), and his attention to the moral welfare and comfortable housing of his charges. As early as 1810, there were some twenty woolen factories in the country and following a brief recession after 1815, it continued to be a profitable industry for many years.

As we have seen, the earliest metal industry revolved largely about the small forge or furnace which produced iron ware for household use, or simple tools needed by craftsman and farmer. Only under the pressure of war did the masters of these forges attempt the more ambitious process of casting large cannon for military use. Rapid advance in the iron industry was not possible in the United States; large capital, a scarce commodity, was required. Furthermore, British iron products made by the puddling and rolling processes easily undersold the American product of similar kind. The preference of American makers for charcoal rather than coal as a fuel, dictated perhaps by the feeling that it produced a superior product or because of the convenience and cheapness of wood, long hindered the introduction of the British, or Cort method of puddling and rolling iron, utilizing coal as fuel. Transportation difficulties hindered the development of large scale iron manufacturing, especially in the South.

In the early nineteenth century, in many of the places where iron was easily obtainable, coal was lacking, or it was too inaccessible to be mined economically. There was no known process of smelting iron with anthracite coal until 1833, when the Geissenhainer process was patented. To a large degree, the extensive development of iron making awaited the demand for rails, which of course did not begin until the 1830's, the first decade of railroad building. The market for large castings was also deferred, until about 1840, when the reproductive metal industry, or in other words, that part of metallic industry which manufactures items of metal from machines made of metal, was well underway. For example, by the latter date, steam engines were coming into wide use. These were generally built of cast iron. The making of stoves was becoming a large industry. Water turbines, formerly constructed of a wooden frame with paddles, began to have cast iron frames. Wooden factory machinery was being replaced by iron machinery, not only in the manufacture of large articles, but even of small items formerly made by craftsmen, such as nails, wire, tacks, screws, files, chains, buttons and the like, as a horde of mechanical inventors devised contrivances to save human hand labor.

Invention fed the industrial revolution and was of course fed by it. When a Rhode Island industrialist, Ichabod Washburn, began to manufacture cards for textile machinery in Worcester, Massachusetts, a city which specialized in that activity, he found his progress hampered by the slow production of wire for the cards. The card-making process was a machine operation, but the wire used in it was prepared by artisans in a slow and inefficient manner. Washburn turned his attention to improving this detail of the business, with the result that by 1830 he had perfected wire-drawing apparatus which revolutionized that metallic industry, and which eventually made of Worcester the greatest center of wire manufacture in the United States. Washburn's work was also of the greatest importance in making possible new uses of wire, as in the telegraph lines which were so soon to appear.

The position of the inventor in the history of American industry is a very important, though a somewhat confusing one. Most intelligent persons are aware of the impossibility of connecting some of the most important devices with a single individual or even, oftimes, to locate their origins in a definite country. That Americans of this revolutionary period were of an inventive turn of mind, seems an unquestioned fact. Roger Burlingame, one of the most interesting writers on the subject, lists more than one hundred devices, mostly of mechanical kind, produced by Americans before 1850, which had far-reaching effects upon the progress and expansion of manufactures.

These included Paul Revere's method of producing cold

rolled copper, in 1801; John Stevens' multitubular boiler, in 1803; Robert Fulton's marine torpedo, in 1808; Daniel Pettibone's heating stove of 1808; John Hall's breech-loading carbine of 1811; Eli Whitney's milling machine of 1818; Jethro Wood's iron plow of 1819; Jacob Perkins' improved stationary steam engine of 1827; Joseph Henry's electro-magnet and telegraph in 1828; Cyrus McCormick's reaper in 1831; Thomas Davenport's electric motor of 1834; Samuel Colt's revolver of 1835; John Ericsson's improved screw propeller of 1837; Charles Goodyear's rubber vulcanizing process of 1839; Charles Thurber's typewriter of 1843; Elias Howe's sewing machine and Robert Hoe's rotary printing press, both in 1846. There were many others; some were original and others constituted hoped-for improvements of devices which had been known for some time.

Many of the new factories did not depend upon steam power, which was for many years too expensive unless a supply of fuel was accessible. The many streams, with their advantageous drops in level, were still sufficient to turn the wheels of many a factory of the 1830's. Even a minor stream like the Blackstone River, between Worcester and Providence, had along its banks in 1840 ninety-four cotton mills, twenty-two woolen mills and thirty-four machine shops and iron works, all run by water power. These companies employed over 10,000 workers. Along the Connecticut, the Hudson, and other eastern rivers were lumber and flour mills, iron works and firearms shops, all of which secured their power from the current of the streams. In the New England area, the force of the streams was increased by constructing dams in such rivers as the Merrimac and the Connecticut, and by the development of the water turbine. By the 1850's, however, the steam engine had invaded the cities of the area, especially where water power was inadequate and fuel costs low.

In other portions of the nation steam power had already become the rule. An American version of the Watt and Boulton engine, designed by Oliver Evans, was popular with many manufacturers. The model was lighter in weight than the English engine, and despite the fact that it used more fuel than the earlier version, it was greatly preferred in early power plants. In the newer communities to the West, as in Pittsburgh and Cincinnati, where water power was not readily available, the new steam engine helped to operate the saw and grist mills which were so vital in the building of the new American West.

The Development of the Industrial System

It took a great deal more than political independence and ordinary human ingenuity to build a full-fledged American industrial system. To furnish adequate supplies of raw materials for the large-scale manufacture of metal products and textiles, agriculture and mining had to be expanded beyond local subsistence levels. In order to deliver such raw materials to points of their manufacture and to deliver the finished products to distribution centers, improvement in transportation was vital. Furthermore, to furnish the necessary capital to expand small, local manufacturing into sizable industries, corporate organization and financial planning were needed. Provision for an adequate labor force must be made. Most of these elements were lacking in great degree in American business organization at the beginning of the nineteenth century.

Basically a lack of requisite capital slowed the progress of the infant manufactures which began, in some such way as has been described, to dot the American countryside soon after the beginning of the century. Alexander Hamilton, in his famous *Report on Manufactures* which appeared in 1791, had been hopeful that much capital, most of which in his day was tied to agriculture and commerce, could be diverted to manufacturing. He believed that by establishing banks, funding the public debt and by attracting to this country more foreign capital, an investing class would be created in the new nation which would turn to manufacturing as a profitable type of venture.

By the time Hamilton published this work, he had already encouraged a project conceived by his assistant secretary, Tench Cox, to develop large-scale manufacturing. A corporation called "The Society for Establishing Useful Manufactures" was created under New Jersey law. The Society, usually called the S.U.M., had grandiose plans. After gathering as much of its authorized capital of a million dollars as could be obtained by public subscription, the S.U.M. intended to establish a wide variety of factories to manufacture, among many items, paper, sail-cloth, stockings, ribbons, tape, thread, carpets, hats, blankets, pottery, brass and iron ware. The company selected a site at the falls of the Passaic River, laid out a town, built factories and equipped them with machinery. A large part of the capital was paid in and operations began. But by 1796 the promoters had to admit failure. The labor supply was undependable; management and engineering were defective. In addition, the investors who had hastily bought shares in the speculative boom of the early 1790's, soon unloaded them.

Nearly twenty years after Hamilton's hopeful report and the premature development efforts of Tench Cox and his S.U.M., another Secretary of the Treasury, Albert Gallatin, lamented that the lack of capital was still the "only powerful obstacle" to the advancement of American manufacturing; this deficiency persisted in large part until after the Civil War. Americans of wealth continued to prefer the older, better-established ways of making money as surer and more profitable. They still placed their prime faith in agriculture, land speculation, new transportation systems such as canals and turnpikes, and foreign commerce, as better ways of seeing their money grow. European investors were no more enthusiastic about American industry than were the Americans. Yet by 1820 the United States Census estimated that $50,000,000 was invested in domestic manufactures; by 1850 this total had grown to $500,000,000, a figure which doubled before the opening of the Civil War.

Capital, even in the small quantities available in the earlier years, flowed into the factory system from a variety of sources. There was a small amount of investing by professional men and farmers, and a larger amount by merchants like the Browns and Almys of Providence, and the Appletons and Lowells of the Boston area. A marked shift of capital from foreign commerce to manufacturing occurred as a result of Jefferson's Embargo and the events of the War of 1812, as already noted. The expansion of the iron industry in Pittsburgh in the 1830's and 1840's was supported by stock purchases by local merchants, commission and forwarding agents, and canal transporters. Many enterprises were prudent combinations of mercantile and manufacturing activity, the store and the mill carrying one another along when dull times affected either. There was also a rise in the rate of profit in even the smaller factories, which induced the owners, as times improved, to reinvest more and more of their funds in the business, instead of scattering them in other ventures.

Organization of a very simple type was the rule in the early business enterprises. Two or more individuals would pool their resources as partners, or if the situation later warranted it, they would enlarge the partnership into a joint-stock arrangement; this was actually an enlarged partnership. Articles fixing the conditions of the partnership or the inter-relations of the stockholders would be adopted; shares could be in cash or in property; the contribution of land, building material, or power rights might entitle a subscriber to a given number of shares. No authorization by law was needed for such a company and until the middle of the century some variety of the partnership or the informal stock company constituted the usual means of organization in the basic industries.

The chartered corporation made its way into industry gradually. In the early years after the adoption of the Constitution, most financial operations had been carried on by Federal and State governments, such as the sale of the large bond issues floated to finance Hamilton's debt-funding program. Government money continued for a time to occupy the main position in support of banks and of internal improvements. Before 1830, almost the only chartered corporations engaged in profit-making were those of a semi-public kind, such as banks, insurance companies, and bridge and turnpike companies. With the beginning of the railroad, however, in the 1830's, this new form of corporation soon dominated all the others. For the first time, the American public was offered a wide variety of bonds and common and preferred stocks. In the same decade the chartering of corporations by states began with the passage by Connecticut in 1837 of a general incorporation act. This trend, imitated by other states, facilitated the gathering of capital funds, and assured, or seemed to assure, more permanence to the investment. In addition the practice gave rise to an entirely new division of law.

With the setting up of the chartered industrial corporation, foreign capital began at last to flow into American industry. In order to distribute the shares of companies more widely, various agencies contributed their services. For a time the Second Bank of the United States became such a channel of distribution. Agents of foreign banking firms, as for instance August Belmont of the Rothschilds, began to locate in the larger cities to deal in American securities. In such centers as New York, Boston, Baltimore and Philadelphia exchanges were set up. Investment banking became an American activity, and much of the buying and selling, the speculation and manipulation which came to characterize financial areas like Wall Street in New York, were well under way by the middle of the century.

Meanwhile the old basic manufactures like the textiles of New England, the iron goods of Pennsylvania, and the clocks and firearms of Connecticut were being supplemented by the manufacture of articles usually imported from Europe, and of new items which stemmed from the active brains of American inventors. Factory methods were applied to watch making in the late 1840's; the copper sheet and wire industries were established in Connecticut by the 1830's; a promising brass button business was operating in Waterbury, Connecticut, as early as 1823. Carpets were made mostly by hand until 1845, but thereafter the industry underwent complete mechanization. The ready-made clothing industry developed after the appearance of Howe's sewing machine in 1846. In the shoe industry, power-driven machinery began to affect the process of manufacture about 1855.

It was Samuel F. B. Morse, a New York portrait painter and teacher of science, who accomplished more in the instituting of electrical communications than any other individual of his time, though he was not responsible for the basic inventions which preceded the inauguration of the telegraph. Joseph Henry of the Smithsonian Institution performed most of the initial research in electro-magnetics, while Alfred Vail and Ezra Cornell made many practical improvements in Morse's instruments before they operated successfully. But it was Morse who attracted the interest of the influential Jacksonian politician, Amos Kendall, who helped Morse to patent his device in 1837. An experimental line was built, with government support, between Baltimore and Washington in 1844. Within six years so many lines were built that news could be wired to all important points east of the Mississippi. The larger industrial phase of the electrical industry, except for some manufacture of telegraphic equipment, belongs to a later period.

A whole flock of inventions tended to revolutionize the agricultural life of the nation. Jethro Wood, John Deere and James Oliver contributed important improvements to the plow; these paved the way for the appearance of the harrow and the seed drill. More important than any of these was the reaper, ascribed usually to Cyrus McCormick, who in 1834 was given a patent for his invention. His idea was not new; it was said that there had been forty-six patented reapers before his, twenty-three of them by Americans. But McCormick, a practical farmer, deserves credit for the solution of many problems presented by mechanical grain cutting which had not been solved by any devices before his own. McCormick became a manufacturer of reapers, but neither his factory nor others like it boomed before the Civil War, possibly because the machines were so expensive and also because the saving in human labor was not then regarded as necessary. Immigrant labor was available in increasing numbers at that time.

Much the same fate, of perfecting a device before the production and marketing facilities were ready for it, befell Richard Hoe's rotary printing press of 1846 and Elias Howe's sewing machine, which was patented in the same year. Neither moved into mass production until the Civil War gave impetus to the clothing and printing industries, although some of Howe's machines were used early in the ready-made clothing field. Goodyear's vulcanizing process, patented in 1844, brought him fame, but the award of the Legion of Honor ribbon was made known to him as he lay in a French debtor's prison. Many of the new devices made short cuts in manufacturing possible in the factories already established, but were often neither outstanding nor strikingly original. The story of Samuel Colt, who patented his revolver in 1835, is in many respects unique; the invention was actually his own and the great arms works at Hartford eventually made him a rich man. Here special circumstances helped, for the weapon arrived just in time to assist the movement of Americans into the Far West. The Texas Rangers adopted it, and American soldiers campaigning in Mexico found it highly useful.

Deficiencies in human comfort were to some extent alleviated by improvements in lighting and heating before the Civil

War. Americans had for many years suffered from a lack of heat in homes and offices. The period of the 1830's and shortly after was marked by the appearance of the cook stove and the heating stove in the northeastern states, a change furthered by the presence nearby of coal, a more convenient fuel than wood. Few fundamental shifts came in the illumination field except a changeover from candles to whale oil and an improvement in lamps. The friction match, invented early in Europe and patented in America by Alonzo Phillips in 1836, soon came into almost universal use. The use of gas as an illuminant was spreading from city to city and was in some instances being piped into private homes before the Civil War. Such a convenience, however, merely marked a chief difference between the country and the more progressive cities, and was by no means common in American homes or offices in those early days.

The United States was growing swiftly in the decades before the Civil War. From a nation of a little over seven millions in 1810, confined mostly to the Atlantic seaboard, the population had increased to more than thirty-one millions in 1860. The old Northwest and Southwest territories had been divided into states. From Mexico the United States had received the vast domain from Texas to California and had divided with Great Britain the Oregon Territory, an empire in itself. The seventeen states of 1810 had grown to thirty-five in the fifty-year period, and all the land to be ultimately admitted as states was already part of the national domain. From 1810 to 1860 the value of manufactures in the United States increased from $199,000,000 to $1,885,862,000. While in the latter years agricultural pursuits occupied more than sixty per cent of the labor force, other industry was growing fast. In the decade from 1850 to 1860 the value of manufactured goods increased eighty-five per cent, while the workers employed in their production grew from 957,059 to 1,311,246.

For purposes of comparison, the years prior to 1800 may be

considered the age of household industry and of manufacturing in small shops and mills. In the generation after 1800, approximately, the factory system with its power-driven machinery, utilizing first water power and then steam power, gradually developed. About 1830 began some of the refinements of the industrial system, including the chartered joint-stock corporation, flotation of securities in the investment market, and the extension of manufacturing into fields which had been hitherto dominated by foreign industry. By the mid-century, with increased capital provided by the large profits of commerce and those of industry re-invested in its own future, and with a labor force swelled by the arrival of millions of foreign-born, the American industrial system was beginning to assume many of the characteristics of an important manufacturing nation. Much remained to be accomplished, however. Capital was still insufficient to link the two oceans by rail; the great age of steel had not yet dawned; the backward state of the physical sciences prevented rapid progress in the electrical, petroleum and rubber specialties. Little headway had been made in mining the non-ferrous metals. The mechanization of agriculture was in an elementary stage.

Yet the American industrialist of 1860 employed hundreds instead of a few helpers. He could count upon a ready supply of skilled and unskilled labor, still insufficiently organized to hamper seriously his decisions as to wages, conditions, and hours of labor. He was in communication with the great city commodity and financial markets by telegraph. Within easy reach, usually, were railroad facilities to forward his raw materials and to carry his finished products to the larger distribution points. From the latter places his agents might even arrange for their shipment by steam vessels or fast clippers to far distant countries.

Thus American industry, after passing in little more than half a century through all the stages which men had experienced from ancient to modern life, arrived at a point ready for its greater expansion. From 1860 on, this phase of American development was to reach gigantic proportions and surpass the wildest dreams of even the most enthusiastic. It was to save the Union from disruption and to emerge as the most complex and successful industrial society in the entire world.

CHAPTER TWO

ENTER THE INDUSTRIAL STATE

For more than two centuries the people of America had been occupied with the early stages of industrialization. Extraction of the products of field, forest and sea had begun early and had proceeded rapidly; mining, fishing and lumbering had started with the primitive means at hand. In the early nineteenth century manufacturing began to change from a handicraft stage to one in which machinery intervened to build production. Workers were being gathered in larger mills and factories. Business men and lawyers had introduced some of the devices which added to the success and permanence of the early companies. In this way corporation finance, investment banking and corporation law began to assist and strengthen industry. Yet, as noted earlier, the industrial revolution was still incomplete. Transportation was slow and confused; the new telegraph still did not cover the western states; more than half of the people of the nation, including most of the older South, were only remotely affected by large-scale industry. Agriculture was still the chief occupation of Americans.

During the remainder of the century the nation was to experience intensification of manufacturing, progress in agriculture, mining and transportation by mechanical means, and the development of new processes and devices which had in many cases been known but not completely utilized in the earlier periods. Production was to double, quadruple and then increase again, many times over. Complexity never dreamed of was to be characteristic of American industry. Mass labor movements, never important earlier, were to emerge as a chief problem. Applied science was about to transform the simplest techniques into processes calling for extensive training. Countless new products were to stream from the industrial adaptation of the ideas of mechanical inventors, chemists, metallurgists, plant and soil experts, oil geologists and electrical scientists. The fiercest kind of competition was to ensue for the access to basic resources; for the control of key transportation routes; for the mastery of production processes. Wealth was to increase enormously to reward the victors in the competitive struggle, particularly in the fields of management and finance. Great periods of depression were to stop or slow up the wheels of progress and to cause many to regard soberly the social and economic damage which such periods caused and to inquire anxiously into what caused them, and what could be done to end or avoid them.

No one with the slightest knowledge of economic history would term the United States after 1860 a country in which industrial peace reigned. Yet it would be a distortion of history to consider the times lacking in social gains. The worker obtained, with great difficulty to be sure, growing recognition of his part in the total industrial process. Enormous gains in human health and well-being were realized from the spread of mass production. The consumer, whoever he might be, was bound to benefit, for it was he whose purchasing power was needed to keep the great industrial machine humming. No industrial system worth the name could possibly have succeeded without careful consideration of human needs and without striving to bring more and more of the products of industry within reach of the masses of people. While short-sighted policies were frequently to dictate the decisions of management, the wisest and most resourceful business men began to realize that it was the customer who was the real dictator of industry. This meant that, by a long and arduous process, American industry would make its greatest advance subject to the desires of the American people.

The Civil War and American Industry

No human experience can be as upsetting or disastrous to orderly existence as war. It follows also that the bigger the war, the more serious are its effects upon political, social and economic life. Moreover, a struggle between one large part of a nation and the other part has some consequences far more deadly than any other kind. At the outset, the unity of the nation is utterly destroyed; people of the same blood, often the same family, are pitted against one another. Almost inevitably, in case one side is completely victorious, the social and economic life of the other is completely disrupted. So it was in our Civil War, and many years were to pass before the South could rise to a healthy and normal economic life, following the destruction of its cities and the wrecking of its agricultural system. The entire nation suffered for a generation from the vast human losses and the incalculable destruction of the conflict.

The effect of war upon economic life is immediate and initially disastrous. The depressed condition, however, is followed usually by a quick upturn. The war boom is a common occurrence in modern life; some of its effects are unhealthy but all of them are extremely important. The United States by 1860 was just beginning to recover from the serious industrial depression of 1857, which had been a natural consequence of the over-expansion and reckless speculation of the 1850's.

After the election of Lincoln in the fall of 1860 and the secession of the Southern states which followed, business men awaited anxiously the result of efforts to prevent an actual outbreak of hostilities. When the South Carolinians fired on Fort Sumter, and thus precipitated a war, the formal economic ties with the South came to an end. Trade with the Confederate States ceased. Chaos ensued in Northern banking and commercial centers, for on their books were debts of hundreds of millions of dollars owed by Southern planters and business men. The debt total was swelled by ownership in the North of millions of dollars worth of the securities of Southern states. How could these ever be paid?

An even more difficult situation developed in the West. There the banks had issued paper money based upon securities which in many cases were of little value. Panic swept the Western cities as their paper money depreciated, and many banks, forced to liquidate their holdings at serious discounts, had to close their doors. Along the Ohio and Mississippi Rivers, almost closed to navigation, cities like St. Louis and Cincinnati felt the pinch of depressed prices as goods destined for the Southern market clogged the docks and warehouses. In agricultural states like Indiana and Illinois, corn was so plentiful that the price fell to ten cents a bushel. In some cases it was not worthwhile to market the crop and it was used as fuel. Efforts to ship goods to the East soon choked all the land and water routes and freight rates rose to prohibitive levels.

By the summer of 1861 the situation began to change, as the Union defeat at Bull Run made it seem likely that a long war was ahead. Financiers became more hopeful of profits as the usual signs of heavy governmental expenditures and inflationary prices appeared. War contracts came to the depressed Western cities. In New York and other great Eastern cities the war boom was soon in full sway. Stocks began to rise. In 1862 they increased an average of forty per cent, and in later years rose still higher. In the year 1863 nearly a billion and a half dollars worth of issues changed hands; the following year the industrial investor entered the market and the turnover in stocks in New York City alone rose to more than a hundred million dollars a day.

This speculative activity which attacked Wall Street was due in part to the abandonment of specie payments by the banks in 1861, and in part to the decision of the United States to finance the war to some degree by the issue of unbacked paper money, or "Greenbacks." The new government notes soon vied with the plentiful issues of similar notes issued by the state banks; since there was no longer any apparent worry about ultimate redemption of these issues in coin, they circulated freely, as demands for more and more money increased. In the words of Jeremiah Best, writing during wartime in *Harper's Monthly,* paper bills "circulated like fertilizing dew throughout the land, generating enterprise, facilitating industry, developing internal trade."

There was an obvious reason for this sudden activity, besides the inevitable and somewhat unpleasant features of speculation and extravagance. Larger armies were being raised than ever before in our history; thousands of men needed shoes, uniforms, overcoats, tents, rifles and ammunition; the cavalry had to secure horses, saddles, harness, fodder, small arms; artillery units needed cannon, ammunition, caissons, wagons and other equipment for their horse-drawn units. The emergency swiftly placed a premium upon all transportation equipment and building materials: locomotives, freight cars, coaches, river vessels and cargo boats, lumber and iron products. Above all, the new armies needed enormous quantities of food, for armies, as Napoleon once remarked, advance on their stomachs.

One of the first industries to show a remarkable build-up was cotton-spinning, which had become highly developed even before the war. For two years the mills of the North operated at high speed by means of the 50,000 bales of surplus stock on hand in 1860, but thereafter were depressed because of the extreme shortage of the fibre supply. Efforts were made by private experimenters to introduce cotton culture into Indiana, Illinois and Missouri, projects aided by government distribution of seed through the various state agricultural societies. These efforts helped to supply local needs, but did not materially affect the shortages which began after 1862. There existed considerable overland trade between the sections at war, in which cotton played an important part; officially this trade extended only to points controlled by the Union Army, licensed by Treasury officials. Foreign sources of cotton, as for instance South America, were not sufficient to change the picture.

While cotton manufacturing sank in importance by the middle of the war, woolen factories expanded; new companies were set up and much of the labor force available from the closed cotton mills was absorbed by the new woolen mills which were speedily built. Wool processing increased from eighty-five million pounds to more than two hundred million pounds during the course of the war. The growing of flax was stimulated but it could not be easily manufactured on a machine basis, as most of the existing machinery was adapted to cotton or wool and not to flax. The ready-made clothing business expanded markedly from the small shops in the Eastern cities to larger establishments, and was devoted largely to the making of soldiers' uniforms; yet the making of clothing was still a matter mostly of home and small shop. While a number of Elias Howe's machines were used in the larger shops, the complete mechanization of the industry had not yet taken place.

In the case of the shoe manufacturing business more steps had been taken towards mechanization by the wartime period. Mechanical devices for cutting and rolling leather, pegging machines, and variations of the sewing machine for stitching the tops, were introduced between 1840 and 1860. During the war these various aids were assembled in factories, mostly in eastern Massachusetts.

Industries not important before became increasingly impressive in size and in profitable possibilities. Sugar refining became the ninth most important industry in the nation. The manufacture of whiskey increased; breweries thrived and more were started under the helpful operation of German-American brew masters. Valuable new salt deposits were discovered along the Saginaw River in Michigan, which helped to relieve shortages caused by the destruction of the Kanawha works in West Virginia during the war.

Probably the most important growth in industry was connected with the production of iron and steel. The older anthracite mines in Pennsylvania increased their yield as coal came into heavy demand for use in locomotives and factories. Exploitation of the bituminous areas in the middle West began as new mines were opened. While Pittsburgh and other Pennsylvania cities remained the chief centers of iron manufacturing, western ore from the Lake Superior district began moving via the new Sault St. Marie canal, as a result of which iron smelting began in Cleveland and the Ohio Valley towns.

The basic increases in iron and coal production had important effects of a military character. While for two years the North was forced to import iron and steel for the making of ordnance, after 1863 it was able to supply its own needs in that direction. In the production of small arms, for example, the North became self-sufficient after two years, and achieved a production rate of a million muskets a year. This was dominantly a government industry, carried on in thirty armories and arsenals, although some dozen or more private arms manufacturers assisted in the task.

Another area in which the superior resources of the North began to tell heavily was in railroad construction. Every year, on the average, eight hundred miles of new tracks were laid; locomotives were manufactured, railroad bridges were constructed and some efforts were begun to make the gauge of the lines standard. Congress, on the supposition that there was a military need for rail connection between the Mississippi Valley and the Far West, authorized in 1862 a project for a transcontinental road, but the actual building of such a line was deferred to the postwar period. The Chicago and Northwestern, having absorbed a number of smaller companies, became for the time, with 1152 miles of track, the longest railroad in the country. Traffic between New York and Philadelphia, which was very heavy, ran mostly over the tracks of the Camden and Perth Amboy. The Baltimore and Ohio, which had extensive trade connections wtih the states of the inland South, suffered from attacks by Confederate troops but received Union military aid to keep open its valuable routes to the interior.

In 1859, on the very eve of the war, oil was discovered in considerable quantities near Titusville, Pennsylvania. The presence of petroleum on the surface of the fields of Venango County had long been observed. "Rock oil," as some called it, had been bottled up for use as a medicine and some farmers had used it to grease their wagon axles. Few suspected its illuminating qualities until a promoter, George H. Bissell, having remembered some experiments on crude oil by his teacher of chemistry, sent samples of the Pennsylvania product to Professor Silliman at Yale College. Silliman soon informed Bissell that an excellent illuminant could be made from it at small cost, and that other products such as naphtha and paraffine could be obtained in the refining process.

Bissell suspected that there might be large pools of the oil beneath the surface. So he hired a driller named E. L. Drake to begin operations in the Titusville area, with the result that by August, 1859, Drake's pumps were bringing up twenty-five barrels or more of the stuff every day. Titusville and the other towns around it at once experienced a boom, as an army of prospectors descended upon them. The refining process began in nearby cities and the kerosene lamp quickly came into its own, as the superior illuminating qualities of kerosene came to be known.

The loyal states likewise demonstrated their ability to outproduce the Confederacy in food products. Such totals were achieved that the North was able soon not only to supply its armies but also to provide large export surpluses. European harvests were bad in 1860-1862, with the result that American exports of foodstuffs mounted. During the war years more than $300,000,000 worth of breadstuffs were purchased from the North by European nations. Farm acreage increased and so did the production totals, despite the drain of military service on the labor supply. It is said that the grain and flour receipts of Chicago alone, in 1861, were nearly double in weight the entire cotton crop of the South. The attractive prices of pork products induced many a corn farmer to feed his crop to the hogs and ship them to slaughtering centers. Chicago climbed steadily to supremacy in meat-packing as well as in the handling of grain and lumber during the war years. With the great midwestern states producing such a vast quantity of pork, which in one year totaled four million carcasses, it was no wonder that Northerners began to scoff at the Southern claim that cottton was king!

Sugar shortages during the war were met in various ways. Maple sugar production was increased but affected the total very little. Efforts to produce a satisfactory granulated sugar from Chinese cane proved a failure, though a company was organized in 1863 to carry on such operations. Sorghum culture was increased in the midwestern states from Ohio to Iowa. The suitability of beet sugar culture was carefully studied in the reports from France and Germany, and some progress made, but the industry did not thrive particularly during the war. Scarcity and high prices continued to rule in the sugar production field.

It is rather remarkable that relatively high production in

most industries was maintained, despite the withdrawal from agriculture, mining and manufacturing of so large a part of the labor force for military duty. This may be explained in part by the increased employment of women and children, by the conversion of many grain farms to stock raising (which required fewer hands), and by governmental policy under which, by the Homestead Act of 1862, vast areas of the public domain were given away to settlers. By the end of the war nearly two and a half million acres had been parceled out. Related closely to the Homestead Act was the Immigration Act of 1864 which authorized the importation of contract labor under the supervision of the Commissioner of Immigration.

Railroads like the Illinois Central encouraged the expansion of agriculture by sales of land at cheap prices. New state agricultural societies were formed and projects like the cotton-culture and sugar-growing schemes were forwarded. Important for the future mainly was the establishing in 1862 of a United States Department of Agriculture, under a commissioner. Helpful to some degree in keeping production at a high level was the use of the new McCormick reapers which increased the speed of cutting grain. Some of the states continued their state fairs throughout the war in order to keep informed on agricultural matters at a time when food products were in such high demand.

As the Union armies advanced, and the plantation system slowly crumbled, thousands of Negroes were added to the ranks of Northern labor. In the manufacturing field, as well as in agriculture, the scarcity of workers was alleviated by an increase in woman and child labor, the advantages of the new Immigration Act, and by the increased use of labor-saving machinery.

Thus it was that the states which remained loyal to the Union more than demonstrated their ability to raise and maintain an army strong enough to defeat a resourceful enemy. Loyal support of the Union was more common, in the North, than the rascality and greed which unfortunately always show themselves in a time of great crisis. But the great upsurge of patriotism would have been vain in such a situation, probably, without the massive production which the North was able in four years to bring to bear. The Confederacy, which had confidently believed that its control of the valuable raw cotton would bring to her side the great industrial powers such as France and Britain, found that there was litttle justification in such extravagant faith. Although large loans were secured abroad in the first part of the war, with future cotton deliveries as collateral, such arrangements were not as influential with European governments as the food shipments which were even more desperately needed and which could be obtained, in the quantities needed, only from the Northern states. Confederate credit abroad, never high, shrank noticeably after Southern military victories terminated in mid-1863.

Having failed to influence foreign policy by her various cotton schemes, which included an embargo and then official crop destruction (both to create artificial scarcity), the Confederacy was compelled to resort to a further remedy to bolster her failing economy. She endeavored to convert her agriculture to a self-supporting one. Heroic efforts were made in this direction, especially in stimulating greater food production, even in the areas formerly devoted to cotton, tobacco and other staples. Although great agricultural states like North Carolina and Georgia assented to this policy, they also found it equally necessary to curtail the distillation of alcoholic beverages which were a regular ration in the Confederate armies and which were also necessary to the Navy and the Medical Department.

The Confederacy found the blockade of its ports, which was begun early in the war, to be an increasing annoyance. While ways were found to evade the ships of the Northern squadrons, the supplies which the blockade runners brought in were more profitable to individual speculators than helpful to the army. The difficulties of blockading so long a coastline and the success of large numbers of ships in getting through to port certainly prolonged the war, but these minor successes could not save the Confederacy.

The slave system, which had fastened upon the South a concentrated agricultural economy, depended for its success upon the free flow of trade with the industrial North and the exchange of its raw materials for the manufactured goods and specie of Europe. Cut off from first one activity and then the other by the grim facts of large-scale war, the Confederate States were doomed unless they could industrialize their economy. They made attempts to do this, but the effort was too small and came too late. A corollary to this condition lay in the fact that, being a predominantly agricultural area, the South had never developed its road and rail transportation sufficiently, with the result that while great stores of goods could be imported through the blockade, they could seldom be moved effectively to the combat areas. As late as February, 1865, considerable shipments of goods were still entering the port of Wilmington, North Carolina. The Confederate Commisary reported, the same month, that en route to Richmond were two and a half million rations of meat and seven hundred thousand rations of bread. Trains of supplies were waiting on railway sidings, or delayed in warehouses, in various parts of the Confederacy. Yet in the spring of 1865 Generals Lee and Johnston, with their armies, were completely cut off and helpless.

Greater manpower, plus a more highly developed economy, had accomplished this result. Production by Northern fields, mines and factories had supplied the Union armies with the means of victory. Man for man, leader for leader, the South had matched the North in fighting ability and in military stamina; with anything like a well-developed industrial and commercial economy behind it, the Confederacy might have won a negotiated peace and its independence from the United States. Lacking such a mechanism, and without the means of forging it, the South could not win. In possession of the means to produce for its armies, the North could hardly fail.

Conversion to Peacetime Industry

Few persons, could they have made the choice, would have selected war as a means of bringing about the fully developed industrialism which emerged in the North after the cessation of hostilities. Such a situation could hardly have developed so soon, however, without the stimulus which war had provided. The South, its economic system in ruins, its currency worthless and most of its credit facilities destroyed, was in no condition to enter any competitive struggle with the triumphant North. Despite the generation of bitterness which ensued between the two sections, however, it has long been an accepted fact that the uprooting of the slave system was in the end beneficial to both sections of the nation.

The essential injustice of human slavery and its economic weakness as a way of life pointed to its eventual disappearance, by whatever method it should be accomplished. While the way it was done will always be deplored by Americans, few can deny that strong logic and reason were on the side of the Union. An economy as dynamic as that of the United States could hardly have tolerated much longer a system so antiquated. The commercial and industrial capitalism of the North, operating under a free labor system, had already, by the end of the war, proved itself immeasurably more successful.

Leadership in manufacturing, commerce and agriculture naturally passed completely into the hands of the business leaders of the North. This was so in the political field as well, though we may ignore for our purposes the severe political arrangements which dominant politicians instituted to "reconstruct" the Southern states. The economic significance of the reunion of the United States is plain to see: the last serious obstacle to nation-wide industrialization had been removed. Possessed of vast material resources and having already made important strides in putting machines to work in order to produce and process the materials which were all around them, there seemed to Northern business men no limits to the possibilities of the future. Within a few years it became apparent to the people of the South that its only salvation lay in adapting their economy to the new order of the day.

The business prosperity which had come to the North early in the war continued after the fighting was over. Capital was plentiful from profits made during 1861-1865; the continued investment of war profits in industry was encouraged by the high protective tariff and by the uninterrupted circulation of the Civil War Greenbacks. Almost two billions in United States bonds were outstanding, the principal and interest of which were payable in gold, regardless of the cheapness of the money with which most of them had been purchased. Gold and silver began to pour in from the new western mines. European investors, chary of risking their funds in wartime projects, now returned to the American market and provided additional capital for railroad promoters, steel and textile mill operators, oil refiners and meat packers. Banks bulged with money and credit was easy. The National Bank Act of 1864 had led to the organizing by the end of the war of over fifteen hundred national banks, whose issues of notes, geared to the banks' purchases of United States bonds, almost equalled in volume the Greenbacks outstanding.

It was favorable conditions such as these, when added to the abundant supply of labor from the returning armies and the increased flow of immigrants from Europe, which provided industry with continued momentum for the postwar years. All the records which had been made earlier were broken. As Allan Nevins well stated in his volume on the postwar period: *The Emergence of Modern America*:

More cotton spindles were set revolving, more iron furnaces were lighted, more steel was made, more coal and copper were mined, more lumber was sawed and hewed, more houses and shops were constructed and more manufactures of different kinds were established, than during any equal term in our history.

The Age of Steel Begins

Iron to feed the great furnaces to supply steel for tools, to furnish raw material for the steel rails with which to span the continent and from which to cast the large machines which were revolutionizing the processes of industry, was discovered in increasing amounts in many parts of the nation. Areas which had been producing small amounts of metal, such as the Chattanooga-Birmingham region, were developed into larger and more profitable fields after the Civil War. Most important of all, the building of railroads into the West and the greater application of steam power to freight vessels, enabled mine owners to bring into the general industrial picture the vast "mountains" of ore already known to lie in the area of Lake Superior. This region had been almost untouched before the Civil War, but extensive mining operations began there in the 1870's. The production of the Marquette, Menominee, Gogebic and Vermillion ranges, which clustered about the western end of Lake Superior and the northwestern shore of Lake Michigan, soon dwarfed all other sources of iron ore. In the 1890's the new Mesabi range, near Duluth, was found to contain as much ore as the four other ranges combined. The entire area, by 1919, was furnishing eighty-five per cent of all the iron ore mined in the United States. Other fields were soon uncovered in the Rocky Mountain states from New Mexico to Montana, as the streams of settlers of the postwar period forked out into the great Far West.

Before the Civil War the use of steel had been confined to

the manufacture of small articles like tools and cutlery, in which quality, rather than low price, was the determining factor. It seemed incredible that items of such great bulk and weight as locomotives and passenger cars could ever be made of such an expensive material. Iron had already served to give added strength to bridges, buildings and machinery, but as greater weight was added to railroad trains, as the growing cities called for increased height for buildings, and as manufacturers demanded greater strength in the industrial machinery which was being planned, amounts of steel were needed which could not be supplied by the old process.

The manufacture of steel before the Civil War was a slow and uncertain process. The purest ore and the finest grade of charcoal were required to produce pig iron suitable for conversion into the harder material. Selected pigs were hammered into bars and packed in containers between layers of charcoal and brought to a high temperature in special retorts. By this slow method the iron bars were converted to steel by the absorption of carbon from the charcoal. Even then, not all the steel so expensively made was suitable for the fine cutting edges required in tools and high grade cutlery. Obviously such a time-consuming process would hardly do for the mass manufacture of girders, rails and heavy rolling stock.

Henry Bessemer, an English industrial inventor, helped solve the problem by a new process which he devised in 1855. A large mass of molten pig iron was poured into a retort or converter; through this mass, blasts of hot air were blown for about twenty minutes. The oxygen of the air combined with the carbon of the iron, and the resultant monoxide gas burned off at the mouth of the converter. When the converter had burned off all the gas, a measured amount of carbon mixed with other materials was restored to the iron. The mass was then poured off as steel. This invention changed the nature of modern industry; steel was made in half an hour instead of three months, stronger than iron and hardly more expensive.

The introduction of the Bessemer process into the United States and was delayed until the war years by the fact that an American, William Kelly, had patented a similar idea and the English process, while superior to Kelly's, could not at first be utilized in the United States. An important feature of Bessemer's process was the use of an alloy composed of iron, manganese and carbon, called "spiegeleisen." This preparation was controlled in Great Britain by a man named Mushet. American promoters allied to Kelly secured the American rights to Mushet's alloy and prevented its use by the Bessemer group in this country, since the latter group could use spiegeleisen only in Great Britain. Kelly, on the other hand, needed the improved methods of Bessemer's process in order to develop his own steel converters but could not use them without infringing the Bessemer patents.

The upshot of this was the merging of the rights and interests of the various groups in an organization called the Pneumatic Steel Company, organized in 1866. Metal for the first American steel rails was poured by the Pennsylvania Steel Company near Harrisburg the following year from the second set of Bessemer converters to be set up in the nation. Thus was inaugurated the age of steel. Steel prices dropped from $106 to $17 per ton, and annual production reached 375,000 tons by 1875, the majority of it steel rails, but large amounts in springs, car axles, plow shares and stove pipes. One steel rail, it was found, would outlast twenty iron ones. As the years advanced, and the life of Bessemer steel was put to heavy tests, it was found to be still too weak for the work it was called on to do. Rails were crushed, torn and ground away by the pounding wheels of the heavy freight trains which they had to carry. Again a better product was needed and was found to be steel prepared more slowly and with greater care in the "open hearth" process, a method known since 1868 but not extensively used at first because of its greater cost. In the new method, laboratory tests were applied to the product in the process of refining. The time and expense which such procedure added were justified by the results, for open-hearth steel grew in popularity, and by the end of the century its output for rails surpassed that made by the Bessemer process.

In later years of steel manufacture, vast improvements were made in methods and in final products. The electric furnace came into use, which made possible the use of scrap, even in the production of high grade steel. The coke ovens which formerly had been located in the coal mines, were installed in the steel mills, which allowed the utilization of the gases produced in the coke process as sources of power, even for the generating of electric current. By the use of alloys, greater perfection in the quality of steel was attained, and a non-corrosive surface was obtained by the invention of "stainless steel" through introducing a large amount of chromium into the mixture. More discussion of present-day steel developments will be carried on in a later chapter.

The small manufacturer, who secured his patent rights from Kelly or Bessemer or Hewitt, gave ground in time to the large corporation with immense capital. The total invested capitalization of the iron and steel business, rated at $121,000,000 in 1869, doubled every ten years and reached three and a half billions fifty years later. Integration of small into large business was well under way by the middle 1880's, as rolling mills absorbed furnaces and fuel producers; as mining companies were drawn in to assure an ore supply; and as transportation companies were purchased to speed the raw material to the processing plants. By the time the various stages of the steel-making process from mining the ore to the finishing of rails, beams and plates were controlled in one organization, there followed the trend of bringing in the makers of secondary products, as wire and tube manufacturers. The final stages are to be seen in the joining of these rather extensive vertical combinations into one or more horizontal organizations. The incorporation of the United States Steel Corporation in 1901 set in motion America's first billion dollar industry.

Tying Together the Loose Ends of a Continent

By the opening of the Civil War the traffic carried by rail had just begun to equal the amount borne by the waterways of the nation, but the value of the goods they carried greatly exceeded that of the internal waterways. As new rail routes were opened, they usually took over the larger part of the existing freight business. Canals were losing their commercial importance, though the Erie Canal, due to its fortunate location near the eastern terminus of the Great Lakes waterway, retained its popularity into the Civil War period.

By 1861 the railroad was making its way everywhere east of the Mississippi River. Even the great river routes like the Connecticut, the Hudson and the Ohio-Mississippi system (so far as its downstream shipments were concerned) suffered from competition with rail lines which paralleled them wholly or in part. The new rail lines which crossed Wisconsin and Illinois to the Mississippi soon secured all the business they could handle. The continued importance of the upper Mississippi, the Missouri, the Great Lakes route and the Erie Canal during the war period was due entirely to the great volume of traffic, not all of which could be carried by the existing railroads. The slow and dangerous water routes were not preferred; they were merely in high demand temporarily. Even in the prebellum South, railroads which connected Chattanooga, Memphis and Nashville with Atlantic ports began to pull to the East much of the cotton which had formerly been floated by water into the Mississippi and to New Orleans. On the very eve of the war, Cincinnati and New Orleans were connected by rail.

The war once disposed of, movement into the West began

in earnest and demands for railroad extension beyond the Mississippi were renewed. Suggested many years before by Asa Whitney and others, stimulated by the gold rush of 1849 and after, the action of Congress in 1862 was taken too late to make possible a completed transcontinental line during the war. The plan, as finally put into operation, was for one company, the Union Pacific, to construct a route westward across the Great Plains, while another company, the Central Pacific, built eastward over the Sierras to meet the first line. The scheme had many difficult angles: the route was longer than any existing lines; the western promoters faced the problem of many thousand feet of steep grades and other engineering problems unknown to eastern promoters; the amount of capital needed was greater than could possibly be secured at that time by private means. Government aid was therefore necessary, and was secured after a vigorous lobbying campaign in Washington.

Space forbids more than a brief summary of the fascinating story of the construction of America's first transcontinental roads. The precedent for government aid existed in the help extended during the canal and turnpike building era. The promoters of the two roads, headed by Oakes Ames of the Union Pacific, and Collis P. Huntington for the Central Pacific group, obtained from the United States Government enormous grants of land, bond issues and cash. The Union Pacific secured 12,000,000 acres and $27,000,000 in bonds payable in thirty years; the Central Pacific was granted 9,000,000 acres and $24,000,000 in bonds. Soon other roads were able to secure similar though smaller amounts of aid. The Northern Pacific was able to push a line across the states near the Canadian border and the Texas Pacific a route near the Mexican frontier, both with federal backing.

Gifts and loans to great railway promoters were but a part of the economic advantages which accrued to all who had anything to do with these enormous projects. Construction companies, often controlled by the railroads, made large profits from building contracts. The disposal of lands for town sites, timber and mining rights, or farming and grazing plots, were means of making huge profits. The methods used by the promoters and stockholders of the new roads were often anything but honest. Congressional investigations and wide public censure were to follow the joining of the Union Pacific and Central Pacific tracks at Promontory Point, Utah in May, 1869.

Yet the completion of railroad transportation across the Far West was a major achievement of American energy, determination and enterprise. The Central Pacific had to climb to the lofty passes of the mountains; tunnels had to be driven, bridges and snow sheds constructed; materials had to be fetched from the East around Cape Horn. The Union Pacific had equally great, though very different obstacles to overcome. As it advanced west through the plains, materials had to be hauled across greater and greater distances. The extremes of climate made it difficult to urge the labor force, made up largely of discharged soldiers, to more strenuous efforts. Hostile Indians and herds of buffalo disputed the right of way of the advancing rails. Yet the goal was reached and the era of the pioneer trader, the frontier trapper and the warlike Indian tribes was ended soon thereafter.

Meanwhile the Northern Pacific was thrusting its way from the Great Lakes to Puget Sound; the Southern Pacific, also projected by the Huntington group, crossed east from southern California into western Texas near the Rio Grande, where it made connection with a road to give it entry into New Orleans by 1882. The Atchison, Topeka and Sante Fe began modestly as a local Kansas line; later its promoters decided to push on through New Mexico along the old wagon trail; finally, to gain a better western terminus, a spur was thrust southward to meet the Southern Pacific, which gave the Atchison access to California by 1881; three years later it completed its own trackage to the West Coast. At almost the same time the

Northern Pacific had emerged through wild and undeveloped territory to the towns of the far Northwest.

These roads were in many ways the highways of the future, but in the East rather different developments were in the making which affected the immediate progress of most of the business and industry of the nation. Here the quick extension and improvement of the existing lines were of vital concern to capitalists, farmers and shippers. A strenuous battle began between railroad men to secure control of the best routes from the East to St. Louis and Chicago. Access by rail from New York, Philadelphia and Baltimore to the middle West was possible, in the Civil War period, only by changes from one road to another. The great task, after 1865, was to extend and consolidate these roads into long, continuous systems.

The most impressive of the old eastern roads was the Erie, which by 1851 had pushed its way through the southern New York counties to Dunkirk on Lake Erie and by wartime had made connections which extended its service even farther. The Baltimore and Ohio, stopped for many years by the stiff grades west of Cumberland, Maryland, slowly built over the mountains to Wheeling, West Virginia by 1852; six years later the Pennsylvania, blocked by similar difficulties, completed its line to Pittsburgh. Spurred by the success of the Erie road, the New York Central was organized in 1853 and began the consolidating of several roads which gave it access to Buffalo and Niagara Falls.

The keenest rivalry was between the Erie and the New York Central; the prize for which each strove was control of the freight business from the rich grain, livestock and mineral regions of the Great Lakes area, which needed transport to the east. A ruthless competition ensued, in the years following the Civil War, between such capitalists as Cornelius Vanderbilt, Daniel Drew, James Fisk and Jay Gould. This warfare centered about a determined effort by Vanderbilt, with the connivance of Drew, to wrest control of the Erie from Gould and Fisk. Foiled in this effort, Vanderbilt bent his energies towards the development of the New York Central Railroad, for with all his eccentricities and ruthless methods, Vanderbilt was a real builder.

The "Commodore," as he was called, had made more than a million dollars by operating a fleet of steamboats up and down the Atlantic coast. From this activity he turned to railroad promotion at the end of the war; in 1867 the stockholders of the rather loosely-organized group of lines which ran from Buffalo to Albany begged him to take over management of their holdings, which badly needed reorganization. Vanderbilt linked the routes across the state with his Hudson River Railroad and by 1873 secured control of the Lake Shore, which ran from Buffalo to Chicago. In a few years he had developed these separate holdings into a smooth, well-operated system.

It was Vanderbilt's skill and vision which lifted him above Drew, Fisk and the others. He ripped up the old rails and replaced them with new ones of steel; the old wooden bridges were torn down and shiny new ones installed; he bought new locomotives, built impressive terminals in the cities and effected economies in operation which cut the running time from New York to Chicago in half. He was quick to provide his trains with the new Westinghouse air brakes and to acquire the new Pullman cars. The Central became a standard gauge system throughout. By 1880 Vanderbilt possessed the best railroad system in the country and had secured the best of the freight business from New York going west. In about twelve years he had run an initial investment of some ten million dollars into a property worth a hundred millions, and had accomplished most of this miracle after his seventieth year.

The railroads had an almost incredible effect upon industrial America. It was due to the transcontinentals, more than to any other factor, that the West was opened, the Indians subdued, and the way prepared for the vast development of mines, farms and towns in the trans-Mississippi area. The new roads provided short cuts between the Atlantic and the Pacific and brought overland across the country a vast amount of the produce of the Orient, and returned to the West, and beyond, the products of eastern United States and Europe. All metallic mining and processing industries were stimulated. Some of the more detailed industrial effects of the further development of the railroads will be the subject of a later section.

Sources of Power, New and Old

The pattern of the new industrial economy of the United States is thus illustrated by the developments in iron and steel and by the evolution of massive railroad systems. None of these extensive changes would have been possible, however, without the presence here of enormous natural sources of power. Water power had been the earliest and most valuable power source, and had served the older seaboard communities well, so long as only a limited amount of power was needed, and wherever seasonal variations did not hamper seriously the manufacturing process. In the larger or more permanent industrial processes the ordinary drop of streams was soon inadequate; until large engineering projects could increase and regularize the flow of water and tap sources difficult of access, water power by itself would become and remain an obsolete element.

Coal, petroleum and natural gas were found soon enough in areas sufficiently close to the thickly-populated sections to carry the industrial load at an increased tempo for almost the entire century after 1800. Steam power, developed from coal, was the principal form of power used in industrial America for two generations or more, or roughly, from about 1840 to 1890. Electrical power, generated from water power, coal, oil or even from gases came into common use by the end of the nineteenth century. While all power development known today tends toward the complete electrification of industry, it is significant to note that we are still dependent upon those natural sources of power which have been present in nature untold centuries before the continent was inhabited.

Of all sources of power, coal was the most important between the Civil War and World War I. It was also the chief item of freight carried by our railroad lines. Anthracite, or hard coal, was at first of principal importance. Its concentration in northeastern Pennsylvania, near to navigable rivers and great centers of population gave it ready outlet to nearby markets in the cities. With the improved methods in combustion science, anthracite came into wide use in factories, steamships and on the railroads. The situation changed soon, due to the increased output of bituminous or soft coal, beginning about 1870, with the result that by the end of the first World War more than six times as much bituminous was produced than of its former rival, anthracite. Soft coal had the added advantage of being a ready source of gas for lighting and heating and its coke became essential in the smelting of iron. Areas yielding various types of bituminous coal were found scattered throughout the nation in the Appalachians, the Great Plains, the Rocky Mountains and on the West Coast. Of these four great areas the most useful to industry have been the Appalachian fields, from near the southern shore of Lake Erie to northern Alabama, because they were nearest to the large industrial cities. The product, also, was superior in quality to the western bituminous coals.

For several years after Drake had succeeded in bringing petroleum to the surface in the Titusville area, the market was mostly supplied by the popularity of kerosene, a component, as a cheap and satisfactory illuminant; no suspicion existed of its vast potentialities as a source of power. Yet such a narrow use was sufficient to boom the oil business, and to set in motion the great refining and shipping activity which was the foundation of the Standard Oil Company. The more advanced stages of the petroleum industry belong properly to a later portion of this volume. It is sufficient for the present to note that this industry profited enormously from the development of mobile power units as applied to motor cars, locomotives and steamships, as soon as engineering had developed the internal combustion engine of the 1870's and the Diesel engine of 1897. As a power fuel, petroleum was found easier to transport, less bulky, and cleaner than coal.

Prior to 1880 the electrical industry had been concerned principally with communication problems. By the time of the Civil War the use of the telegraph was fairly widespread; by 1866 Cyrus Field had succeeded in laying a transatlantic cable. A dozen years later the telephone, after a public demonstration in 1876, actually began to operate commercially. Thereafter, for the next seventy-five years, miracles of lighting, communication, transportation and power were accomplished as this force came to be understood and as the trained scientific minds of America applied themselves to its development. A later chapter will detail a number of these matters.

Science Comes to the Aid of Industry

Characteristic of the years following 1865 was the extensive use of our material resources without much thought as to future supplies. Waste and inefficiency were common in mining operations, in the cutting of timber, and in the haphazard agriculture of the time. Enormous amounts of land were given away to railroads and to settlers, and leased to oil promoters and lumbermen, with little or no thought as to the eventual shortages to which such exploitation would surely lead. Millions of acres were stripped of their forests, with no provision for replacement of the valuable trees. Erosion by wind, water and dust led to the abandonment of great areas of farm land, a condition partly brought about by wholesale destruction of the upland forests, which thus eliminated the best conserver of the soil at the head of the river valleys. Not until the twentieth century did a nation-wide movement get under way to conserve and reclaim, as far as possible, these invaluable assets in forest, mineral and agricultural lands.

Basic to the continuance of a prosperous economy is the scientific use and development of the soil. It is as necessary to apply scientific methods to agriculture as it is to develop an intensified mechanization in manufacturing. Without such methods, it would not have been possible to secure the grain for our great milling industry, the vegetable produce to keep our food processing plants operating, or the grass and fodder crops to sustain our meat and hide industries. The mechanization of the more laborious farm activities had already progressed quite far by the late years of the nineteenth century, but many obstacles remained. A constant warfare had to be carried on against insect pests which consumed the crops. Diseases which attacked plants and animals needed study in order that remedies might be applied.

An important step was taken, as we have seen, toward making agriculture a scientific, instead of a casual activity by the establishment in 1862 of the Department of Agriculture. In the same year the Morrill Act was passed by Congress. By this law there was to be set up in each state at least one college to give instruction in agriculture and the mechanic arts. In order to support the movement a grant of 30,000 acres was made to each state (or territory eventually to become a state) for each Senator and Representative the state had in Congress. After the sale of the lands by the states, the proceeds were to be invested in securities to support the new schools. After many vicissitudes, in which some states profited greatly by sales of desirable lands and others were less fortunate, colleges of agriculture began to dot the land from Florida to the Dakotas. In these institutions, in many cases, were trained the teachers, demonstrators, research workers and organizers who helped to revolutionize American farming.

Later legislation, in the 1880's, set up in each state an agricultural experiment station. A number of states used their grants very wisely by giving their funds to institutions already in existence; the funds would then be used to set up within the organizations new colleges of agriculture; in this way a strong institution like Cornell University secured endowments for agriculture and gave the new school the benefit of coordinating its teaching and research with that of the regular departments of pur and applied science of the university.

From the new colleges, from the United States Experiment Stations, and from the various bureaus of the national Department of Agriculture, came a steady stream of studies and discoveries which have been of extreme value to American agriculture. Improvements in the breeds of cattle, scientific tests of milk, effective methods of dealing with the Texas cattle fever, the hoof and mouth disease of both hogs and cattle, and the relation of bacteria to plant disease—are some of their most valuable studies. Entomologists made substantial progress in the war against the gypsy moth, the brown-tailed moth and the boll weevil by breeding here the natural enemies of such pests as soon as they could be discovered. Soil chemistry evolved and brought forth new types of fertilizer. In the twentieth century, after a vast amount of preparatory groundwork, the results of the new scientific agriculture were placed effectively before the average farmer. Pamphlets, specially prepared, on all kinds of subjects, were distributed by the Department of Agriculture. Special courses were offered in the agricultural schools. By means of the popular farmers' institutes and through the work of hundreds of county agents, necessary farm information was distributed to thousands of farm communities. As time went on, agricultural education was broadened and simplified down to the secondary level. A more detailed account of agricultural matters appears in a later chapter.

The technical education movement was also extended to fields other than agriculture. Before 1850, hardly an institution of high grade offered scientific instruction aside from the United States Military Academy at West Point. After the mid-century a rapid growth began of institutions devoted wholly or partly to the teaching of science. In 1850 Rensselaer Polytechnic Institute became a full-fledged college of engineering. In the decade which followed, approximately one hundred colleges were founded, in most of which some advanced instruction in science was offered. Harvard in 1847, and Yale in 1854, opened schools of applied science, while the Massachusetts Institute of Technology opened its doors the very year the Civil War began. Colorado started a school of mines in 1874, followed by Michigan in 1885, and others. By 1880 some four hundred and fifty colleges and schools were offering courses in chemistry, physics, metallurgy, mining, and mechanical and electrical engineering. The movement to organize business administration on a college level began with the founding of the Wharton School of the University of Pennsylvania in 1881.

In addition to a very impressive build-up of advanced colleges offering training in the fields of engineering and of applied science in general, technology of value to industry expanded down into the secondary education stage and to what we would call today the "adult education" level. Trade institutes offered training to mechanics and young men who had insufficient preparation to enter engineering school, or who lacked the monetary resources to attend them. Cooper Union began its long and useful career in 1857; textile schools were started at Lowell, Massachusetts and in Philadelphia; technical and vocational education was started in the high schools, especially in the larger cities. The number of successful inventors and industrialists who had enjoyed nothing but a common school education was still very large; there was still a place, a very important place often, during the 1870's and 1880's for the poor boy who left school at an early age to find an opportunity in industry, and by hard work and self-denial, to reach a very important post in American business.

But the Carnegies and the Rockefellers were passing out of the technical and industrial picture as the years went on. Technical education of a more formal kind became necessary as the industrial processes became ever more complicated. While management would continue to contain a number of men whose peculiar business abilities and personal force were able to offset their formal educational deficiencies, even these qualities would not long suffice for positions in which highly-developed scientific knowledge of detailed processes was vital.

One way of looking at technology is to regard it as an enormous extension of human strength. A turbine in one of the great electrical generating plants weighs nearly a million pounds; to perform the work done habitually by such an enormous machine would require the concentrated efforts of 400,000 men. In steel plants, heavy steel plates or bars are sheared off as easily as an individual could cut cardboard. In one of the largest automobile plants, entire ships have been cut into scrap for smelting by shearing machines operated by one man working a single lever.

Back of such colossal mechanization lay the formidable advancement of the applied sciences. The steps to make such achievements possible were being taken in the laboratories of the universities and engineering schools, in the scientific agencies of the national government, and in the research departments of great engineering and manufacturing corporations, as the new century began. During the forty years ending in 1900, the age of steel dawned and the railroad steadily forged unbreakable links between the parts of a once-divided land. Old sources of power were utilized more completely and new sources and forms of power were developed. Technical education and research, sponsored by the government, by business, and by the general public were coming to the aid of industry.

In the new century, man's knowledge of the forces of nature was to become even more extensive than before; he would even challenge cosmic forces whose very existence had been unknown in any previous time in history. In a bewildering variety of ways American industrial science was to enable the United States to outproduce any nation which had ever existed. Let us now turn to some of these industrial achievements of our own time.

CHAPTER THREE

TRANSPORTATION IN THE TWENTIETH CENTURY: LAND AND WATER

Perhaps the most remarkable characteristic of the present century, and one which has often been applied to Americans, is speed. The rapid accomplishment of work which at one time occupied many hours or days, has become through mechanized methods a matter of minutes, or even of seconds. In nothing is this business of speed more obvious than in our movements from place to place. Our grandfathers were content if the family horse could make the trip to town and back in an hour, or the long ride to the community picnic in thrice that time or even longer. Train rides of a hundred miles were slow, time-consuming affairs, besides being dirty and uncomfortable experiences. Persons who had journeyed a thousand miles from home, or who had visited other states than their own, were rare exceptions. They had travelled!

In the 1950's the trip to town is a matter of only a few minutes; the long train ride of yesteryear may be over and done in a little more than an hour. A business man may take breakfast in his New York home, board a plane at LaGuardia Field, transact business over luncheon in Chicago, and be home well in time for dinner. The speed of travel thus enjoyed by individuals has worked a revolution in commerce and industry. The motor delivery unit has eliminated the horse in the short hauls which make up so large a part of urban commercial activity. The railroad has vastly improved its freight-forwarding records of fifty years ago, despite the competition with great fleets of motorized freight carriers, even on a transcontinental basis. In the field of air freight and express, extensive progress has also been made, though not to the extent of serious rivalry with other freight carrying agencies.

Between the two extremes of this picture, from the slow passage of individuals and goods through necessary space to their present attainment of speeds of several miles per minute, lies the entire history of twentieth century transportation. Such miracles of travel were not attained easily. Basic problems of power had first to be solved; to a large extent the attainment of supersonic speeds required a complete revolution in mechanical, electrical and chemical technology. In addition to the basic inventions in these fields, hundreds of operational improvements to motor cars, ships and railroad equipment had to take place. Moreover, men had to learn to fly powered devices and develop them to commercial efficiency.

The Railroad Makes Its Major Contribution

At the opening of the twentieth century no form of transportation existed which competed seriously with the railroad. The great net of lines with which all are familiar today was practically complete. All over the United States ran the great trunk roads, connecting the great cities which they had so large a share in building. From the cities, then as now the collecting and distributing centers for large marketing areas, ran smaller subsidiary lines to rural communities, to lake and river ports, to mines, and to junction points near the larger cities. Along two hundred thousand miles of track the steam locomotives puffed their way, hauling long lines of cars laden with raw materials or a myriad of manufactured products. The wealth of a continent moved almost exclusively on rails in 1900, bearing ore to the blast furnaces, wheat to its processing centers, livestock to the great meat packing cities, coal to the factory towns, and oil to the refineries. The freight business, except for the small part borne by the internal waterways, was a railroad monopoly.

A goodly share in such a princely commercial operation was a prize worth having. While it was almost certain that most goods of that day would have to travel by rail, the precise routes and railroads over which they would move were not fixed, and bitter competition ensued for control of the profitable freight between the main shipping centers. Rate wars began, and the most advantageous access to the large eastern, midwestern and Pacific coast cities was contested with great gusto by the larger lines. At immense cost, frequently, companies purchased parallel, competing lines to gain exclusive control of a valuable route or territory. The Boston and Maine originally controlled the major railway freight business of New England, yet the Boston and Albany, controlled by the New York Central, was soon able to acquire a valuable part of the shipping in and out of the "hub" of that area. The powerful Pennsylvania Railroad, eager to share in this wealthy freight territory, acquired the Long Island Railroad in 1900, tunneled the Hudson River at a cost of $100,000,000, and under the East River to Long Island. By its new Hell Gate Bridge it secured valuable connection with the New England roads. To rival its powerful eastern competitors, the Baltimore and Ohio also sought a share of the valuable seaboard business by building trackage parallel to the Pennsylvania from Baltimore to Jersey City, and by establishing a terminal on the Hudson across from New York.

As the older forms of carrying on competition with a rival, such as rebates, pooling the available freight on a percentage basis, and other devices, came into disfavor through government regulation, the consolidation movement became popular. While such a trend had always existed, ever since Commodore Vanderbilt began to buy up lines between New York and Chicago, the new trend dwarfed those operations. When outright sale proved to be impossible, the financial method of buying stock in another road, and thus securing representation on its board of directors, worked well. In this way the Pennsylvania advanced its position by buying heavily of the stock of the Baltimore and Ohio and of the Norfolk and Western. The Reading Railroad came under the joint control of the Baltimore and Ohio and the New York Central; the disturbing competition provided by the Chesapeake and Ohio was ended by its purchase by the Pennsylvania and the New York Central.

This trend is best exemplified by Edward H. Harriman, the master consolidator of them all. Starting as a director of the Illinois Central Railroad, he became its dominating official; he proceeded to build it into a highly efficient and greatly expanded property, possessed of top financial credit. When the Union Pacific was reorganized in the years after 1893, he forced his way into the revamped road and became its controlling force by 1900. The next year the U.P., at Harriman's behest, purchased the Southern Pacific, and four years later, his group, having acquired a large block of Sante Fe stock, moved into directorships of that road.

Meanwhile James J. Hill, with the assistance of J. P. Morgan and Company, had consolidated the northern railroads, including the Great Northern and the Northern Pacific; these two railroads came also to control the Chicago, Burlington and Quincy. Since the Burlington Road, which reached Denver, was a potential invader of the vast Harriman territory, a battle began for control of the great western roads. Harriman had prepared for such a situation, as he thought, by purchase, as early as 1901, of large blocks of Northern Pacific stock. But although he was soon in control of a majority of the preferred shares, he was unable to achieve similar control of the common stock. The preferred stock did Harriman little good in his objective, as it could be retired by vote of the shareholders of the common. In the ensuing turmoil, shares of Northern Pacific, sought by brokers of both Harriman and Hill, were forced up to $1,000 a share, and threw the stock market into confusion.

The way out of the dilemma seemed to be the formation of a holding corporation, the Northern Securities Company, by which the securities of the Northern Pacific, the Great Northern, and the Chicago, Burlington and Quincy were exchanged for stock in the new organization. Since Harriman and two of his associates were also directors of the Northern Securities Company, and since Harriman was a member of the executive committee of the new combination, it is clear that railroad competition west of the Mississippi was thereby nearly eliminated. Furthermore, Harriman, though he had failed to take over control of the Hill-Morgan lines, was given a decisive place in the councils of the northern systems.

As is well known, the Supreme Court, in the famous antitrust case decided in 1904, took a different view of these proceedings than did the artists who put together the Northern Securities Company, and ordered its dissolution. But Harriman fared extremely well, despite the decision; his Union Pacific still held $22,000,000 worth of stock in the Great Northern and the Northern Pacific; from sales of his other holdings in the Hill-Morgan roads he netted $35,000,000 more, which he and his associates proceeded to use in purchases of stock in other roads, both western and eastern. He acquired great blocks of New York Central, Baltimore and Ohio, and the Central of Georgia, among eastern roads. By his death in 1909, Harriman's holdings were scattered from the Atlantic to the Pacific, from the Gulf to the Great Lakes.

While the Supreme Court demolished much of the Harriman empire after his death, the trend which his career illustrated has not been reversed. By 1906 one-third of the stock of American railroads was in the hands of other roads. Interlocking directorates intensified this situation, as did the important influence of the great banking interests upon railroad management. In 1912 the Morgan bank owned stock in nine of the great railroad systems, and partners of the bank were directors in a number of these lines. As the holdings of individuals and family groups declined with the years, the integration of railroad systems through outside ownership, lease, interlocking directors and holding companies increased.

The battle between the railroads and the government of the United States centered largely about the question of whether railroads were purely private enterprises (and hence able to exercise control over their financial operations), or whether their very size and strategic importance to American industry made them public or quasi-public corporations and hence subject to extensive control. In the year 1906, the Hepburn or Interstate Commerce Act enlarged the authority of the Interstate Commerce Commission by giving it power to review rates and determine their fairness or unfairness, and also enlarged the scope of its activities to include supervision of pipe lines, express companies and sleeping-car companies. Through legislation in the ten years following the passage of the Hepburn Act, the railroads practically lost the power to fix their own charges. Legislation by Congress in 1916 established the eight hour day for workers on railroads and other interstate carriers.

Pressure of war conditions led to governmental operation of railroads between 1917 and 1920, a period ended by the passage of the Esch-Cummins Act, which returned the lines to private ownership. The series of circumstances sketched above indicates the pattern of relationship between government and the railroads which has endured, with minor variations, to this time. Government control has become a permanent policy, yet private ownership has been vindicated. The difficulties of railroad management have been infinitely increased, not only by greater government control over wages and rates, but also by the powerful position in recent years of organized labor.

The solution of the problem of competition, in the early years of the century, was its elimination by the methods we have noted. To a large extent this method was successful, though restrained in some extreme instances, as in the Northern Securities Case, by the courts. Consolidation of railroad lines in as large a nation as the United States, if carried on with due regard to the needs and convenience of the American public, seems logical and necessary. Even the methods employed by the founders of the Northern Securities Company were, at the most, controversial; four of the Supreme Court, including the liberal Oliver Wendell Holmes, favored the combination.

Improvements in Railroad Equipment and Operations

The battles of the railroads and their relations with the Federal Government are important here mostly as background of the story of their industrial achievements, despite setbacks. A transportation organization has basically a broader significance than many industrial companies have, in that while it is an industry by itself, with its own peculiar problems, it is in addition a carrier of the materials of all, or almost all, the other industries. This is true whether one deals with extractive industries, such as minerals; agricultural, such as wheat; manufactures like shoes and clothing; or the purely commercial, wherein the railroad acts as a kind of long-distance delivery service for the retailer or jobber. In distinction to most industries, also, its properties are scattered widely, exposed to weather and other damage. The element of repair and maintenance is therefore extremely heavy.

The history of railway equipment is a fascinating drama of the ingenuity of the human mind. In 1951 the United

States commemorated the passage of a century-and-a-quarter since the first railway in America began its operations. Between Gridley Bryant's horse-drawn line, built to haul granite from Quincy, Massachusettts to tidewater on the Neponset River three miles distant, and the Diesel-drawn Meteors and Zephyrs of the 1950's, intervened developments which make a fantastic story.

In the early days of railroading, the locomotive consisted of a vertical steam engine carried on a flat car; the passengers rode in small coaches not unlike those of the highway. Freight cars were mere beds or platforms on which the goods were stacked, covered sometimes with tarpaulins. In the United States, prior to 1870, freight, or "burden" cars were still small, boxlike affairs, almost square in shape, with four wheels, and a capacity of three to four tons. The introduction of the four-wheeled truck, and soon after, the double-truck box car, marked the appearance of a distinctly American type of freight carrier, and added the space needed for the growth of freight business in the 1880's and 1890's.

Shortly after the Civil War, George M. Pullman began the construction of heavier and more luxurious special purpose passenger cars. The Westinghouse air brake, first demonstrated successfully in 1868-69, was improved to such an extent that by 1887 a train of fifty freight cars traveling at a twenty mile per hour speed was brought to a standstill in the space of 171 feet, a feat formerly requiring five hand brakers over a distance of 1500 feet. The decade after 1870 brought forth refrigerator cars with iced compartments, and cars for transporting livestock, equipped with watering troughs and feed bins. Automatic couplers for freight cars were perfected by the 1880's and soon came into general use.

Devices of this sort led to increased speed of freight shipments and improved their handling, but soon made the old wooden frame car obsolete. Lack of timber for the larger cars, and the breakage caused to center sills and draft boards by the increased speed and larger loads necessitated a change to steel-reinforced and finally to all-steel cars. The first steel freight cars were ordered in 1896 by several railroads from C. T. Schoen of Allegheny, Pennsylvania, who had pioneered in making steel parts for wooden cars. These first steel cars were designed to carry coal. Thereafter the railroad equipment industry developed very fast, as dozens of specialized types of freighters began to appear: box cars, ore cars, refrigerator cars, tank cars, and also those designed for logs, poultry, furniture and finally automobiles, to serve the special needs of new types of freight.

As one might readily guess, the appearance of the steel car at the turn of the century, of capacity up to 100 tons, made possible large economies in handling freight, added a safety factor and contributed to the durability and long life of the railroad's rolling stock. But it made necessary large changes in operation and maintenance. Heavier locomotives were needed, heavier rails, stronger bridges, facts which revolutionized the railroad equipment business and the steel industry as well. In line with greater efficiency and speed much improvement was made in the standardization of parts. As a result largely of the recommendations of the Master Car Builders Association more than fifty kinds of journal boxes were reduced to five sizes and one interchangeable type; couplers to one type, axles to five sizes and one interchangeable type; brake shoes to one type; brake heads to three types and four sizes. All brake beams became interchangeable.

The passenger car underwent a similar complete revolution in the period, from the small coaches to the eight-wheeled car, which was better at rounding curves than the earlier ones, but which still had candle light, heat from one stove and still lacked closets, lavatories and water coolers. This was the type of car in which our ancestors of the generation after 1850 still rode. Its length was from thirty to forty feet, the seats were iron-framed and there were no springs. The further progress of the passenger coach followed much the same pattern as the freighter, with greater length, steel reinforcement, better braking and coupling; finally came the all-steel passenger car, about 1904, strongly-built and completely non-inflammable except for the wooden dashes and doors and the cane seats. The American Car and Foundry Company constructed the first of these cars for the London Underground Railway and for the Interborough Subway of New York City. Two years later, in 1906, the New York Central placed an order with the same firm for steel passenger cars, an example soon followed by other railroads. The passenger could now ride in relative comfort and safety, with fair ventilation, window shades, steam heat, hot and cold running water, electric fans, headrolls, footrests and parcel racks.

The evolution of the locomotive to heavier and more powerful types was dictated by the development of the country's business and the increased length and weight of both passenger and freight trains. Great giants produced in the Baldwin and American Locomotive Works kept pace, though with some difficulty, with the increased demands of the nation. Multiple driving wheels, improved boilers and cylinders, better braking power, pushed the steam locomotive to the maximum capacity of power output. Yet steam power was not efficient or flexible enough to keep the railroad abreast of the demands of the times. Despite the fact that the steam locomotive has increased its tractive power by more than fifty per cent since 1923, its peak year of production, the ensuing thirty years have seen a steady decline in its manufacture. In 1937 only 380 units were made.

Better power units than those of steam were badly needed, especially after the rough experience of the railroads during World War I. Electric locomotives had been tried as early as the 1890's, but the vast expense of power transmission was almost prohibitive, except for short runs near or through cities. The smoothness and efficiency of electric power were admitted, however, provided that a more convenient original power source than coal were available and the resultant difficulties of providing several hundred thousand miles of transmission lines or charged rails could be obviated. The answer proved to be the Diesel engine.

The pioneers in building Diesel-powered locomotives were the engineers of the Electro-Motive Engineering Corporation, which began its operations in Cleveland in 1922. Since 1930 this concern has been the Electro-Motive Division of the General Motors Corporation. The replacement of steam locomotives by Diesels, which take fifteen-car trains over the heavy grades of the Raton Pass of the Sante Fe, or pull others through the winding passes and tunnels of the Western Pacific, the Union Pacific or the Great Northern, are familiar to all western travellers. Not only on the western lines, but on those of the East as well, Diesels have become a familiar sight after thirty years of experimentation and improvement. The successful performance of Diesels in yard switching and freight hauling has long been demonstrated.

At the American Locomotive Works at Schenectady two important events took place in 1948. The Company celebrated one hundred years of locomotive manufacture, and began exclusive production of Diesel instead of steam units. The best of the steam locomotives, according to one of the largest manufacturers, spend from ten to twenty per cent of their time in the roundhouse, while Diesels are idle less than four per cent of the time. Brake maintenance is lessened in the latter because traction motors like the Diesels supply braking power when run in reverse. The power delivery factor of the newer types is superior, for the Diesels deliver smooth running power contrasted to the hammer-like thrusts of the steam locomotive's drivers. Thus they contribute to more economical roadbed maintenance. "Dieselization" of American railroads is well underway. It is said that four out of every five crack passenger trains and nine out of ten of the fast freights are Diesel-powered at the present time. In 1948 General Electric and the American Locomotive Company jointly developed a gas turbine locomotive, tested it, and the next year placed it on regular freight service on the Union Pacific. Steam will probably remain for some years as a power source on many of the "coal" lines. It would be poor economy indeed to scrap the 35,000 steam locomotives still in use, for they represent a huge investment. But their replacement eventually by Diesel-electrics or gas turbine units is clearly "in the cards."

The ultimate in railroad transportation is undoubtedly the luxury train, an accomplishment of years of research in the making of railroad equipment. One of the great California Zephyrs, for example, consists of a streamlined Diesel locomotive, a baggage car, three Vista-Dome coaches, a Vista-Dome buffet-lounge car, two six-bedroom and ten-roomette cars, a diner, a sixteen section sleeper, and a Vista-Dome lounge-observation car, having one drawing room and three bedrooms. These magnificent cars are all of stainless steel, built by the Budd Company of Philadelphia; from five of them the passengers may view the mountain scenery from the Vista-Domes or upstairs penthouses built into the roofs. The latest type of fluorescent lighting directs illumination to where it is needed without glare. The trains are equipped with radio and a public address system. Indeed, each berth is a model of comfort and convenience, with its switch panel controlling porter service, radio entertainment, ceiling lights, a four-position fan, heat and air conditioning.

Every provision known has been added to increase safety and comfort in these trains. Structurally the cars have Timken roller bearings, coil bolster springs, high speed control air brakes, copper tubing on brakes, copper steam and water pipes, machine-balanced rolled steel wheels, with rubber pads under the center plates, under the equalizers over the journal boxes, and at the end of the bolsters. Such details of construction smooth the journey as the miles streak by. Together with the effective inter-communication and centralized train control systems, the luxury trains comprise an amazing achievement in smooth and safe travel.

While passenger travel is still a vital part of railroading, it is the movement of vast fleets of freighters which has the greater economic importance. From the great wheat-producing prairie states fast Diesel-drawn trains move the millions of bushels of the harvest to the terminal elevators, and from the terminal elevators to the flour milling centers. As either wheat or flour, the fast freights bear the product to the St. Louis "gateway," whence others carry it to Gulf or Atlantic ports from which it goes to the four corners of the world. Perishables such as the fruits of Florida and California are packed into pre-iced cars (one of which can carry 500 boxes of oranges). After careful bracing of the boxes and sealing of the car, it is ready to be attached to a train such as one of the Florida "Marketers," which pick up additional cars as they move through the great citrus belt. From Florida the Seaboard may run the fruit cars to North Carolina, where the train is re-iced and later may deliver its cars of oranges at Richmond to the Richmond, Fredericksburg and Potomac line, which will run them to Washington. Here the Baltimore and Ohio picks them up and takes them to Jersey City, whence they are lightered across to New York City. This typical movement of fruit is accomplished by the coordinated efforts of three railroads at a total freight cost of less than one dollar per box, including the refrigeration service. Five days from the time the fruit was picked in a Winter Haven grove in central Florida, it is served on the tables of New Yorkers.

Likewise impressive is the movement of coal on roads like the Baltimore and Ohio, the Virginian, the Chicago and Eastern Illinois and the Western Maryland. For example, every day from Webster Springs, West Virginia, an eighty car coal train of the Western Maryland climbs the 105 mountainous miles to Elkins, over steep grades all the way, in just over three hours. Six "helper" locomotives are used to drag the tremendous load of 5,565 tons on its difficult route. It has been found that such a big move of coal is more economical than the dispatch of a number of lighter trains, for it keeps the division clear and makes for quicker shipment. A typical day's movement of coal over the Western Maryland involves the movement of between 600 and 700 loaded cars from the eight coal districts which it serves; on some days as many as 1000 cars are loaded.

Railroading today evidences material progress after more than 125 years of development. In mechanical equipment, safety, efficiency of operation (particularly of Diesel-powered trains), and the matching of any type of transportation in comfort and convenience, the railroad has kept pace with the demands of modern industry. The future should bring still further progress in speed, traction, and in the elimination of marginal lines in the interest of transportation efficiency. So long as seasonal and very heavy movements of raw materials are needed, over long distances to terminals and processing points, the railroad freight train will continue to be the mainstay of the American transportation economy. In the nonperishable and bulky fields especially, the railroad continues to hold its own. In 1948, as near to a normal peacetime year as can be found in recent times, the railroads carried nearly three billion tons of freight.

Water Transportation

The system of internal waterways of the United States, with the exception of the Great Lakes, was decidedly of secondary importance at the opening of the present century. There were good reasons for this. The great rivers of the country run mostly in a southerly direction, while commercial traffic has come to be east-west. Furthermore, the canal systems constructed in the nineteenth century to connect western rivers and lakes with the eastern seaboard were no longer able to meet the competition of the railroads, which had secured large governmental assistance in a period when support for the internal waterways was being cut down. In addition, the railroads by about 1910 had acquired control of nearly a third of the existing canals, which they did not develop, and in many cases stopped using altogether. As for the remainder, the railroads lessened their competitive importance by maintaining freight schedules which the independent canals could not meet, or by acquiring canal terminals and running their own lines of boats. River transportation had likewise declined sharply by the beginning of the century, partly through competitive methods by the railroads, which lowered rates on lines which paralleled rivers and in some cases by establishing their own lines of river steamers.

The Erie Canal carried eighteen per cent of the freight traffic between New York and Buffalo in 1880. By 1906 the percentage sank to three, as the New York Central and Erie Railroads took over the bulk of this business. River traffic on the Mississippi, the Illinois and the Missouri routes fell by equally large margins in approximately the same period. Only the Ohio, of the more important streams, failed to show this trend, as its freight business in 1906 was only three per cent less than in 1889. The vast sums which had once been appropriated for internal waterways seemed to have been wasted, as the century began.

Despite such indications, which seemed to point to the rapid decline of water transportation, as other types of freight forwarding grew to large proportions, there actually was a revived interest in internal waterways during the first half of the twentieth century. A part of this came about because of the wide public excitement over the Panama Canal, which was being constructed between 1904 and 1914. At about the same time the growing interest in conservation of our natural resources gave publicity to waterways which might be improved to provide better water supply, greater power facilities, and cheaper transportation than land transit afforded.

It is also true that the excitement over this subject, which may have been political in origin, did result in several concrete accomplishments, following strong support given by Presidents Theodore Roosevelt and W. H. Taft, both of whom set up waterways commissions and fostered necessary legislation. The Ohio River was canalized at federal expense between 1911 and 1929; improvement of the Missouri River channel began in 1912; substantial improvements were made to Mississippi River navigation between 1906 and 1936. In 1914 the Cape Cod Canal, which cut Atlantic coastal voyages by seventy miles, was opened to navigation. The Erie Canal was thoroughly overhauled and enlarged from 1903 to 1918, a project financed by New York State.

Commerce on the Great Lakes, however, needed no artificial stimulus from politicians or waterway enthusiasts, for this matchless highway of trade, by its position and the direction of its channels, remained a natural and vital part of our transportation system. Continuous voyages by deep-draft vessels were possible through it; there were few problems of narrow channels or cargo changes to make operational charges high, as in the case of rivers and canals. Of course it is true that the greatest usefulness of the Lakes as a long-distance commercial highway was effected partly by the construction of connecting channels, as the Sault St. Marie Canal between Superior and Huron, the Lake St. Clair-Detroit River channels between Huron and Erie, and the Welland Canal which bypasses the Niagara River. But these partially artificial channels were well worth the expense of construction, as they were all relatively short and in each case, opened up several hundred miles of continuous water navigation. Most important of all, the entire Great Lakes route, except for the connecting channels, requires little or no maintenance.

The general pattern of Great Lakes business was established

by 1900 and has not changed materially since that time. The huge iron production of the Mesabi range in the Superior district and much of the great wheat production of Saskatchewan, Alberta and Manitoba, plus a smaller proportion of United States wheat, are carried along this route. The iron ore is dropped off at ports from Calumet on Lake Michigan to Buffalo on Lake Erie. Hard and soft coal are the principal cargoes returned to the ports of the upper lakes, either for the heating and industrial purposes of that region, or for more distant shipment to Canada and the northwestern states. Active competition between shippers has kept freight charges down. Several important railroads, including the New York Central and the Pennsylvania, have developed shipping lines to direct large amounts of the Great Lakes traffic to their connecting roads. A second group, consisting of big industrial concerns, such as the Standard Oil Company and the United States Steel Corporation, maintain their own fleets of tankers and ore boats. In addition a number of smaller companies and individual owners run cargo ships. By 1910 only two nations, Great Britain and Germany, possessed greater merchant fleets than that plying the Great Lakes alone.

The astounding success of the Great Lakes route in an era of general decline in our internal waterways was due largely to its advantageous position between the great productive area of the West and the fast-growing industrial cities from Chicago to Buffalo, as mentioned before. But there were other reasons as well. It was more difficult for the railroads to gain control of the grain traffic of the northwest than of the prairie states, which were crossed and recrossed by rail lines, with no serious water competition. Another strong advantage lay in the improvement and enlargement of cargo carriers on the Lakes. The steam vessel had become the dominant type of vessel there after the 1880's, and the size increased from a few hundred tons burden in the early days to vessels of 7,000 tons by 1920, and up to 12,000 tons at the present time. The newer ore and grain ships were adapted carefully to their cargoes, with long, straight sides, with wheel houses and engines at the fore and aft extremities, and cargo space between, reached by a long series of hatches through the level decks. Docks were constructed at the terminals to fit these craft.

At ports such as Duluth, Superior or Two Harbors, at the northwest extremity of the iron ore "move," ore trains run out over the docks and come to rest with each car above an ore pocket; when the bottoms of the cars are opened, the ore flows by gravity into these pockets. Next the ore boat comes up beside the dock, its hatches open, to correspond to the filled pockets; steel shutes are lowered from the pockets into the hatches and down goes the ore into the ship. A 12,000 ton ship can be filled in this way in four or five hours, an operation which took four times as long in 1905, when the ships were smaller. At Conneaut on Lake Erie, one of the more important ore ports, where the shipment may be delivered, the vessel is docked much the same way as at the other end of the journey. Almost as soon as the ship has stopped, the unloaders go to work. A battery of five of these Hulett machines, each ninety feet high, dips down into the hatches of the ship. The brains of the Hulett unloader are in the bucket leg of the contrivance, which carries its operator along as it dips into the ship, to come up with seventeen tons of ore at one bite. By such means, an ore boat can be unloaded in five hours, compared with thirty-one hours in 1905. The loading and unloading of grain and coal is likewise speeded up by similar mechanical contrivances.

Another survivor of the more prosperous days of the waterways was the coastal trade plying the Atlantic waters, though much less important than the Great Lakes trade. Rate wars went on between ship owners and the railroads, especially with the newer lines such as the Seaboard and the Atlantic Coast Line. By 1912-1913 most of the New England coastal shipping was railroad-owned or controlled. From New York down to

Florida five railroads and one consolidated shipping firm (the Atlantic, Gulf and West Indies Steamship Lines) controlled ninety per cent of the business. In the Gulf, the bulk of the coastal trade remained for some time in the hands of independent shipping lines, with one railroad competing for a share. Coastal shipping registered a fifty per cent increase in the twenty years ending in 1917.

The picture was quite different in the foreign commercial shipping field. At the opening of the century less than a million tons of shipping engaged in foreign trade bore American registry, the lowest figure in our entire history. In 1901 the proportion of American imports and exports carried under the American flag represented only 8.2 per cent of the total value of our foreign trade, whereas a century earlier more than nine-tenths of it was American-borne. We had been slow in adapting steam to ocean vessels because our costs of building and operation were greater than those of foreign shipping firms. We also failed to subsidize our merchant marine and thus lost leadership to nations which did so. Most of all, domestic industry with its quicker profits, plus the rapid development of the West, took the minds of nineteenth century Americans away from the possibilities of a large merchant marine, especially between the Civil War and 1900. We became "isolationist" in this regard.

In the early years of the twentieth century, some voices were raised in favor of governmental mail subsidies and of bounty payments for cargoes carried in American ships. Back in 1891, the Ocean Mail Act had offered small and inadequate subsidies for the carriage of foreign mail, under the terms of which the International Navigation Company started a weekly service to England in 1895; the lean payments helped the Ward Line to maintain service to Cuba and Mexico, the Red D Line to Venezuela, and the Oceanic Steamship Company from San Francisco to Australia. President Theodore Roosevelt, with the strong backing of Senators Frye and Hanna, recommended legislation for establishing an American merchant marine in 1903; a bill which carried mail and cargo subsidies for every type of American steamship in foreign trade, other than to Europe, was defeated through the opposition of representatives of the South and the Middle West. A renewal of efforts by the President in 1907 also failed. As a part of the Panama Canal Act of 1912 and of amendments to it two years later, encouragement was tendered to owners of vessels engaged in foreign trade to bring them under American registration. As a result of these acts, some 170 ships of 580,000 total tonnage were brought under American registration, including ships of the United Fruit Company, the Standard Oil Company and the United States Steel Corporation, but the bulk of vessels in which American capital was invested continued to sail under foreign colors.

To a large extent, as President Roosevelt told Congress in 1903, the question of an American merchant marine was involved in the larger question of American defense, a problem which had become more complicated with the acquisition of overseas possessions and with our responsibilities for protection of the Panama Canal. One of our gravest deficiencies, with the coming of the war in Europe in 1914, was our deplorable lack of shipping. What had formerly been a matter of choice became suddenly a matter of necessity, with Allied shipping being driven from the seas by German submarines, while the United States was still forced to rely substantially on the very nations which were losing so many of their ocean carriers. So Congress established, in 1916, a United States Shipping Board, with authority to construct, purchase, lease and operate merchant vessels in United States service. Organized in January, 1917, the Board began operations just before the declaration of war against Germany in April of the same year.

With the slogans: "ships will win the war" and "America must build a bridge to France," the Shipping Board set up an Emergency Fleet Corporation in April, 1917. Under the effective leadership of Charles M. Schwab, the famous steel maker, three great shipyards were constructed. One of these, at Hog Island near Philadelphia, was constructed in five months; it could handle seventy-eight ships at one time and at its peak of activity could turn out a 7,500 ton ship in three days. The Corporation increased from 256 to 934 the number of shipways in which seagoing craft could be constructed. More than a hundred of these shipways were kept busy assembling "prefabricated" ships, the parts for which were manufactured all over the United States. By the end of the war the nation had a fleet of 2,600 ships with a 10,000,000 ton total. This represented about half of the wartime shipping requirements as originally estimated.

In 1919, after the war had come to a rather sudden end, the Shipping Board was still conducting about 42.7 per cent of American foreign trade. The United States Government wished to discontinue this system because of its "socialistic" aspects but wished to retain a large merchant marine under private United States ownership as a bulwark of national defense. The Jones Merchant Marine Act of 1920 sought to accomplish this double purpose by directing the Shipping Board to sell ships to private American owners or to companies controlled by American citizens. The Act also created a fund from which owners or prospective owners could draw, granted special tariff reductions on goods imported in American ships, and offered generous subsidies to shippers who would carry United States mail. It was specified that all trade to our colonial possessions must be carried on in American ships.

This solution did not work well. Many ships had to be sold for scrap, while hundreds lay idle for years, rusting away, at yards along the Hudson and Delaware Rivers. By 1928 more than 1000 ships had been sold, but the Shipping Board was still operating some 300 of them. Private investors did not take kindly to this kind of shipping business, particularly when they were offered slow, nearly obsolete coal-burners; furthermore, the government offered no money for conversion or new construction. By 1928 the American merchant marine had deteriorated so badly that Congress again came to the rescue, this time by the Jones-White Marine Act, in which $250,000,-000 was appropriated for new construction. The remaining ships were offered at bargain prices, and the mail subsidies were increased. Still the interest of the public lagged. High tariffs had by this time curtailed our foreign trade seriously and shipping costs had mounted to an all-time high.

It is fair to say that world economic conditions, which had deteriorated badly, hindered the progress of all foreign trade, and therefore interfered with the attempts to establish a merchant marine in the late 1920's and early 1930's. The European countries were endeavoring to collect huge sums from Germany, in order to transfer large payments, on account, of their war debts to the United States. Only a vast build-up of trade could have furnished the means of balancing such large indebtedness from nation to nation. Such a trade increase was prevented by widespread economic nationalism by which European nations and the United States sought to preserve their domestic markets through the imposition of high protective tariffs.

Some improvement of the foreign trade situation came through the passage of the Reciprocal Trade Agreements Act in 1934, the fruit of many years of study by our able Secretary of State, Mr. Cordell Hull. By this measure the President was authorized to adjust the statutory rates of tariffs by as much as fifty per cent, upon the negotiation of mutually advantageous bilateral treaties with individual nations. The Secretary had made more than twenty such treaties by 1942, which substantially reduced American tariffs towards the nations concerned by an average of about twenty-nine per cent of the existing rates.

Neutrality legislation in force from 1935 to the outbreak of World War II also had a deterring effect upon our foreign trade. All arms, munitions and implements of war were embargoed; none could be shipped from the United States to any belligerent. The President could also prohibit goods of any description from being carried out of the country to a belligerent, even if the vessels were foreign-owned. Yet, by an arbitrary definition of belligerency, China and Japan were both supplied with goods useful in war, even after the attack on China by Japan in 1937. Of course, this policy did not materially affect the general trade picture.

A further effort to organize and maintain an American merchant marine came in 1936, with the passage of the Merchant Marine Act of that year. The government was authorized to pay the differentials between the foreign and American construction and operation costs. The Act abolished the Shipping Board and created a new Maritime Commission to supervise the shipping activity of the nation; the Commission could own and operate ships when unable to charter or sell them to private operators. No notable improvement resulted from this Act, and until a new war provided the necessary stimulus, the merchant marine continued to languish. Under wartime conditions, between 1940 and 1945, some 8,500 merchant ships were built. According to the Department of Commerce, the American merchant fleet increased from 1379 ships of about eight million gross tons on September 1, 1939, to 3,492 ships of more than twenty-five million gross tons on December 31, 1950.

One noticeable feature of the American merchant fleet, even in the periods of great expansion, as 1917-1920 and 1940-1945, was the lack of ships of large tonnage built in American shipyards. It was perhaps inevitable that war building programs would emphasize rather small cargo ships, which would be easily replaced and which could be indefinitely duplicated by the pre-fabrication method. Understandable as this policy is, as a war measure, it did not help to place us, even by 1945, in a class with the nations which constructed giant liners for peacetime ocean service. Lloyd's Register of Shipping for September 1, 1951 carried data on ninety-two steam and motor ships of more than 17,900 gross tons. Of these, fifty are British and only seventeen of American registry. The addition of the 51,500 ton *United States,* which entered service in 1952, gives us eighteen large liners. Whether this presages a real beginning, under governmental encouragement, of a great period of United States built, owned and operated liners of large size, is a matter for the future to unfold.

When we ask what has been the net effect of the efforts to improve American water transportation, domestic and foreign, it seems clear that in only one area has this type of commercial activity been continuously profitable: that is the Great Lakes region, which continues to provide the necessary link in the transfer of agricultural and mineral wealth from the interior to the Atlantic seaboard. Atlantic coastal shipping has continued on a limited scale, largely subsidiary to the railroads of the East. Transoceanic shipping under the American flag has not been profitable or even self-supporting, despite the determined efforts of several of our leading statesmen and the stimulus of two world conflicts. While we have very large shipyards, such as those of the Newport News Shipbuilding and Drydock Company, which built the *United States,* the great backlogs of orders which they have represent defense construction largely.

The Motor Car Enters the Picture

Labert St. Clair, author of an interesting book on transportation, expresses the opinion that "European engineering genius conceived the gasoline-propelled automobile, but American salesmanship made it a financial success." While there is substantial truth in the generalization, it might be added that it seems difficult, if not impossible, to determine the actual European inventor. The reason for this situation is that the automobile is not a single invention at all, but an aggregate of inventions which have been made, and are still being made, to improve self-propelled transportation vehicles.

By concentrating upon oil-propelled units, the problem becomes a bit simpler, but still confused. After inventors had made a great many attempts at self-propulsion of various sorts, the French inventor Etienne Lenoir, after three years of commercial success with his stationery engine, placed one of them in a vehicle in 1862, and drove the crude contraption from Paris to Joinville-le-Pont, some ten miles up the Seine, using "street gas" as fuel. In 1866 Otto and Langen, German experimenters, invented a gas engine which was a great improvement upon Lenoir's; United States patents were granted for three types of the Otto and Langen engine in 1867; by 1885 Gottlieb Daimler, also a German, evolved a petrol engine which was somewhat similar to those of Otto and Langen. The decade of the 1880's brought a welter of claims by individuals, principally French and German, for the honor of having produced the first motor car propelled by petrol or gasoline. Most important among these seem to have been Carl Benz, Gottlieb Daimler and M. Levassor. Benz developed a gasoline tricycle in 1886; Daimler joined forces with Benz shortly after, and together they produced a vehicle often referred to as the first successful gasoline motor car. Levassor obtained patent rights from Daimler, and adapted the latter's engine to a vehicle of his own design; in his motor car, Levassor discarded the "carriage" design, placed the engine in front, with the axle of the crankshaft parallel to the side members of the frame. The arrangement of clutch, gear box and transmission were established in the so-called Levassor-Panhard car, giving its designers a high rank among developers of the modern motor car. Early English cars were almost exclusively steam-propelled.

Meanwhile, in the United States, there had been much interest in self-propelled vehicles, but little progress, except for a few rather unsuccessful "steam buggies." In 1879 George Selden, a patent attorney of Rochester, New York, who had been experimenting with gasoline engines for several years, applied for a patent on a gasoline-driven automobile, the first such application in American automobile history. He kept his patent application alive until 1895, at which time he was granted sweeping rights, under which he collected hundreds of thousands of dollars from automobile makers until 1911, when his patent rights were narrowly limited by judicial decision.

In 1892 Charles Duryea and his brother Frank built and ran the first successful American-built gasoline car at Springfield, Massachusetts. It was introduced to the public the next year as a "horseless buggy," a type widely imitated in the next few years in the United States, ignoring the more modern designs of Panhard and Levassor, of a forward-placed power plant and a chassis constructed for beauty and convenience. Other inventors followed the Duryeas: Henry Ford in 1893, Elwood Haynes in 1894, R. E. Olds in 1895; the latter's curved-dash "runabout" becoming the first American car to be produced in quantity. At the World's Columbian Exposition at Chicago in 1893 the Benz car was displayed and may have stimulated inventors who were playing with the idea of making gasoline cars. In 1895 a race promoted by the Chicago *Times-Herald* and run off under exceptionally bad road and weather conditions was won by a Duryea "horseless buggy."

The late arrival of the United States as a leader in automobile development may be accounted for in several ways. Highways were poor, the public was hostile, financiers were exceedingly hesitant to risk their funds. Furthermore, there was a long argument in this country over which kind of power—electricity, steam or oil—would be best. After some experimentation in all types it became clear that steam required special handling; electric cars were never popular outside cities because of the constant need of battery charging. Moreover gasoline became the most readily accessible fuel. For twenty years, however, the American automobile industry had a career of continual failure. Hundreds of companies began operations, produced cars, then folded up. It was next to impossible to induce men of wealth to invest their funds to back the original inventors until after 1910. It is interesting to note that most of the pioneers of the industry were not capitalists; the Dodge brothers were bicycle dealers, as was John N. Willys; Walter P. Chrysler was a railroad employee; R. E. Olds was a machinist; Henry Ford a maker of cheap watches.

Ford must be regarded as the great individual genius of the motor car industry, for it was due to him, more than any man, that the industry directed its operations eventually toward mass production. Ford came to Detroit as a poor farm boy with the idea of making and selling cheap watches. Finding that this plan did not seem to succeed, he went to work designing engines, and late in 1893, from junk yards and scrap piles, he assembled the parts of his first car. He fitted four bicycle wheels together with gas pipe hubs; a two-cylinder water-cooled engine was made of cast-off pipe; between the wheels he attached an old buggy seat and dash-board; a tank for gasoline was fastened under the seat. The first Ford car had two speeds ahead, one of ten, one of twenty miles per hour. There was no second speed or reverse. A lever shifted the car from low to high. After a few weeks he sold the machine for $200 and started to build another.

Ford's next problem was to organize a company and produce more of his cars. In 1899 he induced a group to set up the Detroit Automobile Company and to advance him $10,000 to build ten of his cars. This arrangement was upset by quarrels between Ford and Henry M. Leland, a motor builder for the company, which led to Ford's resignation. With the latter out of the way, Leland went ahead and started the Cadillac Motor Car Company, which made and sold more than 2,000 cars its first year. Ford had meanwhile organized another company, only to break with his partners on the issue of high or low-priced cars, Ford holding out for the latter. In 1903 he tried again, and set up the Ford Motor Company, bringing into the firm a coal dealer, A. W. Malcolmson, and James Couzens, Malcolmson's bookkeeper. Both men later left the Ford organization, but Couzens held out longer, and eventually sold his shares for $29,000,000. The Dodge brothers also entered the Ford firm and rendered important service as parts makers. Later they organized a separate company and built their own cars. Their huge fortune was well-started by their early association with Ford.

The Ford Motor Company produced at first three types of cars, with prices up to $2,800. From June, 1903 to September of the same year, they made 1700 machines. Profits from the sale of these amounted to $250,000 but during the second and third years their sales dropped seriously. This seemed to Ford a justification of his belief that automobiles were priced too high, and that a one-model, cheap car was the best idea. Therefore he secured control of the Company by 1907 and began the manufacture of his Model T, which was to sell to the tune of fifteen million units before its discontinuance, at prices which ranged from $850 down to $290. He introduced many innovations on auto construction, most famous of which was his foot-controlled gear shift. He also moved the driving wheel over to the left side for convenience in passing other vehicles. The motor head of the Model T was detachable; cylinders were cast *en bloc* instead of singly; an alloy of vanadium steel was used. After twenty years of manufacture the Model T was abandoned, and Ford began the making of more conventional cars.

The changes in appearance, power development and comfort in the automotive passenger vehicle have been so great since the early days of Duryea, Ford, Haynes and Olds as almost to defy description. From machines of one or two cylinders, capable of developing a few horsepower, European and American engineers have evolved motor systems of eight, twelve and sixteen cylinders, of enormous speed and power. The completely open car of the early 1900's has all but disappeared; in its place came the completely closed car of the 1920's and after, which has gained universal popularity; tires were enlarged from bicycle size to six inches in diameter; the removable head engine was gradually adopted by all makers; lubrication was centralized; impressive improvements were made in spring action, seat and upholstery design; brakes were improved and safety glass was added.

Early automobile plants were to a large extent "assembly" organizations; they put together a large number of parts and accessories which they had purchased from other manufacturers; in this feature they resembled other assembly industries such as bicycle, watch, piano and shoe factories. The first automobiles were largely made by hand. Parts had to be fitted together by making shop alterations, in the early factories. This slow and costly handling made a very expensive product which could not be manufactured in large quantities at a price to fit the pocketbook of the person of average means. While the standardization of parts had long been customary in smaller manufactures such as firearms and watches, this was not a characteristic of the early automobile industry. The new product, which had obviously, by 1910, become a fixture of our industrial economy, was so very large and had so many intricate parts that the industry could not advance markedly until tools and machinery were provided to produce the hundreds and even thousands of identical cylinders, pistons, connecting rods, crankshafts, camshafts, crank-cases, axles, gears and frames to allow setting up production assembly lines to speed up the entire process.

To manufacture a cheap car in great quantity, there had to be a concentration on one or two types. For example, in the Ford plant, at first, only two types of chassis were produced: the pleasure car and the one-ton truck. Each pleasure car chassis was an exact duplicate of every other; the motor was the same for all cars, from sedan to delivery car. Standardized parts could therefore be made by the thousand without changing a pattern, readjusting a lathe, or resetting a machine. Even the testing and measuring of parts to make sure that they conformed to specifications became a machine-like process. Automatic machinery was devised to take care of the many steps to make parts, such as cylinders, alike to within a tiny fraction of absolute identity. In one large motor car factory, for example, 800 operations came to have a permissible range of error of only 0.0005 to 0.00025 of an inch.

Most automobile manufacturers for many years made only a small number of the parts used in their cars, because only a very large and highly mechanized factory can make cheap items profitably. This idea may be illustrated as follows. Suppose, for example, that Ford, Chrysler or General Motors should decide to make their own tires to equip their new cars. To do so, they would have to create vast tire factories to manufacture tires not only for new cars, but also for the thousands of cars already in use, in order to operate profitably. But the business of these firms is to make cars, not tires. Furthermore, the rubber business is itself a highly specialized and very large industry. Therefore, the most efficient practice turns out to be the purchase of tires and many of the other accessories for their cars, and to allow the specialists to replace them and supply the continual outside demand. Such a practice does not destroy the individuality of the finished product, for the parts manufacturer, whether of tires, carburetors, speedometers, or whatnot, has to conform to the specifications as to size, quality and performance set by the automobile manufacturer.

The actual assembling of the finished car requires a plant laid out with the utmost care, so that parts and even pieces of parts, may circulate in an orderly and progressive way through the various departments. This calls for time schedules for the various operations, arranged according to the complexity of each one. At one end of a long conveyor belt, or moving platform, the parts begin to move through the assembly line from station to station, at each of which a new part is added, or several parts fitted together. At successive stations, for example, the partly-assembled engine is lifted from the track or platform, a part affixed, then replaced upon the conveyor and moved to the next station. When the assembled items become too heavy to lift, they may be slid along metal-faced benches; when an engine assembly is complete, it is lifted up by an overhead conveyor and carried to the department where the final assembling of cars is accomplished.

Thus, by standardization of parts and the highly developed assembly line, manufacturers were enabled to produce cheaper and better motor cars. Such mechanized processes did not prevent improvements or freeze, permanently, obsolete designs. Competition took care of that. In the ten years from 1925 to 1935, one manufacturer increased the wheelbase of his product twelve per cent, its maximum horsepower 350 per cent, its weight (in the sedan model) forty-two per cent, its deceleration maximum seventy-five per cent, and its maximum speed 100 per cent. Quantity production, backed by progressive engineering, also gave the car of 1935, as standard equipment, such improvements as anti-rattle devices, braces, dash gauges and meters, door locks, shock absorbers, windshield wipers, safety glass and a counterbalanced crankshaft, items which had been available ten years before, but only as extras at added cost.

After the first period of assembling cars was past, a trend began among the larger manufacturers of making their own engines. By 1914 the percentage of engines made by the automobile makers themselves was 87.5, a figure which rose to 97.8 by 1937. During the first forty years or so of the industry, roughly from 1899 to 1937, the total output of new cars rose 1800 per cent. In 1904 only 22,000 of all types were produced; by 1929 this figure increased to 5,293,000 units. While these increases were mostly in the pleasure car field, the auto truck began to be important by 1919, when 120,000 were produced, a figure which grew to 600,000 by 1937. After the restrictions of 1942 through 1945 were past, automobile production picked up again to reach an all-time high in 1950 of eight million vehicles. Of these, 6,655,000 were passenger cars and 1,337,000 were motor trucks.

A good idea of what has happened more recently in the motor car industry may be illustrated by a glance at one large corporation, such as General Motors. Organized in 1908 as a means of bringing together several small companies which made cars, GMC is now a gigantic combination of many companies, some of which manufacture only motor cars, others engines, still others ignition systems and their parts, transmission units, ball bearings, carburetors, lamps, radiators and so on. One hundred factories scattered in sixteen different states employ almost half a million workers. There are also factories and assembly plants in a number of foreign countries. In the year 1950 the Corporation produced 3,653,358 vehicles.

The products of General Motors are not confined to the automobile industry, but include Diesel locomotives, airplane engines and parts, electric refrigerators, oil burners, gas furnaces and many other items. At the present time a considerable part of its productive capacity has been converted to defense industry, of which more will be said later. In 1950 the Corporation received a gross income of more than seven and a half billion dollars. For its complex peacetime operations, it relies upon more than 12,000 other companies, large and small, for services, materials and parts; to these it pays annually about three and a half billion dollars. Thus the trend toward consolidation, as noted in connection with the railroad industry, has proceeded very far in the motor car industry as well. The great bulk of production today is carried on by the "Big Three": General Motors, the Chrysler Corporation, and the Ford Motor Company, though several large independent firms still manufacture cars. The most important of these are the Nash, Hudson, Packard and Studebaker concerns.

Vital to the development of the motor car as a factor in

transportation has been the evolution of a great system of improved highways. Little was done about this problem until World War I. Some states, as for instance New Jersey and Massachusetts, had begun state highway systems quite early. In 1916 and again in 1921 the Federal Government entered the field, with financial assistance to the states for materials and labor, on the condition that they set up state highway agencies to build and maintain the proposed systems. During the 1930's these grants were supplemented by further aid under the National Industrial Recovery Act, and during the four years from 1933 to 1937 a billion and a half dollars was spent by the United States for roads and grade crossing elimination. The enormous increase in cars in recent years has created serious problems of maintenance and expansion of these road systems, which by 1936 had reached a total of 340,000 miles. The burden has been mostly borne by the users of the highways, in the form of taxes on gasoline by both the Federal Government and the states, which as early as 1930 accounted for nearly four-fifths of the expenditures on state highways. By the end of 1949 the total mileage had increased to 571,753. In addition, the Federal Government controls and maintains some 70,000 miles of U. S. highways.

The motor bus, with its more than nine billion passengers in 1950, has filled the place once occupied by the city and interurban trolley lines and has made serious inroads on the passenger business of the railroads, especially over their shorter runs or branch lines. The motor truck, which had been of such small importance in the early days of the automotive industry, has grown to be a vital part of our transportation system. During World War II, government regulations permitted its manufacture in greater numbers than other cars; by the end of the war there was an incease of nearly fifty per cent in truck registrations compared to the pre-war totals. The bulk of trucking, despite the operation of large transcontinental trucking firms, has remained a business of individual owners covering relatively short distances; in the mid-1940's the 4,500,000 trucks in service were owned by more than three million individuals or small trucking firms.

It is quite clear that during the past twenty years the growing efficiency of passenger cars, buses and trucks, plus the immense increase of well-maintained hard surface roads, has developed a new transportation system. The motor bus, integrated into large long distance lines, runs as many miles as do the railroads. The individual passenger car has become the principal means of individual travel. The motor truck, though carrying in 1931 only about one-twelfth as much freight as the American railroads, has since then made serious inroads upon railroad freight business, due to the greater ease with which it handles small shipments. Government regulation of automotive freighting, beginning in 1935, and the financing of new highways by users, have put a heavy burden on this type of freight transportation. The decline of railroad revenues by nearly a billion dollars in the freight category, and of a hundred million dollars in passenger receipts, between 1948 and 1949, may reflect the growing seriousness of automotive competition.

The most important fact about automotive transportation is the tremendous stimulus it has given to large scale manufacturing. Only second to this has been the importance of providing the individual with greater range and convenience in his personal travel. In the replacement of the truck-horse and the delivery-horse by automotive equipment, there has resulted a decided speeding-up in the distribution of agricultural and industrial produce. The motor truck has supplemented the railroad system, because it can go many places where the railroad cannot run. The application of the internal combustion and Diesel engines to many types of farm machinery has been an important factor in the mechanization of farming and has added greatly to the productive levels of agriculture. Similarly the addition of power units such as bulldozers, graders and scrapers to roadbuilding has speeded the construction of great arterial highways and the de luxe turnpikes of America which enable them to bear a far greater volume of travel than ever before in history.

CHAPTER FOUR

TRANSPORTATION IN THE AIR

The airplane is the greatest foe of isolation and complacency ever devised by man. Its immense range of flight necessitates a familiarity with the places far distant from its home base. The steady advance of powered flights into outer space has suggested some of the most daring and far-reaching of scientific experiments, which even include communication with, or flight to, other planets. No invention prior to the developments in nuclear physics has presented such hopeful possibilities to mankind, on the one hand, or such dire potentialities, on the other, as the planes of the mid-twentieth century. Under skillful international control, the conquest of the air would tend to knit all nations into a single society. Yet if supreme control of the air should be concentrated in the hands of a single nation under selfish and irresponsible leaders, a terrible danger would loom to free peoples everywhere. Moreover, a combination of supreme air power with superiority in atomic weapons (which would probably be air-borne and air-discharged) might lodge in the hands of such a controlling force the political and military dominance of the world.

Thus the airplane, as the ultimate device in the achievement of fast transportation, suggests both the best and the worst of future trends. While prediction, whether of scientific progress or of political change, is always hazardous, it would seem that any such monopoly of both air power and atomic energy by a single great power is extremely unlikely. The history of invention and of human progress suggests that most of the important steps which might lead in that direction have been so widely distributed as to make any such monopoly of destructiveness improbable. Human ingenuity has never been concentrated in one place, and the process of scientific industrialization has been fairly widespread. However, the speed of the airplane to supersonic levels, plus the availability of such planes as atomic carriers, makes an air-atomic war of maximum destructiveness a definite possibility.

Men Attempt to Fly

Human beings have nearly always expected someday to fly. While most of the basic inventions which have made air transportation possible have come in the past fifty years, it is interesting to note that both kinds of air flight, lighter-than-air and heavier-than-air, were current in classical legend and in the writings of the ancient and medieval philosophers. The Roman poet Ovid told the story of Icarus, whose wax wings enabled him to escape from the tyrant of Crete; but the sun melted the apparatus and Icarus fell in the sea and was drowned. Yet theorists were impressed by the legend, and the apparatus of Icarus became a prototype of the "ornithopter," a device by which a man could fly like a bird. Leonardo da Vinci, in his *Flight of Birds,* which appeared in Florence in 1505, described with much detail just how such a device, with bat-like wings, could lift a man from the ground; how he could raise himself higher by more flappings, maintain his altitude by spreading his wings, and eventually glide downward as a bird does.

Despite the fact that none of the early projectors of the ornithopter idea built any models to demonstrate the truth of their beliefs, descriptions of this sort of "birdman" contraption appeared frequently in the semi-scientific writings of Europeans for several centuries. Jacob Degen of Vienna developed the idea further in the nineteenth century, as did Sir George Cayley, a British scientist, Otto Lilienthal of Germany, and several others.

Similar suggestions appeared in the United States. Richard Davidson's eagle-like device, which he described in 1841, was supposed to enable a man inside it to operate the wings by a system of gears, wheels and cranks. While Davidson never actually constructed a model, he held to his notion in later writings, and seems to have inspired a number of United States patents for ornithopters; one was granted to Watson F. Quimby of Wilmington, Delaware in 1869; another to Francis X. Lamboley of New York in 1876; to Melville M. Murrell in 1877; and to Charles F. Myers in 1882. All of these seem to owe much to da Vinci's early drawings, though Murrell's wings, in his patent application, were more like Venetian blinds than birds' wings.

One of the most fantastic flying devices was proposed by a Portuguese friar, Bartholomeo de Guzmao, in the early eighteenth century. His airship was to be built of iron, the hull filled with chopped straw. Across the deck was to be stretched a horizontal sail to catch the wind set in motion by bellows in the hold of the ship. Upon the deck he proposed to place two bales of lodestone, or magnetic iron, to attract upward the iron hull of the deck, above which he would suspend several pieces of amber. The amber, heated by torches, would generate static electricity which would attract the straw upward toward the sail, which would already be pulling the craft upward by the bellows action and the attraction of the lodestones on the deck. Even with all these helpful forces working, Fra Bartholomeo seems never to have made a successful flight, though King

John of Portugal granted him an exclusive patent, the first of its kind, in 1709. But his plans at least gained him notariety, for they landed him before the Holy Tribunal on a witchcraft charge in 1717, as a result of which the poor friar's documents were burned and his person lodged in jail.

Lighter-than-air flight, more properly called "aerostation," has had a longer operational history than aviation proper, or heavier-than-air flight. This is perhaps a very natural circumstance, for ballooning seemed to offer a more obvious and practical means of flying, namely, by attaching a heavy object to a captured mass of gas more buoyant than the surrounding air. It was also natural that aerostatic success should come in the late eighteenth century, contemporary with the understanding of the properties of various gases through the work of Priestley, Scheele, Cavendish and Black. The first practical experiments with balloons began in southern France with the Montgolfier brothers, young students of natural science. After much work with inflated paper bags, they built a hot air balloon which carried aloft a fire-pot filled with ignited shavings, chips and straw; the heated air from the fire-pot carried the balloon more than a mile in the air, it was estimated, before it cooled and lost its buoyancy. This demonstration, held in June, 1783, was witnessed by a large crowd and caused wide discussion. As a result, the Montgolfiers were summoned to Versailles by Louis XVI, before whom they sent up another balloon which carried a basket in which were a sheep, a duck and a chicken. The flight lasted about eight minutes, in which the apparatus rose to about 1500 feet. Besides the King and Queen and their attendants, an interested spectator was Dr. Benjamin Franklin, representative of the United States at the Court of France.

Progress in aerostation became quite rapid thereafter. Parachutes had been demonstrated successfully by the Montgolfiers in 1779, and in 1797 André Garnerin became the first human to parachute to the ground successfully. Professor J. A. C. Charles contributed much to ballooning by introducing hydrogen into the gas bag, as a substitute for the hot air used by the Montgolfiers and other early balloonists. Charles also suggested silk as a stronger material for the bags, evolved a valve to release excess gas, and devised the spreading of netting over the balloon surface to distribute the weight of the basket or carriage below. The carrying of ballast became a common feature of the improved balloons. After these improvements, flights with passengers became quite common. On one occasion Charles himself flew a mechanic named Robert from Paris to Nesle, a distance of twenty-five miles, landed his passenger there and later rose accidentally to a height of two miles, a most hair-raising world record.

The early English balloonists were less successful than the French, though a number made attempts to fly in the 1780's. Several foreigners, however, made important flights wholly or partially in the British Isles. An Italian name Lunardi made a very successful trip, witnessed by the Prince of Wales, in 1785, covering twenty-five miles. A detail added by Lunardi, and used also by Jean Pierre Blanchard, who was about to make aerostatic history, was the placing of a set of aerial oars in the balloon carriage, which could be manipulated to increase horizontal motion. Perhaps the most eventful of the early balloon ascensions was the crossing of the English Channel by Blanchard, a French aeronaut, with Dr. John Jeffries, a Boston physician, in 1785. Starting from Dover, England, they succeeded in reaching a point in France about twelve miles inland in less than three hours. The achievement made a tremendous sensation, and resulted in royal honors for both men. It is also interesting to note that Blanchard had experimented earlier with heavier-than-air devices, and with movable wings and rudders on other balloons. He is also an important individual to Americans, not only because of his collaborator Jeffries, but because he later (1793) began the most extensive series of demonstrations so far carried on in the United States. The first of his trips began in Philadelphia and ended forty-five minutes later across the Delaware, about fifteen miles from his starting place. President Washington witnessed the ascension, and just prior to the take-off presented Blanchard with a passport letter, commending the flyer to the good offices of all American citizens. This certificate came in rather handy a few minutes later, when Blanchard came crashing down in the midst of a crowd of outraged Jersey farm folk, who could not understand a word of his explanations.

Some military use was made of balloons in this period, but at the Battle of Fleurus in 1794 such use was limited to an attempt to reconnoitre the enemy's position. A far more interesting chapter in military ballooning is to be found in the adventures of T. S. C. Lowe, a skilled balloonist, who was attached to General McClellan's Army in the Peninsular Campaign against Richmond, during 1862, in the American Civil War. Several writers, notably General A. W. Greely, at one time Chief Signal Officer of the Army, regarded Lowe's work as extremely helpful. Greely declared that Lowe was able to give vital information about Confederate troop movements at the Battles of Fair Oaks and Gaines' Mill. A modern touch in Lowe's reconnaissance work was his carrying aloft of a telegraph wire and a Morse key, by which signals were sent down to a Signal Corps unit stationed on the ground. However, his attachment to the Union Army did not last long, and for most of that struggle the North, and the South as well, had to do without the advantage of aerial reconnaissance. Some use was made of the balloon in the siege of Paris in 1870-71 and by the British Army in East Africa in the 1880's. But aerostatics did not figure largely in any military operations during the century.

Civilian aerostatics during the early days consisted mostly of exhibitions before large crowds or of some rather desperate than two hundred balloon flights, on one journey in 1859 travelled over 800 miles with three companions from St. Louis to Henderson, New York. This distance record stood until 1900 when Count Henry de la Vaulx almost traversed Europe in a balloon; starting from Paris he reached a point inside Russia almost 1200 miles from his starting point. Attempts to fly the Atlantic, usually from Canada, were frequent and disastrous, many of them fatal. The essential difficulty with the free balloon as a medium of travel was its sensitivity to weather conditions and its lack of controlled motion. The French, notably General Meusnier, provided one improvement upon the older balloons, by devising, even before 1800, an elongated gas bag; manual propulsion of a propeller gave such a balloon a speed, however, of only three miles per hour. Many years after, these designs were greatly improved by experimenters such as the French clockmaker Pierre Julien, and the French engineer, Henri Giffard. Under a patent dated 1851, from the French Government, Giffard flew a cigar-shaped dirigible 143 feet long and thirty-nine feet in diameter. His balloon could not be flown against the wind, however, due to the low power of a coke-burning engine, which was suspended beneath the gas bag. But he had demonstrated the practicability of the dirigible balloon, driven by mechanical power.

After gradual improvement in the design, engine power and speed of dirigibles by French, German and Austrian inventors after 1871, a young German officer, Captain (later Count) Ferdinand von Zeppelin, began what were to be the most important experiments in the history of lighter-than-air craft. Zeppelin had been a balloon observer in both the American Civil War and the Franco-Prussian War. In 1898, after much research, he organized a company to make rigid airships of his own special design. The first of these craft, completed in 1900, was a pencil-shaped balloon, 416 feet in length. Sixteen metal girders ran longitudinally to form a frame, which was then divided by aluminum partitions into sixteen compartments. In all except the fore and aft sections were hydrogen bags; over the hull framework was a fabric cover, with air spaces between the balloonets and the outside coating. Suspended close to the keel were two gondolas, each containing a sixteen horsepower Daimler gasoline engine. The complete balloon weighed nine tons. Despite the difficulties in building and operating these huge ships, Zeppelin flew them many times, attaining speeds of twenty miles, or better, per hour. The German Government, moreover, came to his aid, enabling him to construct larger and faster Zeppelins.

In France, at about the same time, a young Brazilian enthusiast, Alberto Santos-Dumont, was exciting all Paris with his stunt flying. Beginning with an engine taken from a motor bicycle and fitted to a balloon of his own design, he flew it successfully in 1898. In the next few years he built and flew fourteen ships of both rigid and non-rigid types in many spectacular flights. On October 13, 1899, he circled the Eiffel Tower several times in a gas-propelled machine; two years later he accomplished a flight from Saint Cloud to Paris, again circled the Eiffel Tower and flew back to his starting point, winning thereby a prize of $20,000. In 1899 the Lebaudy brothers, sugar refiners, sponsored the building of a curious fish-shaped dirigible, the first to be covered with double rubberized material. Julliot, an aeronautical engineer and Surcouf, a balloonist, built the machine, fitted it with a forty horsepower motor and made twenty-nine successful trips in it within a year. By 1906 both the French and the German Governments had adopted the airship as military equipment and several companies were already engaged in constructing them.

Heavier-than-air Flight

The theoretical basis for the flight of heavy bodies in the air was ably stated in 1809-1810 by Sir George Cayley, the well-known English scientist. He proposed that plane surfaces might be adequate to support the weight of an engine and a pilot by the application of power to the resistance of the air. While Cayley, first of all scientists, suggested early the principles which have governed all successful heavier-than-air flight, it took nearly a century of experimentation to solve the problems of aerodynamics which had to be accomplished before successful power flights could be made possible.

Cayley's principles were used by Stringfellow and Henson in 1843 in an attempt to construct a steam-driven airplane, which they failed to make fly, but Stringfellow is reputed to have flown a power-driven model plane in 1848. Research in aeronautics and aerodynamics was furthered by the published studies of the Royal Aeronautical Society after its establishment in 1866, and by the experimental work of some of its members; Thomas Moy made model planes and Francis Wenham worked out data regarding curved or cambered plane wings. An aeronautical exhibition held in London in 1868, sponsored by the Society, brought out models of seventy-seven airplanes, engines and airships.

Balked by the inability to produce engines of low enough weight per horsepower to fit the needs of a large plane, experimenters during the last third of the century turned to efforts with engineless planes, or gliders. Such attempts were not new, but were pursued with more intensity in the last years of the century. A Frenchman named Le Bris obtained startling results with bird-shaped gliders. Most important of all the glider experts was Otto Lilienthal, of Pomerania. He and his brother Gustav began to experiment with gliders and ornithopters, as boys, in the 1860's. Continuing alone after the Franco-Prussian War, Otto collected information on aerodynamics, designed cambered wing surfaces, studied the flight of birds. He also made more than 2,000 glider flights, some of them of several hundred yards in length. His death in 1896 from a glider accident may have delayed the achievement of power flights, for Lilienthal had already built a plane and was planning such an attempt.

Lilienthal's career was paralleled by another remarkable glider expert, Percy S. Pilcher, who was also a victim of a crash in his huge glider, the *Hawk,* which had a wing span of 180 square feet. Pilcher was likewise preparing to add power to his flights when his death intervened. John J. Montgomery, an American, made even more spectacular glider flights than either of the others. He launched his craft from cables stretched between mountain tops and from balloons. Very valuable also was the work of Octave Chanute, an American of French origin, in the gathering of data as to air currents and velocities. He designed an excellent biplane glider and made many flights along the shores of Lake Michigan. He was the principal founder of the American Aeronautical Society, which was organized in Boston in 1895, and contributed markedly in advice and encouragement to the success of the Wright brothers, in their glider experiments and in their final successful man-borne, power flights.

At the turn of the century, and even for several years earlier, a frantic contest was going on among the flying enthusiasts to be the first to construct and fly a heavier-than-air, power-driven, man-carrying machine. While none ever did so, prior to Wilbur and Orville Wright, several important attempts came close to success before 1903, and collectively had much to do with the success of the Wrights. In 1871 the Frenchman Alphonse Pénaud, after various experiments, flew a small model airplane, powered by rubber bands, before the Société de Navigation Aérienne; he later designed, but did not fly, a full-sized plane of the same type. An Australian, Lawrence Hargreave, performed many valuable experiments with ornithopters, box kites and engines designed for use on planes. With a box kite, which he constructed in 1893, he discovered that curved surface cells, convex side up, pulled almost twice as hard as those with flat surfaces; Hargreave, like Pénaud, used rubber bands for some of his models; for others he employed clock springs and compressed air. Horatio Phillips, an English experimenter, after much research on wing sections in wind tunnels, constructed a steam-propelled plane which actually rose a few feet during tests in 1893. Clément Ader, a Frenchman, constructed several steam-driven planes, attempted to fly them in various trials in the 1890's, and in 1897 actually succeeded in getting his third plane, the *Avion,* to fly clear of the ground, though the distance of flight claimed by his friends does not agree with the report of French military authorities who witnessed the trial.

Sir Hiram Maxim, a most distinguished English scientist, tried out a large plane which measured fifty feet across the wings, driven by two steam engines of some 360 horsepower each; in the first of two tests, both in 1894, the plane, riding on iron rails placed about nine feet apart, was lifted clear of the track; in the second trial the plane rose higher than at first, only to crash to earth after its brief rise. Even more important and nearer to success than Maxim's elaborate attempts, was the long-continued effort of Professor Samuel P. Langley, an American astronomer. As early as 1891 he had published the first systematic treatise on aerodynamics; the basic elements of thrust and air resistance were expressed therein in definite mathematical terms. He built many model planes, which he called "aerodromes" and made several successful tests of them. One of his models, powered by a light steam engine, made a flight of nearly a mile over the Potomac River in 1896. Having attracted the attention of the War Department by his experiments, Professor Langley was officially urged to construct an "aerodrome," driven by power and capable of carrying a man in flight. A government fund of $50,000 was placed at his disposal; other funds were also available for his use, through his official position as Secretary of the Smithsonian Institution.

With an assistant named Manly, Langley worked for five years to construct his machine; late in 1903 the two had finished a plane of steel and fabric, driven by two propellers connected with a fifty horsepower gasoline engine. Two attempts were made to launch the machine by catapult, but both tests failed and the experiment was abandoned. It may have been that the method of launching was the principal fault, for in 1914 the Langley plane, repaired and altered so that it could be launched from floats instead of by catapult, made several successful flights at Hammondsport, New York under the supervision of the aeronautical engineer Glenn Curtis. The particular claims made for this plane, as published in a report by the Smithsonian Institution, gave rise to a long controversy which did not enhance the reputation of that famous scientific foundation.

The background of successful heavier-than-air flight illustrates what has become noticeable in the course of scientific invention from the earliest glimmer of an idea to the point at which it promises to have industrial value. It is impossible to give complete credit to a single individual; furthermore, it is clear that every bit of serious investigation may contribute to the final result, as the import of each test, even though unsuccessful, is comprehended by the next experimenter. From each idea, even Pénaud's with his rubber bands, came a new suggestion which someone would eventually use. For each attempt to achieve sustained flight which failed, there were definite reasons for the failure, whether by Maxim, Ader or Langley. It was almost inevitable that by building on one another's failures, some would eventually acquire enough knowledge of wind resistance, plane design and suitable power delivery to accomplish sustained flight. But without the long period of experimentation and near-success before the Wrights, their own achievement would have been impossible.

The Wright Brothers Turn the Trick

No one can understand the services done to aviation by Wilbur and Orville Wright without delving into the background of their triumph of December 17, 1903. Readers of Fred C. Kelly's recent book on the Wright brothers and his subsequent volume of their correspondence receive a most rewarding experience. The author has been able to convey, in his biography, how it was that these two boys of very little formal education were able to accomplish results which brilliant scientists and engineers had failed to achieve. Despite a lack of formal training in science, the brothers were both of a studious turn of mind. They also possessed mechanical aptitude, and both became confirmed and resourceful "tinkerers." They built printing presses, made and repaired bicycles, and designed wind tunnels and engines. Both were types of individuals who have again and again discovered ideas or devised machines of the greatest value to American industry.

Wilbur and Orville may have received their first inspiration to tinker with flying machines from the gift of one of Pénaud's model planes from their father. The accidental death of Otto Lilienthal in 1896 seems to have turned their interest to the possibilities of engineless planes, for soon thereafter they began to read absorbingly all that they could secure on the subject of gliding and flying. They read Octave Chanute's *Progress in Flying Machines,* Langley's *Story of Experiments in Mechanical Flight* and his *Experiments in Aerodynamics.* They obtained and absorbed the *Aeronautical Annuals,* for 1895-1897, edited by James Means. They sent for the reports of the Smithsonian Institution and obtained from Washington a number of papers by Lilienthal in which he described his glider experiences. Besides their avid study of these materials, Wilbur and Orville began to experiment with kites, in order to learn as much as possible about aerial equilibrium. In 1900 Wilbur Wright wrote Octave Chanute in Chicago of their plan to experiment with a man-carrying kite which should have warped wings for lateral control, and an upper surface which could be shifted backward and forward for longitudinal control, principles which they were already observing in the experimental kites they were flying near their Dayton bicycle shop. Chanute was enthusiastic and became from this time a strong supporter of the Wrights in all their experiments.

In the same year, 1900, they constructed their first glider. It had two wing spans trussed with wire; each wing was eighteen feet long and five feet wide. A small plane in front could be shifted up and down to control the plane's movements. Upper and lower wings were curved at the edges, were made of white spruce and covered with white sateen. At the suggestion of the Weather Bureau, after inquiries about a place where wind velocities would be at least eighteen miles per hour, they chose a remote stretch of beach near Kitty Hawk, on the eastern coast of North Carolina, for their first flying attempts.

The first glider experiments at Kitty Hawk were not a striking success. Winds were either too strong or did not exist. It was also found, contrary to their calculations, that while winds of seventeen to twenty-one miles per hour might be sufficient to support a glider with a pilot aboard, it would need a far stronger wind to raise it in the first place. They first tried to fly the glider as a kite, with about fifty pounds of chain aboard in place of a pilot; though results of this test were promising, they felt that far more should be accomplished. Their attempts to glide down a hill resulted in flights of a few seconds only. Thus ended the experiments of 1900.

The Wrights returned to Dayton ready to construct a bigger and better machine. This plane, when ready, had a lifting area of 290 square feet compared to 165 in the first one. The curvature of the wings was increased, the span lengthened to twenty-two feet, and the front "elevator" was enlarged to eighteen square feet. At the suggestion of Octave Chanute, several other amateurs in aeronautics, including Chanute himself, attended the 1901 tests. At Kitty Hawk, after a few tuning-up trials, they made a glide of 315 feet in nineteen seconds. Though in most ways their new apparatus was better than the other, it did not move in as level a manner while gliding along a slope, perhaps because of the excessive camber, or convexity of the wings.

With still more alterations, they returned to Kitty Hawk again in 1902 with a machine with which they made more than a thousand glider flights, some of them more than six hundred feet in length, and in some cases against a thirty-six mile per hour wind. With such impressive results, the Wrights now felt that with changes in the controls, including connection of the mechanisms which controlled the wing-warping with those which moved the tail, they were ready to build a successful power flyer.

The next year, 1903, they returned to Kitty Hawk with a glider biplane and an engine of their own design. They had tried without success to induce several automobile makers to build them an engine which would weigh not more than twenty pounds per horsepower. Only one engine existed which seemed to meet their requirements; when it proved to be of insufficient power, they decided to make one themselves. Their engine weighed only seven pounds per horspower, at a time when automobile engines weighed several times that ratio. Their success in this particular was therefore, in itself, an engineering achievement of great importance.

After various misfortunes, including a broken propeller shaft, which necessitated a return to Dayton, their first power flight was attempted on December 14, 1903, with Wilbur at the controls, while Orville remained on the ground to keep the wings in balance. The motor was started, the restraining wire slipped, and the plane began to run down a monorail track which they had provided to help the take-off. After about forty feet of ground travel, the plane lifted from the ground a few feet, then stalled, swinging around and digging into the sand. After two days spent in making repairs of the damage caused by this misfortune, they were ready for a further test on the 17th.

This time Orville was at the controls, with Wilbur ready to run alongside the wing. Unlike the previous trial, made in calm weather, this start was made in a twenty-seven mile wind. As before, the machine lifted into the air after about a forty foot run; the course of the flight was erratic, due to irregularity of the wind, lack of experience by the pilot in handling the machine, and the apparent unbalance of the rudder. While the actual flight was very short, the brothers computed that, by adding the speed of the machine against the wind to the speed of the wind, Orville's flight of twelve seconds was equivalent to one of 540 feet made in still air. As Orville Wright himself stated afterward, regarding this test:

> This flight lasted only twelve seconds, but it was nevertheless the first in the history of the world in which a machine carrying a man had raised itself by its own power into the air in full flight, had sailed forward without reduction in speed, and had finally landed at a point as high as that from which it started.

A second flight on that day carried the machine a greater distance for about the same duration. The third trial carried the plane over 200 feet in about fifteen seconds. A fourth, and most successful of all, lasted for fifty-nine seconds and covered 852 feet. While the Wrights and several of their companions were standing near the machine, discussing plans for further flights after some minor adjustments and repairs, a sudden gust of wind turned the plane over and damaged it seriously, so that further tests proved impossible that year.

One of the strangest things about the tremendous event of December 17, 1903 is the small amount of public attention it received. The Dayton papers, though informed of it, at first ignored the matter altogether. After a Cincinnati paper came out on the day following the flight with a full page notice, the Dayton afternoon paper was stirred sufficiently to run a piece on a back page, headed: "Dayton Boys Emulate Great Santos-Dumont." Santos-Dumont, as one recalls, was a balloonist, and had never attempted heavier-than-air flight! Nowhere did anyone intimate that the flight was the first of its kind in history. It may be that the quiet and unassuming nature of both the Wrights deprived them for many years of the proper attention to their achievement. They did not seek publicity. To them it was perfectly apparent, as they returned to Dayton for more years of experimentation, that the age of airplanes had not yet arrived. The curious set of circumstances by which Professor Langley, deserving though he was, came to be regarded as the true American pioneer of heavier-than-air flight, may have contributed to the general ignorance shown by Americans and the aeronautical world in general, of the priority of the Wrights' achievement.

Back in Dayton the brothers spent the winter building an improved plane to demonstrate flying to the public of their home area in the spring of 1904. But the demonstration was a failure, and scepticism as to their reputed flights increased. Yet in 1905 in a meadow near town they made repeated flights, all visible from trolley cars which ran by the field. One of these flights was twenty-four miles in length. Their speeds at this period averaged about thirty-eight miles per hour.

On May 14, 1908 the Wrights flew a plane at Kitty Hawk which represented five years of continual experimentation and improvement, and for the first time, Orville Wright, who piloted the machine, carried with him a passenger. Later in the same year Orville conducted trial flights at Fort Myer, Virginia; the United States Army, after some delay, became interested in what the Wrights could do. The Army trials were successful, though in one of these flights, on September 27th, Orville crashed his plane. Lieutenant Selfridge, an Army observer who rode with him, was killed and Orville Wright was seriously injured. His brother Wilbur had meanwhile gone to France to further a possible contact with the French Government, but was meeting with disappointing delays.

Patents on the Wright machine had been applied for on March 23, 1903, but were not granted until May 22, 1906. When Wilbur reached France in 1908 he found there tremendous interest in flying. Louis Blériot had begun making flights in 1907, and so had his close rivals, Henry Farman and Leon Delagrange. It appeared that Octave Chanute, after sharing the confidence of the Wrights, had gone to France and had divulged rather freely the precise ways in which the Wrights had achieved balance with their plane. This information was eagerly picked up by French aeronautical enthusiasts and shortly thereafter they began to make real progress in heavier-than-air flight, though the name of the Wrights was scarcely known. Wilbur sensed that if he were to secure government contracts and recoup the family fortunes, he would have to move quickly. With a plane ready for flying tests, and a field provided by Leon Bollee, an automobile manufacturer, near his factory at Le Mans, Wilbur Wright took to the air and very quickly disposed of the incredulity of the French. On September 21, 1908, after a number of shorter flights, Wilbur Wright smashed all distance records by covering 60.85 miles in an hour and thirty-one minutes. Later he ascended with a passenger and remained in the air for an hour and ten minutes.

France at once acclaimed Wilbur Wright and so did high governmental authorities of most of the European countries. Prizes for altitude records and contracts for planes were soon forthcoming. Later exhibition flights at Pau, in southern France, were witnessed by hosts of notables, including the King of Spain and the King of Great Britain, Edward VII. Later flights were made in Italy and in Germany; in one of the latter the German Crown Prince was carried as a passenger.

Between the earlier and later European tests, Wilbur returned to the United States, met Orville, who had meantime recovered from his injuries, and the two successfully conducted the final Fort Myer trials. On June 30, 1909, Orville, accompanied by Lieutenant Foulois of the Army, flew a plane to Alexandria and back, meeting the conditions of the test and qualifying for a bonus of ten per cent for each mile per hour over forty. The United States Government formally accepted the Wright plane on August 2nd, 1909, at a price, including the speed bonus, of $30,000.

As a result of the signal success of the Wrights in their various demonstrations, companies for the manufacture of the Wright plane were established in France and Germany. On November 22, 1909, the Wright Company of the United States was incorporated, with a capital of $200,000. By the articles of incorporation, the Wright brothers received stock and cash for rights to their patents in this country and a ten per cent royalty on all planes sold. The Company would henceforth bear all expenses arising from suits against patent infringers. With an office in New York and a factory ready by November, 1910 in Dayton, a new and very important American industry was born.

The Development of Commercial Aviation

Even with the basic achievements of the Wrights between 1903 and 1909, the place of the airplane in modern economy was at first extremely uncertain. Its military possibilities appealed at once to William II of Germany, as they surely did to the French. The United States Army, having acquired a Wright plane in 1909, presumably intended to institute a training and experimental program within the Signal Corps. Beyond such developments as these, all in their infancy, of what use could airplanes be to the general public?

The Wrights, who by now had entered the manufacturing field, after intensive work in designing and flying the new contraptions, must have pondered this question deeply. They came to the conclusion, apparently, that they could not operate even a small factory profitably by selling planes for private use; they felt rather that they could only show a profit by sponsoring public exhibitions and demonstrations. Yet in order to have pilots to fly the exhibition planes they planned to make, it was necessary to train men for this work. So it was that they instituted, first at Montgomery, Alabama, and later at Huffman's Field near Dayton, training schools for prospective pilots. The factory made the planes, the pilots flew them in exhibitions, and the Wright Company collected a rental for each plane used. Flying demonstrations were given at Indianapolis, where five planes were used; at Atlantic City, where wheels were first used on the machines for take-offs and landings; along the Lake front near Chicago in 1910; and at Belmont Park, New York, in the same year. At the latter place both foreign planes and those made by the Wrights participated. In 1912 Wilbur Wright died very suddenly; in 1915 Orville, who had succeeded his brother as head of the Company, sold his holdings to a group of capitalists headed by William Boyce Thompson and Frank Manville, and lived thereafter in retirement until his death in 1938.

The commercialization of flying exhibitions, begun by the Wrights, was carried still further in the years just prior to World War I, not only by pilots trained by the Wrights, but of course by many other flyers, both European and American. In 1909 an aviation meet at Rheims, France, called forth thirty-eight contestants. In the same year Louis Blériot spanned the English Channel and Count de Lambert flew from Juvisy Field around the Eiffel Tower and back again to his starting point. Henry Farman attempted to match Blériot's Channel feat, but crashed into the water. Back in the United States, Glenn Curtis won $10,000 by flying, non-stop, from New York to Albany in 1910. Lord Northcliffe, the famous British newspaper owner, offered a prize of $50,000 to the flyer who would first cross the Atlantic, while the Directors of the Panama-Pacific Exposition put up $150,000 for the first round-the-world trip. Neither of these offers was taken up before World War I.

The continent of North America was first spanned, from New York to California, in 1911 by a flyer named Rogers, who had been trained in the Wright School and flew one of the

Wright planes. Speeds in excess of 100 miles per hour were soon attained: Prévost, in France, flew 124.5 miles per hour. Oelrich, in Germany, reached an altitude of over 25,000 feet, and another German, Boehm, set a non-stop record of 1,178 miles. It was clear from various records such as these that planes were beginning to show possibilities of endurance and dependability which might soon have important commercial applications.

From the first there were hints of the immediate usefulness of planes. In 1910 a department store in Columbus, Ohio, arranged to have a bolt of silk carried by air from Huffman Field near Dayton to a park near Columbus. The distance of more than sixty miles was covered in about an hour and the "air express" charges were $5,000, or about $71.42 per pound. But the store made money on the transaction, for it sold the pieces of silk for souvenirs at a stiff price. An airplane demonstration of carrying government mail was carried on for a week between the Nassau Boulevard Airdrome on Long Island and the village of Mineola nearby. Early in the history of the Wright factory, probably in 1911, the first private plane was sold to Mr. Robert J. Collier.

Most of these developments were of minor importance, but they did provide suggestions of future commercial applications. The speed, altitude and endurance records, moreover, demonstrated that planes were becoming better. The first real stimulus to the airplane industry and also to the airship industry, came from the outbreak of war in 1914. While engine design both in the United States and in Europe had made marked advances before that time, it was the war which stimulated every phase of plane and dirigible construction and which likewise produced thousands of pilots and navigators trained in the hard school of combat. The biplane gave way to the monoplane, and the pusher type of engine, placed in the rear, to the tractor type, set forward. Twin engines and multi-cylinder types of great variety were introduced. In Europe the rotary engine was favored, while in the United States the water-cooled V-type predominated. Planes became specialized, whether for bombing, reconnaissance, or pursuit. In Germany particularly, the dirigible type of airship, already advanced so far by Count Zeppelin, was perfected and given extensive military use. The German Navy used large "Zeppelins" for reconnaissance and bombing raids, especially over London. An important German achievement in lighter-than-air flight was the dispatch of a Zeppelin from Bulgaria to a point near Lake Victoria Nyanza, in German East Africa, to deliver supplies to a German military garrison marooned there. The distance of 5,500 miles was flown non-stop, in four days.

The significance of all this to commercial aviation was very great. Every new design or discovery made in the travail of war might prove of great importance to the development of safe, fast and comfortable commercial flying. During the war, the United States gained considerable experience in the manufacture of aviation motors. One such product, the Liberty Motor, was the result mainly of research performed by the Packard Motor Car Company; it was a twelve cylinder job and several thousand were completed. The abrupt end of the war precluded any final judgment of its effectiveness, however, since very few Liberties were used in combat. To some extent American engines, of the Curtis-Wright types, were used in planes of European manufacture.

During the decade following the war, the developments in the heavier-than-air field were extensive in the United States. One phase of this activity was the so-called "barn-storming" by many individual flyers, often combat-trained, who acquired war surplus planes and performed with them at country fairs, or in some cases acquired airstrips and conducted a limited and quite hazardous passenger sight-seeing service. These barnstormers did some aerial photography, carried out advertising stunts for commercial firms, engaged in races, or gave instruction to would-be flyers. The value of this sort of thing to scientific aviation was probably small. Many of the flyers were poorly trained, had defective equipment and by their recklessness did little to inspire public confidence in the future of aviation.

More important were the many flights by experienced flyers, undertaken to demonstrate the speed or distance capabilities of the airplane, and the possibility of establishing regular air mail or passenger routes. A non-stop flight from New York to San Diego, a distance of 2,516 miles was made by Lieutenants Kelly and Macready, in 1923, in twenty-six hours and fifty minutes. Records across the continent in both directions were constantly improved in the years following. Transoceanic flights were begun in 1919 by United States Navy flyers in the NC-4, with two stops, on a route from Newfoundland to Plymouth, England. The first non-stop flight across the Atlantic was accomplished by two British flyers, John Alcock and Arthur Brown, about a month after the crossing by the NC-4. Their course ran from St. John's, Newfoundland, to Clifden, Ireland and was covered in sixteen hours; this feat gained them the *London Daily Mail* prize of $50,000. The first westward crossing of the Atlantic was made on April 12-13, 1928 by two German aviators, Captain Koehl and Baron von Huenefeld, from Dublin, Ireland to Greeny Island, Labrador.

Polar flights were another feature of this most important decade. On May 8-9, 1926, Lieutenant Commander Richard E. Byrd of the United States Navy, with Floyd Bennett as pilot, flew from King's Bay, Spitzbergen, to the North Pole and return, in fifteen hours and thirty minutes. Three years later, Commander Byrd, with Bernt Balchen in the pilot's seat, flew over the South Pole. Altitude records were raised to 41,794 feet by 1929 and speed records to nearly 300 miles per hour in 1927. Impressive records of continuous flight in the air were made in 1929; two flyers, Jackson and O'Brien, flew continuously for over 420 hours, refueling while in flight.

It must not be forgotten that the period was a great one,

also, for lighter-than-air craft of the dirigible type. A British-built ship, the R-34, made the first transatlantic non-stop flight, from Edinburgh to New York, a distance of 3,120 miles; after a few days the airship returned to Pelham, England, also non-stop, a distance of 3,200 miles. In 1923 a German built dirigible, the ZR-3, commanded by Dr. Hugo Eckener, made an epochal non-stop flight from Friedrichshafen, Germany, to Lakehurst, New Jersey. The great dirigible covered the 5,066 miles in eighty-one hours and seventeen minutes. This airship was then delivered to the United States Navy and christened the *Los Angeles.* Dr. Eckener followed up this achievement by establishing a regular commercial service from Germany to the United States in 1928; in that year a new ship, the *Graf Zeppelin,* arrived by the same route as the other, carrying twenty passengers and a crew of forty. The most impressive feat by the Germans, who had assumed the lead in commercial lighter-than-air aviation, was the world-circling trip of the *Graf* in 1929 from Friedrichshafen, which was made in twenty days, fifteen hours and seventeen minutes. The *Graf* and later the *Hindenburg,* a sister ship, made many crossings from Germany to the United States, and from Germany to South America.

It was unfortunate for the aviation industry, both lighter and heavier-than-air, which was making such impressive technical progress and was arousing such wide public interest by its transcontinental, transoceanic and polar flights, that these developments were most promising on the eve of the depression. Aviation, like all other rapidly growing enterprises, suffered by too rapid expansion and had attracted speculative capital. The industry needed careful, conservative management; the results of the experimental period, which included the war just past, had to be carefully analyzed. Planes had to be designed especially for specific purposes. To carry passengers in large enough numbers safely and in reasonable comfort would require very large and powerful types. To raise large loads of freight from the ground and deliver them successfully at a distance presented further problems which would have to be solved, if a profitable industry were to develop and enter the competitive transportation field.

Efforts to apply the experience of what might be called the "sport" or "spectacle" period of aviation to sound business began quite early after the war. As we have seen, the Germans made the most impressive beginning in the commercial airship field, and they had no serious competitors. Commercial progress in aviation in this country has been almost exclusively related to lighter-than-air craft. After the war, as early as 1920, air-lines were established for mail and passenger service between Key West and Havana, and between Seattle and Vancouver, British Columbia. Four years later the United States Post Office Department inaugurated transcontinental mail service from New York to San Francisco. The Kelly Air Mail Act of 1925 authorized the Post Office Department to contract with air transport operators for the carrying of mail; as a result there were fourteen domestic air mail lines begun in 1926, and the service was soon extended to Canada and Latin America.

While American airlines were seeking mail contracts and making early beginnings in passenger service, new speed and distance records were made and broken so frequently as to make a recital of them impossible. The transcontinental record was reduced to less than eight hours by Ben Kelsey in 1939. Speed records reached 400 miles per hour in 1934. Round-the-world and transoceanic records were made and broken with dizzy frequency. An important event took place in November, 1937, when, after several years of experimentation, regular transpacific air mail service was established between the United States and China. Transatlantic air mail, passenger and express service was begun in 1933 by the Deutsche Lufthansa; mail service by heavier-than-air craft followed the same route the following year. In the spring of 1939 regular passenger service was started between this country and Europe; on May 20th of that year the first mail was flown from the United States to

Europe, and five weeks later the first passengers took to the air.

Aside from the constant making and breaking of all kinds of records by planes, the 1930's beheld a new type of scientific research into the upper air. Since the airplane is limited in its altitude possibilities because of the effects of oxygen deficiency on the mechanical and human elements involved, several scientists, notably Professor Auguste Piccard of Belgium, began experiments in high altitudes by means of air-tight gondolas attached to balloons. Piccard's ascent to 51,775 feet in 1931 was soon exceeded by his own subsequent flights and those of several others. Captains Stevens and Anderson of the United States Army reached an altitude of 72,395 feet over Rapid City, South Dakota in 1935.

Meanwhile the development of lighter-than-air transportation for commercial and military purposes suffered a crippling blow in a series of airship disasters between 1933 and 1937. The largest dirigible in the world, the *Akron,* which was capable of carrying over 200 persons, crashed during a storm off the Jersey shore in April, 1934, a fate which was shared by its sister ship, the *Macon,* two years later near the West Coast. Both the Army and the Navy had begun an extensive dirigible program, assisted materially by the transfer of the Zeppelin patents to the Goodyear Company in 1924; the latter firm thereafter had constructed a number of airships, including both the *Akron* and the *Macon,* both of which were attached to the Navy. The loss of these ships, coming so close to one another, discouraged further construction of large airships by United States Army and Navy authorities, though smaller airships continued to perform valuable service, in reconnaissance and aircraft barrage service, during World War II. Most tragic of all, in loss of life, was the burning of the *Hindenburg,* just after completing a successful crossing from Friedrichshafen on May 6, 1937. The high inflammability of hydrogen, commonly used in all but American dirigibles, and the unmanageability of the great gas ships in severe weather, threw serious doubts upon the future of the larger lighter-than-air craft.

The development of air transport in all countries has depended upon governmental assistance. Municipalities in the United States and in other countries have established airports. The Federal Government has made appropriations for weather service aids to flying, has aided in the laying out of airways across the continent, has helped to perfect air equipment through its testing service, and has subsidized mail transportation. The mail contract system set up in 1924 proved unsatisfactory; the government lost money, companies without contracts were discontented, and passenger service was neglected. Congress passed the McNary-Watres Act in 1930 with the objective of developing a unified passenger and air-mail service. In 1938 aviation was placed under a new body, the Civil Aeronautics Authority, a step which signalized the coming of age of the airplane as a feature of the nation's transportation system. In a dozen years the growth of passenger transport was enormous. In 1925 only 5,782 passengers were carried on the domestic airlines of the country; by 1937 this figure grew to over a million. In 1948 over thirteen million passengers took to the air.

The Air Mail Act of 1934 authorized the award of contracts for the transportation of mail over designated routes to bidders for initial periods of not more than three years. Maximum rates paid were thirty-three and-a-third cents per plane-mile for loads of not more than 300 pounds, and one-tenth of the base rate for each additional 100 pounds or fraction thereof. The Act creating the CAA in 1938 contained provisions for changes in administering air-mail compensation. With the passage of new legislation in 1940, extending the authority of the CAA and the Civil Aeronautics Board, those agencies may revise and fix rates of compensation for the carrying of mail upon complaint of an airline, and may also initiate investigations to determine the justice of the rates charged by the lines. This method of determining rates offers an opportunity of considering financial and operational factors before fixing the rates to be paid an airline for mail service.

In the early days of air mail, the income derived from its delivery amounted in many cases to from half to two-thirds of the carriers' total revenues. In recent years, however, the percentage has decreased in comparison to other revenue such as passenger, air express and other income. In 1948, for example, air mail revenues accounted for only 13.7 per cent of the operating income of the lines having contracts, compared to 86.3 per cent for other types of income.

After the famous shipment of the bolt of silk from Dayton to Columbus aboard a Wright biplane in 1919, the American Railway Express attempted the shipment of 1000 pounds of merchandise in a large military plane from New York to Chicago. Bad weather and fuel difficulties necessitating two special landings, together with damage done to the plane on this trip, put an end to the experiment for the time. In the six or seven years following there was some small package freight business, but not until 1926 was a regular air freight service undertaken. In that year 1,733,090 pounds of freight were carried. The next year the Ford-Stout Air Services, operating between Buffalo, Cleveland, Detroit and Chicago, carried over two million pounds, all of it for the Ford Motor Company. The same year the American Railway Express established an air express service by contract with airlines carrying passengers and mail. By 1928 a number of airlines were engaging in this activity, the largest being the Boeing Air Transport, Western Air Express, Mutual Aircraft Corporation and National Air Transport. As air express developed and embraced the entire nation, its services were extended from twenty-six cities in 1927 to 375 cities in all parts of the United States. In the year 1947 the volume of goods carried by air reached 188,850,000 pounds. In the same year the volume of air express carried between the United States and foreign countries reached a total of 33,406,000 pounds.

The range of air express and its limitations are interesting. Emergency goods are carried, since situations develop, as for example the urgent need for a precision part of a machine, which justify a high transport rate. Style goods, as millinary and other wearing apparel, are favored items carried by air; so are jewelry, scientific instruments and other light items of high value. Motion picture film, photographs, and publishers' electrotypes are frequently shipped by air. The increase in the carrying capacity of modern transport planes is indicated by the fact that in 1940, 17.6 per cent of all air express shipments included items carried for the heavy industries, including machinery, hardware, automobile parts, rubber goods, and electrical and aviation products. Inflammable and explosive articles, corrosive acids and most live animals, are excluded from air transport. Limits as to weight, size and value are also prescribed.

A distinction is usually made, though it is not a fine one, between air express and air freight. In the former case the shipment is expedited, is a relatively small one, and of course bears a premium rate. Air freight is in larger shipments, may be carried in planes especially designed for cargo carrying and may also move in unscheduled flights after a load is gathered at a collection point. A great stimulus was given large-scale air freight shipment by the formation in 1941 of Air Cargo, Incorporated, by agreement between four of the largest airlines: American, Eastern, Transcontinental and Western, and United. By 1942 nearly all the other certificated domestic carriers had joined in sponsoring Air Cargo. American Airlines pioneered in 1944 in providing extensive cargo service under a regular system of class and commodity tariffs; other lines later adopted this system of tariffs, though each company's rates show detailed variations.

In 1947 certificated domestic carriers transported nearly thirty-six million ton-miles of air freight, compared to less than fifteen million the year before. In the following year, air freight came to about seventy million ton-miles and air express to about thirty-one million. Subsequent years again showed substantial increases. Types of goods which move by air freight are similar to, but not identical with, those carried by air express, as already shown, but air freight has also moved plant and nursery stock long distances, has carried live seafood to markets in the interior, and has borne choice fruits and vegetables to distant markets. It is likely, from recent estimates, that the greater increase in air cargo will be in perishable items and in wearing apparel. Limitations on goods acceptable for air cargo are practically the same as those for air express; here, as in express, dangerous shipments, those too large or too heavy, or too valuable to risk air travel, are refused by the carriers. In the opinion of leaders in commercial aviation, as for example President W. A. Patterson of United Air Lines, the future development of air cargo will make that item the most profitable type of air revenue. This will of course depend upon the construction of larger and more efficient planes, improved handling of cargo on planes, reduced costs of operation, and improved coordination with water and ground transport.

The Impact of the Airplane upon American Industry

The airplane business is very different from all other kinds of manufacturing enterprise. While the plane is related to locomotives, ships and motor cars as a type of transportation equipment, any other resemblance to them, industrially, is lacking. In the first place the others are built to travel in relatively stable media, while planes must tour the highly changeable conditions of the upper air. Mass production of railway equipment and of motor cars has proceeded so far that the highly efficient assembly line methods can be applied, since both the railroad equipment and automotive industries are assured of enormous volumes of sales for products which have become more standardized in design. It is very difficult to apply such methods to airplane manufacture, unless assured, as during 1940-1945, of immediate and almost unlimited markets. Even during such a period, when the pressure of the world's greatest war gave the airplane business its biggest "shot in the arm," there were still present the difficulties of designing a bewildering variety of models, only to have requirements change abruptly, as new military needs arose and as more powerful models were called for.

Few if any of the advantages which have been gained by the older and more firmly established manufacturing industries are possessed by the airplane business. It developed rapidly through its "sport" and public spectacle stages into a highly unstable commercial stage. The years of development under normal conditions have been very few, hardly more than a dozen. In 1939, the last normal year before its great wartime expansion, the aviation industry was comparatively small. There were then 125 plants of all sizes, which employed some 49,000 workers; the industry produced in that year some $280,000,000 worth of planes and parts. It was this industry which was suddenly asked, in the emergency of 1940, to expand its capacity from a possible 5,000 planes yearly to what seemed the utterly fantastic figure of 50,000 planes annually. While most everyone knows that the industry accomplished this miracle, and did reach the productive totals asked for by the President, that fact did not solve the vexing problems which beset the makers of commercial planes, nor has it made the postwar development of the industry any easier.

Certain goals had been reached, however, in the twelve years or so before Pearl Harbor which it is worthwhile to note. The industry contained, by 1940, about thirty major companies, a few of which might be called "integrated companies," including the United Aircraft Corporation and the Curtis-Wright Corporation, which made the complete product—the frames, the engines, the propellers and so forth. Certain other companies assembled planes of their own design by making use of ready-made units, as propellers or engines, made by the other companies or by one or more of the aircraft supply concerns, of which there were about 200.

Various specialized activities were already in evidence by the 1930's. Curtis-Wright (in its airplane division) specialized in military planes, as did the newer Grumman, North American, Glenn L. Martin and Consolidated Aircraft concerns. Douglas, Boeing and Lockheed were the leading producers of commercial transport planes. Piper, Taylorcraft and the Aeronautical Corporation of America manufactured only small private and business planes. The Allison Company, engine manufacturing subsidiary of General Motors, was showing much activity in the aviation engine field; the two leading supply manufacturers were the Sperry Corporation and the Bendix Corporation.

In order to indicate what has happened in the aircraft industry in recent years, it is necessary to sketch the schedule under which planes were being manufactured before the war. In 1938, a fairly good year, some 3,675 planes were produced; these must be broken down into 1800 military planes, 1425 light commercial planes, 300 private and business machines, and 150 airline transports. These could be broken down still further into sub-classes. In all cases the designs of these varied-purpose machines were changed frequently, and such changes were very expensive to make. For example, in designing the Douglas DC-4 transport, 500,000 hours of engineering time and 100,000 hours of testing time were required.

Wages paid in the pre-war aircraft industry reached a total of 27.5 per cent of the value of the product, compared to sixteen per cent in the automobile industry. The reason for this is to be found in the fact that fewer but more versatile multi-purpose machine tools were used in the aviation manufacturing process, and these required great skill in manipulation; the personnel training factor was large. In the larger factories, semi-assembly line methods were beginning to be used, in which conveyors were used to carry the partly assembled units, but did not move them continuously or as fast as in the automobile assembly lines.

Profits in making aircraft, prior to 1941, were small and exceedingly irregular. For example, during the years 1929 through 1940, nine of the leading companies had annual deficits for one or more years; one for eight years, another for five, a third for four, and so on. Only one company, among the large manufacturers, showed no deficits for the entire twelve years, and the profits of this firm varied from 9/10ths of one per cent to better than 48 per cent. This irregularity of return on capital investment may be traced to the constant change in plane design, the short life of the average plane, and the relatively small volume of business. In 1933 United Airlines spent $2,500,000 for fifty-five new Boeing transport planes. Within six months, however, the Douglas firm brought out a new plane which was faster and had greater carrying capacity than the Boeing model. TWA immediately bought a fleet of the new Douglases, and United was forced to spend another $1,500,000 to remodel its Boeings. This series of events led both Douglas and Boeing to bring out still newer, better models. The effect of this sort of thing upon manufacturers' costs and stockholders' dividends is easy to comprehend.

Net costs to the manufacturer could be cut down if, first of all, designs could be frozen long enough to permit the setting up of complete mass production methods and thus avoid the heavy engineering and plant costs of new designs; secondly, if plane production could be increased indefinitely, and thus increase gross income, such engineering costs would not be so ruinous to the manufacturer. It is difficult to see how the first condition can be met, for change and improvement constitute the very lifeblood of the airplane business. Furthermore, conditions of national security make it vital that improvement in aircraft be constant. As to stimulating the production totals, such a result is dependent upon an expansion of freight and passenger revenue far beyond anything predictable in the near future, or, upon the coming of another war, which most business men feel is an utterly wrong way to stimulate manufacturing, if it can be avoided.

It is quite possible that the aircraft manufacturing industry will show great strength in the future, as the various commercial uses of planes register increases similar to those which have rendered the railroad equipment business and the automobile manufacturing industry so profitable. It is logical that a comparatively new industry should have extraordinary difficulties, especially one which presents so many technical and production problems as does airplane manufacture. Indeed, recent statistics of commercial airlines operations give reason to hope that the industry will yet prove a thriving peacetime activity. In the first six months of 1951, air express showed an increase of 44.6 per cent over the corresponding period in 1950; passenger traffic registered a gain of 37.2 per cent in the same period. The volume of air mail increased 37 per cent at the same time, due partly to the fact that rates paid for carriage of air mail were decreased over 66 per cent, an item which saved the United States Post Office Department $13,000,000 during the first six months of 1951. Freight transport dropped in the same six months period, but only slightly, and that small decrease (1.3 per cent) may be readily accounted for by the diversion of many cargo planes to the Korean airlift. The fifteen major domestic airlines earned thirty per cent more in the first nine months of 1951 than in the corresponding period of the year before. Domestic and international United States scheduled air carriers attained a safety record for 1950 of only 1.4 passenger fatalities per 100,000,000 passenger miles, the second lowest fatality record in any single year.

It is figures such as these which please the manufacturer, for only great expansion in commercial aviation can place the manufacturer in a position to enjoy a steady and permanent demand for his product. Viewing the entire matter from beginning to end, the vital fact of all probably is, that the United

States must continue to encourage and aid by all means in its power, the continuance and success of the aviation manufacturing industry and its principal customer, the commercial airline, for a well-developed and progressive aviation industry is most vital in this troubled period in our history. The year 1954 promises even greater successes.

CHAPTER FIVE

THE NEW WORLD OF CHEMICALS

During the past fifty years, those industries which we consider chemical have come to have immense importance in the economy of the United States. While chemical companies existed from the early days of the Republic, the scope of their activities was small and their products were few. Early processes were devised for the making of paper, glass and soap; heavy chemicals were manufactured for use in the tanning of leather, and later in the refining of petroleum. A sizable explosives industry emerged in the early nineteenth century, which served the military needs of the nation and also provided the means for mining coal more easily, for clearing land, and for blasting away large obstructions which lay in the paths of the new railroads. Research and production methods for the infant chemical industry originated basically in Europe. The United States possessed priceless deposits of salt, lime, sulphur and coal; there were immense timber resources as well, but the full industrial use of such valuable chemical raw materials did not get under way until the twentieth century.

Some Early Beginnings

Back in 1802 Éleuthère Irénée du Pont de Nemours, a talented young Frenchman, started a powder mill on the banks of the Brandywine in Delaware. The young man was the son of a famous revolutionary leader, the Sieur du Pont, who had come near death by the guillotine through his determined opposition to the excesses of the Reign of Terror. Young Irénée had distinguished himself earlier as a student of, and assistant to, the great chemist Lavoisier, who was later to fall victim of Robespierre's tyranny. At length, in 1799, after enduring continual persecutions for their political views, the du Pont family, headed by the venerable Sieur, took up their residence in America, as so many French had done during the revolutionary period.

The young nation was in need of many things, among them a good grade of explosive powder for hunting, mining and military purposes. Encouraged by President Jefferson, good friend of the du Ponts, Irénée obtained moderate financial support from his friends and constructed his mill, which after many vicissitudes became a thriving business. The first, and for many years the only product of the du Pont firm, was black powder of several degrees of fineness, made from the old formula: charcoal, sulphur and potassium nitrate (or saltpeter). Government orders during the Mexican and Civil Wars stimulated the progress of the du Pont powder business. Industrial use became more important following the Civil War, especially after the introduction of a high grade blasting powder by Lammot du Pont, grandson of the founder.

In 1880 the manufacture of nitroglycerine and dynamite was begun in the du Pont mills. These two new chemical explosives, of infinitely greater force than any earlier types, came to be of great importance in the transportation and mining industries. Ten years after the decision to make the new high explosives, Pierre S. du Pont, the new head of the company, undertook the manufacture of smokeless powder, based on nitrocellulose. This was a significant move, for in a few years this new raw material, nitrocellulose, served not only as a base for its new powder, but provided the firm with a number of new products, among them lacquers, belt cement and leather finishes. By 1910 an artificial leather was developed, made by coating fabrics with a cellulose preparation. At the opening of the first World War, the du Ponts were producing nitrocellulose plastics and rubber-coated textiles, as well as enormous quantities of explosives.

In recalling a few of the early du Pont industrial successes, a relation may be seen between what was at the outset a simple (though very dangerous) manufacturing process, and chemical industry as it was about to develop. In their search to improve a product, new properties of their raw materials were discovered; so the du Ponts learned, in their investigations of the possibilities of nitrocellulose as a base for smokeless powder, that a whole new world of chemical products was awaiting discovery. Their experience was to be similar to that of many others.

Their experience also illustrated another trend in the development of industrial chemistry, namely the dependence of the United States upon the research of European chemists, for many years. The early career of Irénée du Pont, who received his great inspiration from Lavoisier, was an example of that. But there were others. Back in 1845, Professor Schoenbein of the University of Basle produced a nitrated cotton which he termed "guncotton;" the same year another chemist, Professor Sobrero of Turin, first prepared nitroglycerin; both were explosives of terrific power. No way was found to control nitroglycerin until Alfred Nobel, a Swedish inventor, discovered that a powdered earth called kieselguhr would absorb three times its weight of the chemical. Thus by 1867 a way was found to handle and ship the dangerous explosive, molded into sticks of protective earth, without fear of premature explosion.

Meanwhile there seemed to be no way to use or handle the equally dangerous guncotton. Scientists worked upon the problem for many years without success. It so happened that a small portion of it came into the hands of a young Boston medical student, J. Parker Maynard. One day he put some of the stuff in solution with ether and alcohol, and produced what came to be termed "collodion." This harmless solution, placed over a wound, dried quickly into a hard, protective film; millions of bottles of the preparation were sold and were soon to be found in medicine cabinets the world over. A bottle of the collodion lay on a shelf nearby Alfred Nobel one night as he was kept awake by the pain of a sliced finger. As he applied the remedy over the wound he pondered, as the collodion hardened into a dry and smooth coating. An idea suddenly occurred to him, and he hurried to his laboratory. By morning he had solved a problem which had vexed him for years: how to protect explosive charges when used under water, as in mine shafts and harbor and river deepening. And the compound which he had prepared was the forerunner of Nobel's blasting gelatine. In time the du Ponts became manufacturers of dynamite; and later, of Nobel's newer blasting gelatine.

In a print shop in Albany, New York, a young employee named John Wesley Hyatt was looking over a printed circular sent out by a new concern, Phelan and Collander; they would pay $10,000 to anyone who would produce a new material for billiard balls. He and a friend named Brown started a project to make artificial ivory. After trying many substances without success, Hyatt noticed one day that a bottle of collodion which had been standing on the shelf at the print shop had overturned; the stuff had run out of the bottle and had hardened into a smooth, ivory-like substance. Hyatt and his friend began investigating this material, and learned that an English chemist, Parkes, had already produced a plastic by combining dried collodion (or pyroxylin) with camphor. This "Parkesine," as it was called, was too costly to be commercially valuable. The two printers went to work; in an old shed behind their boarding house they built a press, and by substituting heat and pressure for the costly solvents which Parkes had used, they succeeded in producing a plastic. This was the first cellulose plastic suitable for large-scale manufacture. The inventors called it "celluloid" and patented it in 1870.

Over in France a young graduate of L'École Polytechnique, Hilaire de Chardonnet, was also attracted by the smooth texture of pyroxylin. Inspired perhaps by the great Pasteur, who was toiling to eliminate the silkworm disease which threatened one of France's largest industries, Chardonnet cast about for some way in which silk might be replaced. Many others had worked on this problem, but in vain. The Frenchman constructed a machine in which a metal disk, or "spinneret" would turn about, throwing whatever liquid might be inside through tiny holes in the sides. Into the chamber of the machine he poured some pyroxylin. As the disk revolved, the syrupy liquid squirted through the holes into a stream of hot air, which evaporated the ether and alcohol in which the pyroxylin had been dissolved and left the cellulose material in the form of light, fluffy filaments. In this way, in the mid-1880's, was born the essential idea of rayon and many other man-made fibers. Too costly in those days to be developed commercially, another cellulose product was added to those which the new century would see striding into vast production.

Explosives, acids and cellulose products were not the only laboratory substances to be developed early in chemical history. Dyes were also produced chemically in the mid-nineteenth century. W. H. Perkin, an English chemist, while working on a process to produce quinine from coal tar, stumbled upon a substance with a purplish or mauve color. This was in 1856. Perkin was so impressed with the possibilities of his synthetic dye that he built a factory and began its manufacture. This neophyte industry added new dyes such as magenta and "Turkey" red, but the real development of the industry was soon transferred to Germany where most of the synthetic dyes of the late nineteenth and early twentieth centuries were developed. Although other countries tried to compete in this respect, the Germans soon secured such a grip upon the processes of manufacture that they could undersell any competitors in the field.

World War I caused serious interruption to German supremacy in the production of coal-tar dyestuffs and pharmaceuticals, which they also had developed to a high degree. Since a large amount of German chemical business was conducted abroad, in the countries with which Germany was at war, seizures of German chemical patents and properties began after 1914. In the United States the Alien Property Custodian took over some 4,500 German dye and medicine patents and in 1919 sold them to a corporation, organized to aid American chemical industry, called the Chemical Foundation. This organization in turn leased these patents to American firms, which began the manufacture of many of the products for which they had bargained. A very important medicine, salvarsan, of great value in the treatment of syphilis, came by such means into American hands.

One of the earliest chemicals known to man is soda. It is said that the early Egyptians used it to melt sand in order to extract gold from that very common substance. They didn't get the gold, but in the fusion process they may have hit upon a method of making glass, a material of which they made very effective use. Soda was also used by the ancient alchemists in Egypt and elsewhere in making dye and soap, and as a medicine. Throughout the years soda, prepared usually from alkali-bearing plants, was a common item of commerce, but was a costly one. Chemists of the eighteenth and nineteenth centuries set about finding a better means of its preparation than by recovery from plant and wood ash. In the 1860's two Belgians, Ernest and Alfred Solvay, developed a process for making soda ash by the reaction of ammonium bicarbonate with salt brine.

They began commercial operations in Couillet, Belgium, in 1865. This development, called the ammonia-soda method, became the standard process and furnishes most of the commercial soda ash used at the present time.

In the United States, the need for alkaline products was met for many years by household manufacture or by importation of potash and soda from Europe. Patents were granted early for "improved" methods of preparing potash, but the product remained scarce and expensive here until the late nineteenth century. Two Americans, Rowland Hazard and W. B. Cogswell, obtained American rights of the Belgian owners of the Solvay process, and established the first American alkali plant at Syracuse, New York, in 1884. The reason for selecting this site for the Solvay Process Company, as it was called, was the presence in the area of large deposits of salt and limestone. Solvay soon was producing bicarbonate of soda, caustic soda and many other alkaline products in great demand.

So it was that by the end of the nineteenth century a broad basis was laid for a very large industrial development. Synthetic dyes, cellulose plastics, high explosives and alkaline products were all on the way; the building of these industries into sizable ones, following the rather random discoveries, required a long time. Few of the early experimenters dreamed that chemical industry would one day reach the size and importance which it now possesses.

The Nature of Modern Chemical Industry

In certain directions chemical companies developed in much the way as did the other principal industries. The making of chemicals is, after all, a very old human activity, as we have seen. Many of the fundamental raw materials and products have been familiar to science and human industry for centuries. Ancient peoples mixed vegetable dyes, compounded herbs for medicinal use, utilized earths in making pottery and glass. Another strong resemblance which it bears to other industries is its close dependence upon basic scientific discoveries. Our earliest knowledge of the nature of many of the chemical elements came only with the fundamental discoveries of chemists such as Boyle, Dalton, Priestley, Cavendish, Scheele, Lavoisier and others. Until these men had contributed their efforts we could prepare no pure chemicals. They could not be analyzed or measured. Just as in other industries, too, there was often a time lag between a discovery of importance, such as Cavendish's isolation of hydrogen in 1776, and the use of the discovery industrially. This is another way of saying that chemistry became industrially important only after industry itself had undergone drastic change. In still another way, chemical industry has followed a conventional pattern; it has developed to large size the way industries have, by the elimination of competition through combinations of many companies into a few large firms. In this particular it shows strong resemblance to the American steel, electrical and railroad transportation industries.

Yet chemical industry in this country has some features which are unique, which do not follow a standard industrial pattern. There is, for example, more diversity in chemical industry than in most industries. At the start, it is sometimes quite difficult to say whether a given firm is chemical or not. A rayon company, such as the American Enka Corporation, manufactures a textile from cellulose derived from bleached wood pulp or cotton linters. A number of chemical processes turn this material into yarn; thereafter, in the textile departments of the company, mechanical processes are used.

There are many other forms of business which use chemical processes in part, as in the making of rubber, glass, paper and steel, yet these are not usually classified as among the chemical group. Of course there are many about which no doubt at all need be entertained: the makers of heavy chemicals, as soda and sulphuric acid; the drug and dye (or "fine" chemical) manufacturers; and the makers of the large group of organic and inorganic products such as alcohols, plastics and tanning materials. Then there are other firms which process natural raw materials by chemical means into another form which is commercially useful, as the hydrogenation of cottonseed oil into an edible fat.

A characteristic of American chemical companies, as contrasted with foreign organizations, is that while consolidation had gone far in this country, no such control over chemical manufacturing by a single group has ever existed comparable to that exerted in the British market by Imperial Chemical Industries, Limited, or as that once maintained over German and much other chemical business by I. G. Farbenindustrie A. G. In this country, as early as the mid 1930's, the four largest firms controlled thirty-seven per cent, and the eight largest, forty-eight and a half per cent, of the American market. Although this was true, there were still hundreds of smaller companies. Even the group with assets of less than a million dollars each accounted at that time for twenty-one per cent of the total chemical output.

Yet the trend in more recent years has been toward dominance of the American chemical market by a few of the large companies. These are: E. I. du Pont de Nemours, Union Carbide and Carbon, Allied Chemical and Dye, Dow Chemical, American Cyanamid, Monsanto Chemical, Air Reduction, and Hercules Powder. Five of these companies together made sales in 1951 of more than three billion dollars. In the value added by manufacture, the products of the chemical and allied producers ranked fifth among American industries in 1951.

The trend among the larger chemical companies has been to enter a number of fields. The du Ponts confined themselves mostly to explosives before World War I, but thereafter entered many other lines of manufacture: dyestuffs, plastics, solvents, heavy chemicals, germicides, rayon; they expanded their interest in paints and lacquers, which they had begun to manufacture earlier. The Allied Chemical and Dye Corporation was formed in 1920 by the consolidation of five companies, each from a different chemical field. These firms were: the Barrett Company, which had specialized in coal-tar products; the General Chemical Company, acid manufacturers; the National Aniline and Chemical Company, which had engaged in dyestuff manufacture; the Semet-Solvay Company, makers of coke and its by-products; and the Solvay Process Company, which as we have noted, was the oldest alkali firm in the nation. Since this notable consolidation, which gave Allied Chemical a widely distributed interest in various fields of chemical industry, the five companies, now divisions of the larger firm, have expanded into many associated lines of manufacture which cover almost the entire chemical horizon. Similar trends of combination, diversification and expansion are true of the other major companies.

There are excellent reasons why chemical companies, more than those of some other industrial groups, tend to expand their activities so widely, and why, from a rather simple form of chemical manufacture, such as producing black powder or soda ash, they expand and add dozens of items to their lists of products. Modern chemistry is a field of rapid and continual change; chemical industry is geared very closely to the research laboratory and is affected immediately by the new discoveries which emerge from that source. It has become almost impossible, commercially, to manufacture a single product; in even a comparatively simple process there emerge not only the product, but several by-products. Good sense demands that a commercial outlet be found for the latter. New products are constantly being discovered. Competition continually forces companies to engage in research to find better lacquers, stronger plastics, purer pharmaceuticals.

Such a dynamic sort of industry presents serious problems to administrators. Very large budget allowances must be made for research. It is estimated that, in 1937, the chemical industries required 300 research personnel for every 10,000 employees, a percentage greatly in excess of that for other industries. The du Ponts spent more than forty million dollars and carried on five years of research in setting up a dyestuffs division before realizing a cent of profit. The organization of this main function of chemical companies is of such importance that a section of this chapter will be devoted to it.

Another high cost of chemical making arises from the rapid outmoding of plant and equipment. Tanks, pipes and stills wear out quickly by subjection to high temperatures and pressures. Much of the plant must therefore be replaced frequently, in some cases every two or three years. Careful calculations have to be made as to the probable disappearance of certain products from the market and as to the time to undertake new research programs. In 1928-1932 a survey of fifteen chemical companies showed that less than two per cent of their capital was in the form of funded debt, a very small quota for large industry. This apparent avoidance of long-term commitments may have been to keep their future financial structure unimpeded, in order to secure large sums for quick expansion.

Another quite common practice in chemical industry is to make arrangements with foreign chemical manufacturers as to patents and processes; such agreements usually secure, for the company making them, domestic rights to certain inventions or processes, and a share in the foreign market of the oversea company. In this way an American company may buy a license to use a foreign process here, and a share in the foreign firm's formerly exclusive market.

Chemical manufacturing is also unusual in that the major raw materials which are used are themselves products of chemical reaction. The most important market of the industry is therefore the industry itself, for chemical companies furnish one another with essential products which are later used as intermediates in some further chemical operation. Among the larger firms, as chemical technology develops, there is a move backward in their production schedule to prepare their own raw materials, laterally into associated or related processes, and forward to carry products to new combinations. It is thus inevitable that larger plants expand, that the number of processes keeps increasing, and that the lists of products become extended almost indefinitely.

The integration of small companies into large ones, or the acquisition by large companies of many smaller specialized firms, has had a definite effect upon the industry as a whole. New inventions or processes, even though other companies have begun to use them commercially, are apt to gravitate into the hands of the big firms; the prospects for a newcomer of developing and marketing a new process are not very bright, especially if it competes with a process already being developed by a larger firm. The result has often been the sale of the small company's patents or processes, and perhaps its plant, to the larger one. Bigness in chemical industry begets greater size. Thus the trend has been towards what some economists call "oligopoly," with competition limited in large degree to fewer, larger companies. In 1945 a survey showed that of the 238 principal industrial chemicals, nearly one-half of them were made by four or fewer companies; of the remainder, the four leading companies accounted for seventy per cent of the production of 100 items. Present studies show a similar situation.

The Place of Research in Chemical Industry.

Laboratory research occupies a high place in the activities of all chemical companies, for it is the principal means by which a corporation can expand into new lines of business; it is also vital to carry on, by research methods, continual efforts to improve existing processes and products. Products which were once in high favor are frequently superseded by others which can be manufactured more cheaply or which may give better satisfaction. For instance in the field of chemical dyes, the vat dyes, which show superior fastness and durability, have gradually replaced the older azo dyes. In the field of insecticides, recent preparations such as DDT and methoxychlor have come to be preferred, in place of the arsenical group. Large companies frequently choose to carry on general or "fundamental" research, with the object of discovering new scientific facts, rather than to reach specific commercial objectives. Such a policy, which was adopted by the du Pont firm in 1927, is justified when one considers the vast potentialities of entirely new lines of research. According to this company, a number of outstanding products were born in its fundamental research division, including nylon fiber, neoprene synthetic rubber, orlon acrylic fiber, and dacron.

The ambitious research program of the du Pont Company includes scientific activity on many levels and in all manufacturing departments. Throughout the Company's thirteen departments, three on the staff level, ten on the manufacturing level, there are research agencies with their special laboratories. In addition to all these, some fifty outstanding scientists, many of them from our leading universities, act as consultants to the research divisions in which their main interests lie. By their visits to the Company's laboratories they maintain a liaison between the outside chemical world and its wide industrial applications.

The impressive scientific organization for research requires an annual expenditure of some $38,000,000 by the du Ponts, according to recent figures, a sum more than twelve times as great as they spent for that purpose in the years before 1914. Thirteen different budgets must be reviewed each year by the executive committee of the Company, and many conferences have to take place in order to coordinate properly the developmental work of their forty-two laboratories all over the United States. The risks of such a program are very great; there is no certainty that any one of the many projects under way at any given time will be commercially successful. Indeed, out of 136 laboratory projects which were going on in one research unit of the du Pont Company in 1948, only eight had become commercially successful by 1951. Of the others, fifty-eight were terminated as probably not worth continuing, while research continued on the remaining seventy. Losses are many, but in the case of the du Pont organization they are kept down by its extremely broad base of research, a careful selection of projects, a willingness to continue them over a long period, and by finding out when to abandon them. It is quite obvious that without such large scope and careful management the program could not be feasible. Only a very large corporation with ample resources could even venture upon such a program.

Chemical companies show some variation in organizing their research facilities. The American Cyanamid Company maintains at Stamford, Connecticut a large central laboratory to serve the thirty-eight plants and mines which compose the various interests of the Company. The organization of this laboratory is highly diversified. In one wing are sections for physical and organic chemistry, pharmacology, antibiotics, and bateriology; near these is an area for live animals used in experimental research. In another part of the group of buildings are extensive areas where many sorts of plastics are studied; adjoining these are sections for work on rubber, paper and leather chemicals. In still another area work goes on in developing processes for recovering minerals and metals from their natural ores. Insecticides, detergents, emulsions, fertilizers, acids, explosives, and numerous other chemical products originate or are perfected in this effective organization.

About 1000 men and women compose the staff of the Stamford Laboratories. Another thousand scientifically trained persons function in its Lederle Laboratories, in its Bound Brook, New Jersey plant, and in the research laboratories of its affiliated companies. The annual research budget of American Cyanamid in a recent year was about twelve million dollars. It is interesting to note that nearly half of this sum was spent on medical research. The Company has made a distinguished record in its development of sulfa drugs, antibiotics, hormones and vitamins.

Not all scientific personnel in chemical industry are chemists. One company, for example, has a research staff of 550 chemists, sixty physicists, eighty engineers, seventy-five biologists, twenty metallurgists, five physicians, 450 technicians and 400 non-technical employees, a total of 1,640. These people are all employed in the laboratory portion of the business, and do not include many scientific personnel engaged in administrative, plant operative or other staff duties.

Mr. M. C. Whitaker, a distinguished research director in chemical industry, has some interesting things to say about the industry and about his own specialty, which is administering the research work of a large company. He emphasizes the difficulty of securing the right type of research persons, and states that perhaps only five per cent of a research organization possesses the "divine spark" needed to bring projects to a successful termination. A project, according to this experienced director, may be brief or quite extended in duration. For example, a curious substance called "melamine" was a chemical puzzle for a hundred years. A research project was started on it by a company to see if it might not serve as a component of molded plastics; for dishes, table tops and so on. The results exceeded all expectations, for this long-standing puzzle of chemistry suddenly became one of the most useful of plastic materials, not only for those expected, but also for others which had not been thought of. It came to be used in a wide variety of heat-setting coatings and enamels, as for refrigerators and washing machines; in another form it added strength to paper towels, wrapping papers and containers; another variety served the leather tanning industry; it also became an ingredient of water-repellents and crease resistants in the rayon and cotton industries.

It is difficult, when one analyzes the recent triumphs of

industrial chemical research, to exaggerate the value of the activity. Cooperation of medical men with organic chemists has produced the sulfa drugs; biochemists have brought the antibiotics into the picture. Continued cooperative research with the medical men may bring solutions to many hitherto unsolved medical problems. Chemists and metallurgists, Mr. Whitaker reminds us, developed alloy steels for high-speed cutting tools; they accomplished wonders in adding strength and anti-corrosion qualities to other metals. Enormous improvements have been rendered to the glass industry by chemists: optical glass, heat resistant glass, chemical glassware, and fiberglass, all in common use, owe much to the industrial chemist. The making of coatings, paints, and varnishes has been revolutionized in the chemical laboratory, particularly in their durability and quick-drying qualities. At one time it required twenty-eight days and five coats of paint and varnish to finish an automobile; today it may be done by the use of heat-setting resins which dry in one hour and last for the lifetime of the car. The soap business has gone chemical to a large extent, in the great growth and popularity of chemical detergents.

Thus it becomes clear that not only has successful research expanded the purely chemical industries themselves, but has effectively permeated many other industries. Petroleum refining, glass manufacture, the making of steel, paper, soap, textiles, paint and many other products could hardly be carried on at all without the active assistance of the chemist in his laboratory.

Some Recent Chemical Industrial Developments

One of the most successful of all chemical achievements is the production of synthetic fibers, the earliest being rayon and acetate; in a single generation since these products were first produced on a large scale they have risen to second place in the textile field, surpassed only by cotton. These fibers are the result of one of three processes, each using the same or similar raw material, cellulose derived from wood pulp or cotton linters. The raw material is treated with caustic soda in the "viscose" process to produce rayon; with acetic acid to produce acetate; and with a copper ammonium solution in the cuprammonium rayon process.

The pioneer firm in the successful commercial development of rayon was American Viscose Corporation which set up a plant at Marcus Hook, Pennsylvania, and began production late in 1910; from a modest output of 390,000 pounds of rayon in 1911, production by its six rayon mills rose to more than 400,000,000 pounds in 1950, an eloquent reminder of the widespread popularity of the fabric as a satisfactory material for clothing and its extensive use in the manufacture of automobile tires. This one company makes about a fifth of all the man-made fibers produced in the country.

Another very interesting and by now quite familiar product of American Viscose and similar companies is cellophane, a product which like rayon, was foreshadowed by chemical experiments in France. Cellophane has over 5,000 uses, but its progress to its present commercial position was neither swift nor easy. First manufactured in a rayon plant in 1924, it required more than ten years of experimental research to give the film its many desirable qualities. The high usefulness of cellophane as a wrapper for fatty, moist or frozen products, its protection of drugs, toys, writing paper and drinking cups, and its use in surgery to render items completely sterile, are so familiar to the public as to need no special emphasis.

The manufacture of rayon and other synthetic fibers is widely distributed among many firms. Besides the American Viscose Corporation, we may note American Enka Corporation, American Bemberg, Industrial Rayon Corporation and du Pont de Nemours. These firms and others have developed several important variations of the original synthetic fiber processes, and a number of new fibers have been produced. Chemical companies which do not engage in the textile business directly, find a large market for their acids and anhydrides among the companies which have gone into synthetic fiber manufacture.

* * *

An entire series of important developments followed the discovery by the French chemist Le Châtelier in 1895 that a flame of great intensity may be produced from a mixture of oxygen with acetylene gas. For some twenty years this discovery was not commercially valuable due to the lack of pure oxygen and acetylene in large enough quantities. At the opening of the first World War, an American firm, Air Reduction Company, was incorporated to attempt the production of oxygen in commercial quantities. The commercial oxygen process begins with the compression of ordinary air to 475 pounds per square inch. Next it is freed of all traces of oil, moisture and carbon dioxide, then it is cooled to an extremely low temperature. Later, under various temperatures, the components of the processed air are "tapped off" into oxygen and other products.

Commercial oxygen is useful in the welding and cutting of metals, in the manufacture of steel, and in medical treatment. Other products of Air Reduction include xenon and crypton, gases used in the electronics industry; argon, which assists in the arc welding and flourescent lighting operations; nitrogen, helpful for processing coaxial cables and in food preservation; and neon, for tubular advertising lighting. Other valuable commercial gases prepared by Air Reduction Co. are hydrogen, by an electrolytic process or from natural gas; and helium, a valuable non-explosive volatile gas for dirigibles, which is obtained from gas wells owned by the United States Bureau of Mines.

In 1924 Air Reduction Company acquired the calcium carbide plant and limestone quarries of the National Carbide Corporation; many prophesied the move a foolish one, as calcium carbide, which had furnished fuel for gas illumination of homes and for auto headlights, was being replaced by electric illumination. This proved to be a short-sighted view, as Air Reduction was soon able to demonstrate. The acetylene, prepared from calcium carbide, became an indispensable aid to industry, for with oxygen, temperatures up to 6,000 degrees Fahrenheit were possible. In addition calcium carbide was found to be an important raw material in the making of neoprene for shoe soles, hose, printing rollers and insulation materials; it also took its place in the manufacture of vinyl plastic materials for umbrellas, shower curtains and furniture coverings. Besides these extensive developments, Air Reduction added to its list the manufacture of welding equipment. Moreover, it acquired in 1940 a large Ohio company which manufactured surgical supplies and thus cemented a contact with the medical profession, already well developed in Air Reduction's production of oxygen, nitrous oxide, cyclopropane and ethylene for surgical purposes.

* * *

The chemical industry has performed many services to scientific agriculture in its production of insecticides, fertilizers and soil conditioners. One of the most interesting of these projects is the production of krilium, result of a ten year research effort begun in 1941 by the Monsanto Chemical Company. Officials of the firm became interested in the problem of why two adjoining pieces of land, having the same care, and with similar rainfall and crop nutrients, should show such utterly different results after planting. Was the difficulty one of good, versus bad soil structure? While soil may have good texture, or a satisfactory proportion of sand, silt and clay, it still may not have, say the chemists, good soil structure. This means that its particles may not be so arranged in clusters which will hold up well under rainfall, or remain porous and granular, and thus be suitable for good crop production. Organic materials already in the soil, or placed in them by soil experts, have the function of securing a better soil structure. Yet even if such improvement is made, it is important that the better structure be retained or no permanent improvement will be realized.

Soil specialists had made hundreds of experiments to give back to the soil the identical "gums" or polyuronides which it had lost, but the Monsanto chemists decided to see what synthetic chemicals would do regardless of their resemblance to the materials chemically akin to those originally found in nature. Hundreds of specimens of poor soils were treated with a wide variety of chemicals, after which comparative growing trials were made. After many failures, a "synthetic polyelectric" which bore the code name of CRD-189, appeared to be so successful that a massive field experiment was carried on with it in all parts of the country, aided by state and federal authorities, commercial nurseries and university consultants. By June, 1951, the verdict was favorable.

Amazing things are claimed for krilium. It is said to resist bacterial decomposition, and to be highly effective in resisting erosion. The effect upon the soil of one pound of the new material is said to equal that of 200 pounds of peat moss or 500 pounds of commercial compost. Soil aggregation and aeration are improved, as are the infiltration and percolation of water to the growing plants, say the Monsanto experts. Large crop increases have been accomplished in root crops, as beets and carrots, and excellent results have followed the treating of land planted to corn. While the conditioner has already been put on the market, precise information as to just how it will work under all conditions is still incomplete.

* * *

The du Pont Company has announced recently the development of mylar, a new film similar to cellophane, which is expected to fill commercial uses for which the older product was unsuitable, such as electric insulation, as a component of laminated products and perhaps for packaging, for collapsible tubes and light weight storm windows. Chemically the new product is similar to dacron, one of the latest du Pont textile fibers, except that the latter is spun, while mylar is cast in continuous sheets. Some other du Pont products which have seen recent wartime use include alathon, as an insulating material for infantry field wire, and lucite, which had importance as a glass substitute in the aircraft industry. Lucite met the exacting military requirements of lightness, transparency, and resistance to exposure and sunlight extremely well.

The Future of Chemical Industry

Nothing could be more hazardous than to venture a guess as to the final form, extent or importance of the chemical industry in the United States. Fifty years ago chemical industry was growing, but was quite small compared to steel, textiles or transportation. Yet while steel production by 1950 had increased to four times its 1900 level, some lines of chemical production, as for instance that of sulphuric acid, advanced at a much faster rate than did steel. In 1950 the nation was producing twelve times as much sulphuric acid as in 1900. Even as we write or speak about industrial chemistry, it continues to increase in general economic importance. Achievements which may seem outstanding as we go to press may well be outmoded by new processes or discoveries which may be made in the intervening months. There is no reason to believe that progress in this field of industry science will taper off or suddenly cease. The du Pont Company confidently expects that sixty per cent of its sales in the year 1970 will be in lines unknown or still undeveloped in the year 1950. It makes this estimate on the basis of its 1950 record, as compared to that of two decades before.

One great chemical industrialist visualizes a time, perhaps

fifty years hence, in which scientists and engineers will infiltrate and direct almost every great industry. That has been the trend in chemical industry, in which the work of trained scientists has received recognition, not merely in conducting the research necessary to advance the interests of their companies, but also in high positions of policy-making importance. It may well be that the scientific investigator will move to the top positions in all industry, and thus parallel a trend already characteristic of the chemical business, and which is increasingly true in the petroleum, rubber, electrical, and paper industries.

A possible clue as to the future of chemicals in human living may be found in the present position of the synthetic textile business, which is such a fast-developing part of the industry. In the forty years during which production of synthetic fibers has been industrially important, the consumption of rayon, nylon and other chemically manufactured textile fiber has risen from almost nothing to a billion and a half pounds in 1950; "man-made" fibers have become a serious competitor of the natural fibers; they may even replace them entirely, except as components of the newer textiles.

To illustrate just what this has meant in the life of the average family, a leading retail department store recently assembled an exhibit in Wilmington, Delaware of all the textile articles owned by a family of four persons, from small pocket handkerchiefs to large room-size rugs. Two pictures were taken, one of the family surrounded by all its textile belongings; and another of the same four persons, with the synthetic products blanked out by white spaces. The result was astounding. Many of the husband's suits, his wife's dresses, the children's toys, along with rugs, mattresses and pillow-ticking, all of rayon, had disappeared. Objects of nylon, as shower curtains, luggage, underwear, hosiery, sweaters, dresses and swim suits were also missing, as were a number of items of acrylic orlon fiber, including awnings, garden umbrellas, sweaters, dresses and window curtains. As more and more items are developed and as improvements of the existing synthetic fibers are made, it is possible that the entire textile industry will eventually become chemical in at least some part of its manufacturing processes.

Scientifically, it seems quite clear that all manufacturing of the future will become more and more chemical, less and less mechanical. While there will probably always be some heavy industries in which a desired change in the form of the raw materials will require large mechanical force, the intervention of chemistry into industry has been too thorough to be dismissed as merely auxiliary or supplementary. Even the metals used in much heavy industry are products, in many cases, of metallurgical research carried on in the laboratories of chemical companies. The Electro Metallurgical Company, a division of the Union Carbide and Carbon Corporation, has produced over fifty metals and metallic alloys; these include many ferro-alloys for metallic products which call for unusual strength, resistance to high temperatures, durability and anti-corrosive qualities.

In whichever direction one may look, modern life is seen to be more and more dependent upon chemistry. A large portion of our clothing is of synthetic materials; a substantial portion of our house furnishings, our lacquers, paints and varnishes, represent new chemical processes; insecticides, soaps, cleansing materials, kitchen utensils, drugs and medicines have the same origin, in many cases. The glass in our windows, the paper and binding of our books, the tires on our automobiles, all bespeak the skill of the chemist.

The automobile of today probably owes more to the industrial chemist than to anyone else. Powered by gasoline containing anti-knock chemicals, it rolls on tires of synthetic rubber, strengthened by synthetic fibers. If the car has automatic transmission, it is because of research in perfecting oils of great density to render them effective in hydraulic pressure. To a large extent the upholstery of the automobile is of synthetic materials. Plastics of all kinds are used throughout the car: in the safety glass of the windows, in the steering wheel, in the knobs and the decorative panels. An entirely new field of chemistry is underway looking to the possible replacement of welds, rivets, bolts, screws and solder with chemical adhesives.

That is only a part of the story. The car is chemically painted and chemically cleaned; it is sealed and cushioned in dozens of places by rubber bushings, pads and weather-stripping. Its fuel, ignition and heating systems are furnished with many varieties of synthetics and plastics. More than 250 chemical products are directly or indirectly used in the automobile industry, according to a recent survey by the General Motors Corporation. Even without the arrival of an all-plastic automobile body, which has already been constructed experimentally, the motor car as it stands is actually a "portable chemical factory, with storage tanks, mixing chambers, pipe lines, chemical reaction chambers, and waste product disposal." Thus it is a perfect masterpiece of the chemical age. As improvements in motor transportation are made, they will come largely through the chemical industry, whether in transmission, fuels, body structure, riding qualities, durability, lubrication or general appearance.

One of the most curious types of human conservatism is that which exhibits strong dislike for things which seem "unnatural." Persons who hold such views set up strong resistance to the use of materials which have origins other than in nature. They like wool and cotton clothes, the fresh produce of the earth as food; they fear and detest the products of chemical industry as artificial or inferior imitations. This is of course an absurd attitude, and one which has little basis in actual fact, for the most elementary of our many industries have always needed some chemical aid before consumption, even though the process may have been as simple as the thorough washing of an article before eating it. It is also a fact that chemical processes as we know them today are just as natural and just as clean as those which utilize the natural raw materials. Carbon dioxide may have a strong and unpleasant smell, but it is just as natural and common as the more pleasing odor of a flower, and it has been with us just as long.

The real truth is that nature, which has placed so many and diverse materials within the reach of men, has not necessarily produced them in a form which is most advantageous. Chemistry has the arduous task of making over these raw materials into forms useful for industry; this is a long, difficult and often unpleasant process. But the results, as we have viewed them in extremely brief space, have begun to be quite satisfactory. We are getting more durable and useful chemical goods than ever before; in quality, appearance, lightness and wearing qualities the whole range of chemical plastics bids fair to transform our world and to help remove from everyday life a large portion of the drabness and waste which purely mechanical products, especially in their later stages, have always bequeathed to us.

As we proceed from a consideration of chemical to other types of industrial science, it will be perceived that chemical methods underlie many of them. Not only in the textile and automotive industries, which we have considered briefly in this connection, but in the paper, rubber, mining, petroleum, and food processing industries, which will be considered in greater detail, chemical methods are assuming greater and greater importance with the passage of time. The chemist is an indispensable man in the advance of modern industrial science.

CHAPTER SIX

THE ELECTRICAL AND COMMUNICATIONS INDUSTRIES

The Civil War was long past and the bitter controversies of the Reconstruction period nearly over before serious progress was resumed in applying to industrial use the mysterious force of electricity. A generation earlier Joseph Henry and Samuel F. B. Morse had developed the electro-magnetic telegraph; even before the Civil War began the Western Union Telegraph Company was already supplying service over a wide area. By 1866, after some failures, Cyrus W. Field and his associates had succeeded in opening permanent cable communication with Europe. Some small industrial development resulted; a few firms were making wire, cable and telegraphic instruments and supplies.

Except for the telegraph, however, electrical invention lagged for some time and the vast possibilities of this force which had intrigued scientists for centuries were still relatively unknown. In the seventy-five years which followed the great Centennial Exposition at Philadelphia the complete picture has been changed. Inventors have gone to work and opened up an entire new world of lighting, power and communications. The telephone, the radio, hydro-electric and turbo-electric power, television and household gadgets of many kinds have become almost commonplace. Great industrial companies arose to build equipment, establish power plants, and to engage in many new kinds of manufacturing; hordes of scientific researchers devised the intricate apparatus of the electrical age, and opened to view the wonders of electronics. By the year 1954, an American of the 1870's, if suddenly brought to life, would hardly recognize his home town, so far has daily living been revolutionized by electrical science.

Telephonic Communication

At the Centennial Exposition in Philadelphia in 1876 considerable interest was aroused by the exhibits in Machinery Hall. President Grant personally turned on the power which set in motion a massive Corliss steam engine; a dynamo built by Moses G. Farmer provided electrical lighting for the building; most interesting of all was a device which had been created by Alexander Graham Bell. The Emperor of Brazil, one of the distinguished visitors at the Fair, held the instrument to his ear, then exclaimed in astonishment: "My God, it talks!"

The year before, Bell and an assistant named Watson were carrying on a series of experiments in a Boston garret. They had been given financial backing by two Massachusetts business men, Gardiner Hubbard and Thomas Sanders, to perfect what was called a "harmonic telegraph." During June of the same year, while the two were tinkering with a device which they hoped would enable them to send a number of messages simultaneously over the same wire, a strange thing occurred. While they were adjusting a transmitter spring to secure an unbroken current in place of the interrupted one which they had been using, Bell caught the sound of a vibration at the receiver end of his line. A sound had been transmitted, as he thought, over a wire, and the key to transmission of the human voice was at last known, for if one sound could be sent over a wire, why not another?

After building his crude telephone instrument, Bell obtained a patent for it in 1876. His application was filed just two hours before another for a similar device, submitted by Elisha Gray, an electrician who had been working for years on the transmission of musical tones by electrical means. Bell got the patent, and Gray's application was denied. In October Bell succeeded in transmitting sounds over a two-mile telegraph line between Boston and Cambridge. As soon as the invention appeared to be successful, Bell's backer, Hubbard, offered it to the Western Union Telegraph Company for $100,000, but they turned the offer down, much to their regret later, when its commercial success seemed to be assured.

But the money was running out; Hubbard and Sanders had received no return on their investment thus far, and it proved to be up to Bell to find the revenue elsewhere to finance his device. Luckily Bell had been a teacher of speech, and was accustomed to giving lectures before an audience. So he decided to present the principles to audiences who would pay an admission for their entertainment. A lecture given at Salem, which he illustrated with messages sent by his assistant Watson from a point several miles away, so amused his hearers that these performances became popular, and soon Bell was besieged with offers to repeat them in other towns. This bit of showmanship probably saved him from financial disaster. Certain it is that the income from Bell's lectures, given to capacity audiences, provided the first capital to finance the telephone. By 1877 a line was built in Boston for private telephone service; the next year a central switchboard was installed in New Haven, and licenses were soon sold to local companies for the right to use Bell telephones.

But this early success proved to be too easy and was not to last, for the Western Union, already regretting its earlier decision, decided to enter the field; it formed an organization, the American Speaking Telephone Company, and engaged Thomas A. Edison to devise for it a transmitter, which he proceeded to do. For a time the new concern threatened to engulf Bell, because the new Edison equipment was at first superior to his, but Bell did not give up so easily. He engaged an engineer, Francis Blake, to make improvements on his telephonic equipment, and also entered suit against the Western Union for infringement of his patents. In 1879 he won this case and effected the departure of Western Union from the telephone manufacturing field.

Bell's own manufacturing was carried on under an informal agreement with his original backers, who had again entered the picture. Hubbard and Sanders were to advance the cash, and all three were to share in any profits which might result; later Watson, Bell's competent assistant, was admitted to this "Bell Patent Association." This arrangement was succeeded by the Bell Telephone Company, in which the patents were assigned to Hubbard; of the five thousand shares of stock which were issued, the original three associates received about thirty per cent each of the shares, with Watson taking the remainder. Hubbard, as the principal supplier of capital, insisted on a policy which seemed at first unwise, but which later proved to be extremely profitable. No instruments were to be sold; they were to be supplied only under a leasing agreement.

More capital was needed, to expand the telephone business. This could hardly be furnished by the few individuals who had spent most of their funds in getting the invention started. Sanders, the other principal backer of Bell, now came forward with the answer to the problem. A number of regional companies were started, the first of which was the New England Telephone Company, with capital stock of $200,000. The Bell group kept the larger part of the stock of the new company, in return for the exclusive right to operate in New England under the Bell patents. Several reorganizations of the parent company took place, but when it proved necessary to increase the company's capital beyond ten million dollars in order to expand further, the State of Massachusetts, under whose laws the Bell firm had incorporated, refused to allow it. Since the plans of the Bell group included a scheme to connect the many local lines into a single system, it was now regarded as necessary to form a new company under more favorable incorporation laws. A new company, chartered by New York State, was called the American Telephone and Telegraph Company and was formed with the object of building long interconnecting lines. Even this device, of a subsidiary of the Bell firm, with a more favorable charter, did not meet the financial needs of the times, with the result that, in 1900, an even more elaborate reorganization took place. Under the new plan, the AT & T, as it has long been called, took over the parent Bell firm by exchange of stock. As fully developed, the AT & T took care of the long lines (those of fifty miles or more) and assumed control of, or made agreements with, the companies which performed local services. A manufacturing corporation, the Western Electric Company, took over the production of telephonic equipment. The Bell Telephone Laboratories undertook to perform the necessary research.

An interesting feature of the telephone industry, even after this consolidation, was the continued existence of many independent companies, which numbered as many as 6,000 at one time. In many communities, competition was keen between rival companies. In one small city, for example, two companies maintained service for a total of not more than two or three thousand subscribers. One firm used Bell instruments with which one had to call a number to the central operator; the other provided a dial instrument by which one called through directly. Since business and residential subscribers were divided in their opinions as to the advantage of having a supposedly "private" telephone, the great majority of them were forced to have both phones in order to obtain complete local coverage. "We have both phones" ran the advertisements of the local merchants. Thus telephone costs were large and the small companies were unable to expand their services and improve their operative efficiency. The passage of the Graham Act by Congress in 1921 allowed consolidation of these small competing companies without fear of prosecution under the anti-trust laws. Thus one form of useless competition was virtually ended.

So far as local service was concerned, the equipment which came into use was not so different from that which was used in the early days of the telephone, except for the elimination of batteries and the increased use of dial instruments. The latter development necessitated, of course, the addition of automatic switching equipment. Transmission lines have changed from the overhead forests of wires to small bundles in lead-jacketed cables, placed underground. Telephone service has improved immeasurably and so has the clarity of telephone signals. All this suggests a tremendous advance in operating equipment.

In the case of long lines, however, new devices were needed to extend telephone communication from coast to coast. Under the traditional method of transmission, telephone signals could not be sent over wire through distances as great as from New York to San Francisco. To secure such continuous operative service, electronic amplification had to be used. In regard to electronics, more will be said of its connection with the development of radio, later in this chapter. The American Telephone and Telegraph Company was planning in 1912 a long line to reach San Francisco in time for the opening of the exposition in that city in 1915. The trouble was that the electric currents to carry the signals over wire were too weak to be audible by the time the line reached Denver. Luckily there was available a device which might, if improved, secure the required amplification needed to traverse the added distance.

This item was the de Forest audion tube, invented in 1905, and acquired by the Company a few years later. Bell laboratory experts went to work on the de Forest tube and were able to develop it into a workable telephonic amplifier; transcontinental service was opened on July 29, 1914. The almost prohibitive cost of providing wire circuits for the additional long lines was obviated during the 1920's by the devising of the carrier system, another application of electronics, by which long distance wire circuits were enabled to carry four, sixteen, and later, by means of the new coaxial cable, as many as 480 separate conversations over each pair of wires. Repeater stations, at which the signals are amplified in each direction, have been set up to handle the many carrier conversations on each pair of wires. The carrier-repeater system, with its many improvements and variations, is operated by electronic equipment, which though expensive, is far cheaper and infinitely more durable than would be the enormous quantities of wire and other equipment which would otherwise be needed to carry so many extra circuits.

This was not all which the newer science of electronics could do for the long distance telephone industry. Wire circuits may be exchanged for radio links to bridge water barriers or areas through which wire could not be strung or laid conveniently. For example, at Norfolk, Virginia, an ultrahigh frequency radio link carries the telephone traffic across Chesapeake Bay to Cape Charles, twenty-five miles away. To connect the two points by wire would require 400 miles of circuits around the inland route. This radio link takes the traffic coming into Norfolk by a twelve-channel carrier, broadcasts it over a single transmitter at Cape Charles; at the latter point the radio receiver feeds the signal, carrying the twelve carrier channels into a carrier cable. Telephone service proceeds in this way between the two points just as if no water barrier divided them. This is a rather simple illustration of what has become standard procedure in long distance telephony.

In line with such amazing developments are the two-way radio telephonic systems now in wide use in police patrol cars, de luxe railway trains, specially licensed taxicabs, and steamships. Great progress has been made in linking distant parts of the world by radiotelephone. From the New York end huge transmitters provide direct vocal communication with Paris, London and other European cities; from the Miami area links reach out to Havana, Rio de Janeiro, Buenos Aires and other American cities to the south; from San Francisco the messages are broadcast to Manila, Chungking, and Sidney. The United States continental telephone system, by carrier-repeater means through coaxial cable, renders these radio telephonic services available to millions of Americans. That is, it does if they have the price, for it is still relatively expensive to speak directly to London, Rio or Manila, mainly because no way has yet been devised to send more than two or three voice signals simultaneously through the same transmitter over the radio links of the world-wide telephonic system.

The telegraph and telephone system of the United States has become one of the most successful, and certainly one of the most useful, of all our industries. By 1947 the nation had thirty million telephones. In 1953 the telephone and telegraph companies received a gross income of more than four billion dollars, paid out more than two billions in wages and salaries, and handed to about a million stockholders some $300,000,-000 in dividends. The modest enterprise begun by tinkerers like Gray, Bell and Watson, financed courageously by Hubbard and Sanders, has grown into an indispensable giant serving most of the cities of the civilized world.

Electric Power and Lighting

While Bell was beginning his experiments with the telephone, two other important developments were under way. Thomas A. Edison was experimenting with electrical discharges in his laboratory at Menlo Park, New Jersey; Charles F. Brush was building a dynamo to supply current for an arc light in Cleveland, Ohio. Both were successful in providing new types of illumination.

The principle of the electric arc was not new. In the early years of the century Sir Humphry Davy had noticed a luminous discharge in the space between two electrodes, but neither Davy nor any of the other experimenters of the early days could generate enough power to make practical use of this effect. In the 1830's Michael Faraday discovered the secret of the induced electric current, which led to the development of the modern generator and motor. Various others built small magnetos and dynamos, including F. W. Holmes, Antonio Pacinotti, and Z. T. Gramme. One of the first practical applications of the electric generator was in making Davy's arc discharge commercially valuable.

In 1876 young Charles F. Brush, an employe of the Cleveland Telegraph Supply Company, was encouraged by his employer, George W. Stockly, to build a dynamo which would generate enough current to illuminate an arc light. Mr. Stockly planned, if the experiments were successful, to add arc lights to his company's line of electric bells and telegraph supplies. Within a few months Brush completed his dynamo and his arc light, and later devised a way, by use of a shunt circuit, whereby he could assure continued operation of the remainder of a series of these lamps, even if some should fail or become disabled. It was in this manner that electrical street lighting first appeared, from power furnished by a dynamo in a central station. Brush's device was improved upon by other inventors, among them Elihu Thomson, James J. Wood, and Edwin J. Houston. In 1878 John Wanamaker's store in Philadelphia featured a display of arc lamps; the next year a California company began the installation of Brush's arcs for street lighting; in 1880 the Brush Electric Light and Power Company installed three-quarters of a mile of arc lights along Broadway in New York City.

Meanwhile, Edison, working at Menlo Park, was developing a different method of lighting. Backed by a New York lawyer, Grosvenor P. Lowery, the inventor was investigating "incandescent" electrical discharges. After many failures, he discovered that carbon would stand high temperatures if contained in a vacuum. Working beyond this first step, Edison and his workmen were able to produce a lamp with a filament of carbonized cotton which would burn for forty hours. Armed with this outstanding discovery, one of the greatest of the modern age, the Edison Illuminating Company of New York was organized in 1880 to install incandescent electric lights in lower Manhattan. His firm made its own dynamos, conduits, bulb sockets and meters. On Pearl Street a generating station was prepared for operation. After a list of some fifty-nine subscribers had been gathered, all within the vicinity of the station, power was turned on early in September, 1882. In the offices of Drexel, Morgan and Company, of the New York Times and the New York Herald, in Sweet's Restaurant on Fulton Street, and in a number of other stores and offices, the fragile little lamps began to glow. The commercial success of the venture seemed assured. At least so it seemed to the very important banks who had subscribed to the service, for the Morgans furnished half of the million dollars needed to start a company to serve the uptown section of New York, and later backed Edison in organizing a company in Boston.

In New England a group of business men became interested in electric lighting, and brought together several other investors and two important electrical inventors, Professors Thomson and Houston. They organized the Thomson-Houston Company, bought a small factory in New Britain, Connecticut, and moved it to Lynn near Boston. They then proceeded to manufacture electric arc lighting equipment. The new company has historical importance in that it became one of the principal ancestors of the General Electric Company.

Both incandescent and arc lighting became popular throughout the cities of the nation, and spread very quickly to all areas. In both systems, direct current of low voltage, and hence of limited range of transmission, was utilized. Generating plants could therefore serve only the immediate neighborhood where they were located; costs of building innumerable stations were soon prohibitive. This obstructive situation was capably remedied by another inventor, already famous for his achievements in the transportation industry. George Westinghouse had acquired the American rights to an alternating current transformer invented in France by Lucian Gaulard; in cooperation with a brilliant young scientist, William Stanley, improvements were made on the Frenchman's machine.

Alternating current, of which little was known, was believed by some to be of greater flexibility than direct current. It could probably be raised to a higher voltage and used to transmit current for lighting purposes through greater distances. This was the belief which led Westinghouse and Stanley to set up their experimental apparatus at Great Barrington, Massachusetts, in 1885. Early in 1886 they successfully transformed direct to alternating current, sent it over a line at 3,000 volts, then reduced it to a lower voltage for lighting. The demonstration, which was eventually to revolutionize the transmission of electric power, did not at first meet with favor from other electrical experimenters. A strong group, including the Edison interests, objected that the new current, with such high voltages, was unsafe and not fit for common usage. In this particular they were wrong, and alternating current, with greater range and flexibility, became the preferred type. As a direct result of this daring innovation in power transmission, the Westinghouse Electric and Manufacturing Company was to grow and become one of the leading electrical firms of the next century.

A battle soon began between the two leading electrical lighting systems. There was to be an exposition at Chicago to celebrate the four hundredth anniversary of the discovery of America; the illumination of the great fair by electrical means was opened to bidders; both Westinghouse and Edison placed their bids, and since the bid of the former was lower, Westinghouse's company was granted the contract on May 23, 1892. This posed a problem, however, for although Westinghouse could run his current into the grounds where the fair was to be held, and transform it to a voltage suitable for lighting, the Edison Company held the rights to the only practical electric light bulb in existence. In haste Westinghouse began the development of an electric lamp of his own; though the lamp produced was inferior to Edison's, it did stand up sufficiently to light the World's Fair. Moreover, the fact that the new alternating current was used to supply the necessary power strengthened the position of those who held it to be the better type for future development. Despite the hardships, Westinghouse made a profit from his lighting contract.

Thus far the production of electrical current for lighting purposes had been the main concern of inventors, manufacturers and the public in general. Soon, however, they were to note the possibilities of using this power for other purposes. One of Edison's engineers had discovered an interesting characteristic of the dynamo. The ordinary dynamo armature was turned by outside mechanical power to produce electric current, but Frank J. Sprague found that if a current of electricity was introduced into the armature of a dynamo from another unit, mechanical motion would result. Thus the operation of the dynamo could be reversed and a new first rate source of mechanical power could be made available. Sprague became much interested in this phase of electrical power and formed a company to introduce the electric motor into the New England textile factories. The great possibilities of using electric current for manufacturing and transportation purposes now burst upon the minds of inventors and industrialists. In the search for power sources of sufficient magnitude to serve large factories, steam and water power were utilized. In 1877 Jacob Schoellkopf bought riparian rights at Niagara Falls for $76,-000, and construction work on hydro-electric installations began there two years later. In 1882 the Western Edison Company installed a plant in Chicago for generating electric current by steam. The same year, at Appleton, Wisconsin, the first hydro-electric generating plant was put in operation, with a capacity, it is said, of 2,500 sixteen candle power lamps!

Meanwhile inventors were already at work applying the electric motor to transportation. Edward M. Bentley, Walter H. Knight and C. J. Van Depoele were the principal American inventors of the trolley car, which appeared in Kansas City in 1884, in Richmond in 1888 and in most American cities by 1900. While the trolley car is now obsolete, a casualty of the motor age, it should not be forgotten that for a generation it was the principal means of transportation in large towns. It had a brief period, in the early 1900's, of interurban activity, in which high-powered cars built on the lines of railway coaches, furnished fast transportation between cities; systems of these lines were even integrated to furnish connected transport over several hundred miles. Today the electric-driven car survives mostly in city traction lines, most of them operating underground.

During the late 1880's there ensued a series of litigations over the numerous patent rights under which the various electric companies were operating. One of the most important of these was brought by Thomas A. Edison against the United States Electric Lighting Company in 1886. Edison took the position that his patent on the incandescent lamp was exclusive; that therefore all others who were manufacturing such lamps were infringing his rights. After much argument and delay the United States Circuit Court of Appeals in 1892 confirmed Edison's patent on the incandescent lamp as original and exclusive. Yet other vexing legal complications continued to arise. It had become evident that none of the large manufacturing companies, whether Edison General Electric, Thomson-Houston, or Westinghouse could prepare complete electric installations without running the danger of patent infringements of some kind. Consolidation seemed to offer the best way out of such trouble. After a year or so of negotiating, two of the largest companies, Edison General Electric and Thomson-Houston, were taken over by a new corporation in 1892, the General Electric Company, which was soon doing an annual business of more than twenty million dollars.

The Panic of 1893 greeted the new concern with threats of immediate disaster, for the original companies had over-extended credit to the many local operators who were eager to obtain the new equipment quickly; many of these loans went by default, and General Electric became ultimately responsible to the banks for some $3,000,000 worth of endorsed paper with only a little over a million in cash to meet these obligations. The president of the firm, Charles A. Coffin, worked out an ingenious and effective way of meeting the difficulty. Some twelve millions of the securities of the local companies had been accepted in part payment for electrical equipment; these were offered to GE stockholders at one-third of their stated value, an offer which was backed by the company's bankers with an immediate loan of $4,000,000. This arrangement enabled the General Electric Company to weather the serious depression. Thereafter the company's finances were managed to match more nearly the customers' ability to pay.

In the 1890's a significant power development took place in the field of hydro-electric power generation and transmission. Since the late 1870's, when the project was launched for harnessing Niagara for this purpose, almost no progress had been made. A company called the Cataract Construction Company was still puzzling over the problem by 1890, when the directors decided to initiate an international competition for the best plan of making use of the torrent of water. It was not sure at first that a hydro-electric method would be used; some thought that the force of the Falls might be transmitted by compressed air, by water piped under pressure, or by some kind of electrical means. In 1891, an International Niagara Commission, sitting at London to decide on the merits of any plans proposed, made several awards for generating power from the Falls, but none at first for the transmission of such power from its origin to users. The general opinion of the Commission was for some sort of electrical power, and in general favored direct, as opposed to alternating current. In this opinion the Commission received the support of the distinguished Lord Kelvin, and of course from the equally well-known Thomas A. Edison. On the other side of the controversy were George Westinghouse and the Italian inventor Nicola Tesla. Just before opinion veered to the side of alternating current, Westinghouse informed the Cataract Company that he would undertake the electrification project.

By early 1893, Lord Kelvin changed his opinion, and influenced largely by such a famous scientist, who in addition had become the president of the Commission, the latter body decided to utilize alternating current in developing the power resources of the Falls. In October of the same year the Commission contracted with Westinghouse to furnish three 5,000 horsepower generators, which were installed by 1895. In April of that year the power was turned on before a distinguished audience, including Lord Kelvin himself. Of course Niagara Falls, then only a small village of 3,500, could use only a small fraction of the power which would soon be available. Twenty-six miles away, however, was Buffalo, a city of a quarter of a million inhabitants. Westinghouse's generators were capable of developing power for this large industrial center. By November, 1896, the General Electric Company had completed transmission lines to Buffalo, and the electric current, generated at 2,000 volts at the Falls, was stepped up to 10,000 over the Buffalo lines, then reduced in voltage for lighting, trolley car operation and factory power in the large city.

A further evolution in electric power took place almost as soon, in the application of the steam turbine, invented in England in 1884, to the driving of electric generators. American rights to this invention were acquired by Westinghouse in 1896, and the development of it began immediately. The introduction of the steam turbine provided a more efficient power element than earlier steam units. In 1899 the New York Subway system had installed in its power plant seventeen 10,000 horsepower reciprocating steam engines; built to last, as was supposed, they were replaced within three years by steam turbines of only one-tenth their size.

As the century neared its close, substantial beginnings had been made in the fields of electric power and lighting. The impressive achievements of Westinghouse and General Electric were to be carried to even greater lengths by developments about to begin. The relations to each other of these giants of the electrical industry were meanwhile smoothed by the completion of an agreement in 1896 which pooled the patent rights held by both concerns. Together they soon controlled nearly ninety per cent of the electrical supply business.

Radio and Electronics

Electronics may be defined as the science and the industry built around the use of vacuum tubes which control in a countless variety of ways the flow of electric current. These tubes may vary from the small ones in tiny radio sets to immense types which perform important industrial work. Hundreds of them may be assembled in a single piece of apparatus. For example, at the Ordnance Proving Ground at Aberdeen, Maryland, a great electronic "brain" utilizes 18,000 electronic tubes to solve mathematical problems which would otherwise require the services of several hundred computors and operatives.

Electronics serves our large entertainment industry by providing the means by which millions of radio sets convert radio impulses into music and speech, having earlier helped to transform the sounds into radio waves and to send them out into space. Industry makes extensive use of the new science in apparatus which applies heat to the fixing of plywood sheets; by similar means lumps of plastic materials are rendered suitable for molding. Electronic-welding timers control accurately the amount of current needed to fasten together metal sheets in the airplane industry. The mysterious photo-electric cell makes talking movies possible, opens doors, counts oranges, and matches colors.

As one of the newer applications of electronics, television makes use of tubes which can render details photographic without employing more than a candle light. The television receiver itself is a large electronic tube, on the outer end of which the impulses recorded in the studio are converted back into duplicates, or pictures, of what happened there. Telephonic messages from great distances are, as we have seen, amplified electronically. We measure smoothness and roughness of metals by devices based upon these remarkable tubes. Directions can be determined by electronic compasses; human tissues are examined by electronic microscopes. The military use of the new principles are almost endless. It is likely that we are only on the outer edges of complete electronic knowledge.

A Scottish physicist, James Clerk Maxwell, predicted the existence of radio waves as early as 1864; he suggested further that these waves travelled with the speed of light. Rudolph Hertz, in the 1880's, proved Maxwell's predictions to have been sound by actual experiments in projecting and detecting electromagnetic waves. He showed that they could be reflected, refracted (or bent), and focused to one point. In 1892 Sir William Crookes suggested that these waves might be used for communication purposes. Then an Italian, Guglielmo Marconi, did just that. He followed up the experimental work of

Hertz, who had used a small battery and coil to produce a spark, which was detected and reproduced by a small loop of wire several feet away. With a device similar to this, Marconi was already sending messages in 1895 through distances of a mile or more, by adding to the Hertzian apparatus a Morse telegraphic key. Wireless telegraphy was thus born.

Marconi was an unusual combination of daring inventor and successful business man. In 1896 he took out patents for his invention in England and with the encouragement of Sir Oliver Lodge and others established a British company to manufacture and market his device. Three years later Marconi visited the United States and was able to demonstrate to technical experts of the United States Navy the effectiveness of his system over distances of about thirty miles. Here he organized the Marconi Wireless Telegraph Company of America, in which the British Company retained a large interest. Two years later, in 1901, the brilliant scientist successfully detected wireless signals initiated in Cornwall, England, at a station set up near St. Johns, Newfoundland, a distance of 1800 miles from the British station.

Marconi's success stimulated many other inventors and scientists to investigate wireless telegraphy. Three of these were to have a very important part in the development of radio and electronics; they were Professor R. A. Fessenden of the University of Pittsburgh, E. F. W. Alexanderson of the General Electric Company, and Lee de Forest, a young student at Yale. Fessenden, who had been for several years an engineer in the Westinghouse plant at Pittsburgh, began to experiment in 1897, and about two years later brought out a detector, or receiver, which was the first of a number of achievements which led to the development of wireless telephony. In 1900 Fessenden began an association with the United States Weather Bureau to conduct experiments in wireless. Meanwhile young de Forest had graduated from college, gone to the Spanish War, and by 1901 had already invented a new type of detector with which he was able to attain a range of four or five miles. The same year he secured a contract from a small news service to report the International Yacht races off the Atlantic coast; with a thousand dollars of borrowed funds he outfitted a tug with equipment, including a transmitter designed by a friend named Freeman, who went along. The tug followed the yachts and tried to send signals to the shore. In another tug was a Marconi party, engaged by the Associated Press to send in news of the race. Neither of the news services received any signals from the tugs; to begin with, the Freeman transmitter burned out; a coil which de Forest had with him was connected up, but the spark sets of the two floating stations jammed each other, so that no coherent messages were received.

The advantages of wireless telegraphy were discerned very early. Messages could be sent from ship to ship, and from ships to the shore; weather reports, transmitted in this way, would be of tremendous value. Moreover, naval authorities were definitely interested in placing the equipment on their vessels, as soon as its effectiveness should be demonstrated. In making the new devices commercially valuable, the two Marconi firms, British and American, had an important head start on all others. Competition in Europe with a French company was ended by its purchase by British Marconi; the American affiliate absorbed the United Wireless Company and made traffic arrangements with Western Union. Only the Federal Telegraph Company, which had secured rights to the Poulsen arc transmitter, was competing seriously with Marconi in the United States. In fact, by 1913, the Marconi interests were envisioning world-wide control of wireless by agreements made with some twenty foreign companies.

Meanwhile, however, though wireless telegraphy seemed to be approaching the monopoly stage in the hands of the Marconi group, the minds of many inventors were turning toward the more ambitious project of sound reproduction by wireless means. Alexander Bell had experimented with a wireless telephone in the 1880's. Professor Fessenden, who had obtained the backing of several wealthy men in organizing the National Electric Signalling Company, continued to work on the problem of modulating sound and transmitting it through space. With the aid of Alexanderson of the General Electric Company, an alternator was developed, capable of producing wave frequencies of 50,000 cycles per second. Having set up this apparatus at Brant Rock in Massachusetts, he was able to transmit the human voice over the air; audible sounds from Brant Rock station were picked up by wireless operators along the Atlantic coast on Christmas Eve, 1906.

During the same year H. C. Dunwoody, a retired Army officer, with the aid of an inventor, G. W. Pickard, produced the crystal radio detector. These detectors were inexpensive and easy to operate; their introduction led many amateurs to build home-made sets. The age of the radio "ham" had begun. In 1909 Alexanderson's alternator was improved, with a capacity of 100,000 cycles; by 1916 he had developed a multiple-tuned antenna; he was in this way prepared to develop a very effective system of long distance communication. Marconi himself was much impressed with Alexanderson's latest achievements, and came to the General Electric laboratories to examine them. He announced his willingness to negotiate for the use of the alternator throughout his vast telegraphic system, reserving to the General Electric the exclusive right to manufacture it. The involvement of the United States in the first World War prevented the completion of these arrangements.

In 1905 an assistant in the British Marconi organization, reminded perhaps of a phenomenon observed and reported many years before by Edison, that a leakage of current took place across a space between two electrodes in an evacuated glass tube, devised the so-called Fleming "valve." From a small glass tube, Fleming exhausted the air, having placed therein a wire filament and around it, with a space between, a metallic plate. Wires were connected to the filament and to the plate.

Fleming had undertaken research on this bulb in the belief that some such device would reduce the frequencies of antenna currents and thus render them more capable of detection. He found that his valve would pass current when the plate was positively charged, but none when negative, and that it would change the oscillating current to one following only one direction. This one-directional flow of current, emerging from the Fleming Valve, could operate a telephone receiver and make a signal audible and clear. The term "valve" originated in England where Fleming devised the tube; it signified that, by regulating the flow of electrons, the tube acted much as does a valve with a liquid or a gas.

The appearance of the "diode" tube, as the Fleming device is usually termed, led to another invention by Lee de Forest, about two years later. De Forest had been experimenting along lines similar to Fleming's, but had come to the conclusion that the "Edison effect" would be more useful to wireless if a grid of fine wire were placed between the two electrodes in such a tube as Fleming's, activated by a special current carried by a wire from outside the tube. This proved to be a better detector than Fleming's, for it would not only detect the radio signals, but would also make the weak signals stronger. This was the beginning of the "audion" or three-electrode tube, which was to prove probably the greatest single invention in the development of the radio industry.

It was only after the audion came into the hands of the American Telephone and Telegraph Company, and had undergone extensive improvement at the hands of such engineers as Langmuir and Arnold, that the greater possibilities of de Forest's invention came to be realized. A higher vacuum was added to the tube, the tantalum filament was discarded for one of greater electronic capacity; changes were also made in the spacing and arrangement of the various elements. As mentioned earlier, the improved tube was of immediate value to

AT & T as an amplifier in its new transcontinental telephone line.

The general importance of this device can hardly be overestimated. At last radio waves could be detected sensitively and effectively. They could be propagated successfully; the tube could transform alternating into pulsating direct current; it could be used to extend wireless telephony to greater distances through its power of amplification. By employing a series of tubes this amplification could be run up almost indefinitely.

The entry of the United States into the war in 1917 did very little to advance the radio industry. Technically the new device was developed to a point of commercial use; broadcasts had been made from the Eiffel Tower in Paris in 1908 which were heard as far away as Marseilles; two years later de Forest, who had engineered the Eiffel Tower experiments, sent Caruso's voice over the air from the Metropolitan Opera House in New York. In 1915 John J. Carty of AT & T carried on a number of experiments in radiotelephony which resulted in conversation for a few minutes between the operators at Arlington, Virginia, and their colleagues at the Eiffel Tower.

Yet with victory over the ether waves well in hand, it turned out to be a "wire" war after all, with the older cables and telegraph and telephone lines carrying the communications from headquarters to stations in the combat zone. Meanwhile the United States Government took over the commercial transmitting stations operated by Marconi and Federal Telegraph; wireless telegraphy was used effectively by the Navy, as were some of the new submarine detecting devices, many of which were electrical. At the war's end, there was some disposition by officials to retain control of wireless communications, but at the insistence of Congress, they were returned to private ownership in 1920.

Within the United States there now arose a battle royal over patents, many of which involved the manufacture and use of the new vacuum tubes; three corporations: General Electric, Marconi Corporation of America, and the American Telephone and Telegraph Company all needed rights to the basic patents of de Forest, Arnold, Fleming, Langmuir and others. What was called for was a pooling agreement or better still, the creation of a single corporation to take over American wireless facilities. A suggestion pointing to the latter alternative came from the Navy Department, with the interest and backing of President Wilson; American naval authorities were concerned over the reports that the two Marconi Companies were attempting to secure exclusive rights to the Alexanderson alternator, which was considered the most effective device known for the propagation of radio waves over great distances. The upshot of all this was a series of conferences between a number of leading electrical industrialists and Navy men, which culminated in the organization in 1919 of the Radio Corporation of America, with Owen D. Young of General Electric as chairman of the board; Edward J. Nally, a former vice-president of American Marconi, as president of the firm; and David Sarnoff, also a former Marconi man, as commercial manager. An important feature of this organization was the transfer to it of the assets of the American Marconi firm, including the British-held shares. One of the wireless stations which came into its hands was the former Marconi station at New Brunswick, New Jersey, over which President Wilson had broadcast his Fourteen Points to Germany.

The new company established commercial transatlantic radio communication to foreign countries by March 1, 1920, when the first messages were sent from New York to London. Before the end of the year, radio contact was also made with France, Norway, Germany, Hawaii and Japan. This involved the setting up of a new transmitting center on Long Island, and the development of five transoceanic radiotelegraph circuits from the United States to Europe and the Orient. Thus was begun a movement to make the United States a center of world-wide communications.

While in this way the new Radio Corporation of America had begun a very promising message system of long range, it was also true that some confusion existed as to the extension of its usefulness. Radiotelephony was not satisfactory as a confidential form of communication; in fact it was not confidential at all. This fact, which of course suggested the need for further research and development, led some to believe that it had very little value. Of what good, they asked, would it be to the Army and the Navy, if the enemy could listen in on their plans, or to business men, who might in this manner disclose their designs to business rivals?

Some radio engineers, however, were beginning to speculate about another possibility; if the voice, in ordinary conversation, could be heard at such great distances, why could not music and other entertainment also be sent out via the waves? So too could news and educational programs. The difficulty was that there seemed to be no commercial value in such activity. Anyone with a receiver could hear and enjoy the entertainment, but the fellow with the expensive transmitter who had initiated the programs would have nothing but the trouble and expense. David Sarnoff, while still a traffic official with the American Marconi firm, had an idea about this in 1916, when he wrote a letter to his superior, Mr. E. J. Nally.

Why could not wireless be used, asked Sarnoff, to bring music into the household? A transmitter could be set up which might have an effective range of twenty-five or more miles; from one point musical programs could be produced, and transmitted over the air to all within the given area who might possess a specially designed receiver, or "music box." The box could be arranged for several wave lengths and operated, Sarnoff suggested, by throwing a switch. In the same way, lectures, baseball scores and important news could be put on the air. He added the astonishing prediction that if such music boxes could be made to sell at $75 each, a million of them could be sold in the following three years. Later, from 1922-1924, RCA actually did make and sell instruments of a total value of $83,500,000.

Sarnoff's idea thus paid off and started the radio manufacturing business on its way. For the development of the broadcasting station and its organization into an effective medium of mass information and entertainment, considerable credit is due to Frank Conrad, an engineer of the Westinghouse Electric and Manufacturing Company. Conrad was a pioneer "ham;" he had built radio sets as a hobby before World War I; after the war he sent out musical recordings over his sets, becoming the first of the "disk jockeys"; his broadcasts, sent out for the amusement of himself and his friends, became so popular that Harry P. Davis, vice-president of Westinghouse, suggested that Conrad expand his station to larger size and put on regularly scheduled news and entertainment. The result was the construction of a larger transmitter on top of the Westinghouse plant at East Pittsburgh in 1920. This became Station KDKA, which broadcast the returns of the Harding-Cox election campaign of that year.

From this point on, the progress of radio became so swift as to defy adequate description. During 1921, General Electric and Westinghouse manufactured 5,000 radio tubes per month for RCA; by 1922 the national sales of radio sets reached almost $100,000,000. Stations sprang up everywhere. By June, 1923, multiple-station networks began operating, with the setting up of a link between WEAF New York, WGY Schenectady, KDKA Pittsburgh, and KYW Chicago. The same year the superheterodyne circuit was applied to home receivers, as superior to the "regenerative" circuit of the earlier radios. Radio broadened its listener range by broadcasting, from their source, prize fights and political conventions. Radio pictures were sent across the Atlantic. New sets were introduced which dispensed with the older batteries and would plug into the electric light sockets. Commercial advertising, at first avoided, began when, for ten minutes in 1922, the advantages of owning Long Island real estate were promoted over the AT & T station WEAF.

Large corporate management was instituted with the organization of the National Broadcasting Company in 1926. Coast-to-coast communication was initiated the following year by broadcasting the Rose Bowl football game at Pasadena on New Year's Day. At the very beginning of the business depression, television appeared. Dr. V. K. Zworykin of the RCA laboratories invented the iconoscope or "eye" and the kinescope or "screen" in the mid-1920's. After four years of laboratory tests, Dr. Zworykin demonstrated his television pictures at a meeting of the Institute of Radio Engineers at Rochester, New York, in November, 1929. A few sets were produced, but the massive development of television was deferred until after World War II. Meanwhile, as a symbol of the huge growth of the radio industry, the impressive Radio City was planned in mid-town New York, housing at its center the headquarters of RCA and NBC. Radio went to sea in ships, flew aloft into the air, went to war during 1941-1945 and was ready thereafter to meet and conquer new frontiers of communication.

Recent Progress in Electricity and Electronics.

Most Americans are familiar with the results of the enormous build-up in electrical and electronic directions, even though they may know little or nothing at all of the steps by which it has been accomplished. The average family makes constant use of electric power; more people here receive constant benefit from advanced telegraphic and telephonic communications than in any other nation; electrical equipment of wide variety is coming more and more into use, as fast as the American people secure the money with which to buy it. Powered by the current available to millions, householders own and operate refrigerators, vacuum cleaners, washing machines, dish washers and dryers, waste disposal units, air conditioners, toasters, heaters, fans, flourescent lights, hair curlers, even electrically heated sheets and blankets. New appliances and labor savers are being developed constantly by the electrical appliance firms. The average American householder would be lost without electric current; it has become an item upon which we depend for basic living.

At the basis of what everyone has long accepted as an electrical age, and has lately come to be one of electronics also, lie the steady advances in experimental electricity. First came the theorists, such as Volta, Galvani, Faraday and Franklin; later, the more practical inventors: Henry, Morse, Edison, Brush, Marconi and the rest. The third step was the appearance of a vast and highly complicated industry which has developed and put into mass production the results of the work of the theorists and inventors. Here again, as has been demonstrated in several other types of American industry, it has been the practical genius of business men which has placed so many of the material benefits of applied science within the reach of the public. Scientific originality seemed frequently to emerge in European centers, in which a long process of education, begun at an earlier period, brought to completion many of the basic scientific concepts before the United States had developed more than scientific apprenticeship to Europe. It was so in the fields of chemistry, of medicine, of power and of transportation. Even in the electrical field, most of the initial discoveries were European.

In the development of electricity to meet human needs, however, an entirely different story must be told. In the speeding up of electrical development, the United States rather quickly pushed to the forefront, with the magnetic telegraph and its many improvements, the telephone, the electric light, and in developing electrical power. To whatever extent the newer science of electronics has come to be one of the great wonders in an age of wonders, the supremacy of American progress has come to be unquestioned. It is fair to say, however, that a considerable part of American industrial success in this direction, as in others, has been due to the fortunate position which capitalism has attained and maintained in this country in relation to the rest of the world. Wars and political anarchy have not damaged the American economy as they have that of European nations.

In the field of communications, we have seen how skillfully the patents of many inventors were pooled, and the competing interests of large companies were coordinated by the organization of the American Telephone and Telegraph Company and its affiliated companies. In the international communications field an equally impressive organization of the early 1920's, the International Telephone and Telegraph Company, succeeded in building, in less than a dozen years, a great combination of world-wide telephonic and telegraphic service. This concern, cemented together with great skill by two brothers, Hernand and Sosthenes Behn, acquired cable companies and European equipment factories; for a time it was the only company which utilized all known methods of electrical communication: land telegraph and telephone, submarine cable, wireless telegraph and radiotelephone. It is interesting to note that during all the impressive building of the IT&T communications empire, the strong position of the Western Union Telegraph Company in domestic communications continued unchanged.

In the more specifically radio-electric portion of electrical communications, the Radio Corporation of America, formed, as we have seen, to coordinate the highly promising but very chaotic field of engineering and manufacturing, has become one of the greatest companies in the world. It pioneered in the development of radio broadcasting, in the manufacture and sale of radio equipment, in the development of domestic and international communications. The total of its sales of products and services during 1953 was about $750,000,000.

The General Electric Company, after nearly sixty years of manufacture of electrical and related equipment, sold goods and services during 1953 of a value of $2,500,000,000. These included testing and measuring instruments, power equipment for transportation and manufacturing companies, electronic equipment for the radio and television industries, lighting equipment of great variety, materials useful to the chemical and metallurgical industries, x-ray and stereoscopic apparatus for the medical profession, and a wide variety of household appliances.

Westinghouse, a large competitor of General Electric, sold about a billion dollars worth of equipment in the first nine months of 1953. Its list of products totals some 8,000 different items. One of the smallest is a "grain of wheat" lamp used in the doctor's bronchiscope in locating foreign objects swallowed by little Johnny. Its largest is a generator "as big as a house" and weighing two million pounds. It manufactures a motor only 5/8 of an inch long for domestic electric current meters which develops but two-millionths of one horsepower; from this point Westinghouse runs its power units up to alternating current motors of 65,000 horsepower; four of these, when connected up with the huge pumps of the Grand Coulee Dam, would develop enough power to furnish every person in the United States with a glassful of water every three minutes! This firm is planning even larger motors for use in wind tunnels to test jet engines.

The Westinghouse firm, like General Electric, goes far afield from purely electrical manufacturing. It makes locomotives and driving mechanisms for them, both the Diesel-electric and steam-turbine types; elevators, including the very fast types in Radio City; micarta plastic timing gears for the motor car industry; miners' helmets; electrical varnishes; porcelainware insulation. It built the gigantic mounting for the 200 inch Mt. Palomar telescope, the mounting alone weighing a million pounds. In the field of lighting, Westinghouse manufactures 10,000 different sizes and types of lamps, from the smallest up to those of three billion candle power. The company is also one of the great figures in the manufacture of radio equipment and in the operation of radio broadcasting stations.

In the electrical industry, as in the chemical field, research laboratories form a vital part of the industrial organization. In this particular these two types of industry show the closest resemblance. From the Bell Telephone Laboratories and from the research divisions of General Electric, Westinghouse and Radio Corporation of America flow improvements in materials and processes which are quickly reflected in the manufacturing policies of the various companies. As in the chemical industry, the numbers and types of products are very large and are increasing every day. One can hardly talk about the largest or the smallest or the best piece of equipment, or about the newest or most ingenious electrical or electronic process. While one writes about electricity and its many applications, the subject itself moves from under the author and reaches new levels of accomplishment not possible a year or two before.

At the present time electrical and electronic research appears to be proceeding along several principal lines. One important interest is obviously the steady improvement of equipment to serve the older, standard electrical markets: better lighting, more compact power units (like the "packaged power," featured by General Electric), superior household appliances. For instance, much attention has been given to the re-designing of generators and transformers in order to reduce their size; this trend, which is not restricted to power equipment alone, has the very important objectives of saving scarce materials and of providing adequate operative efficiency in smaller space.

The mobilization effort since 1950, despite the Korean peace in 1953, has stimulated a second important area of research, one which has always been a concern of the electrical industry, particularly in crucial times. Materials which are in critical supply must be conserved. Substitutes may be found for the scarcer ones; in some cases alloys or plastics provide the answer. This is an immense field of activity and only a few examples can be mentioned here. To replace brass, plastics and plastic-coated steel are being used; shortages in aluminum have been met by the use of plastics reinforced with fiber-glass; cast iron is used in place of die castings. A magnetic latch for refrigerator doors has been discontinued by General Electric because its manufacture involved the use of cobalt which is needed in defense industry; the same company found that by reducing the size of the heating unit in an electric iron, large amounts of aluminum and magnesium oxide could be saved. In this type of research, which is dictated partly by defense requirements, the principal theme is that there must be no drop in equipment performance. At the same time the appearance of products may be less attractive, their size altered, and costs may rise slightly. These, however, are minor defects when economy of materials is effected.

A third trend in the research carried on in the laboratories of electrical companies is the broadening of the scope of activity, outside the conventional limits of electrical manufacturing and service. This trend is a common one in all present-day industry, but is especially noticeable in the electrical field. For many years electrical furnaces have been used in the metal refining industries; in the last ten years electric-arc furnaces have come into wide use. A new device has emerged from the electrical laboratory which digs coal without previous undercutting or blasting. Ships may now be steered by "electric helmsmen," portable remote-control stations from which signals, magnetically amplified, control the delivery of power from the ship's motors.

Electrical companies have been instrumental in the performance of much research for the United States Armed Forces, both in the last war and during the present defense build-up. They have had an important place in the development of atomic energy. As early as 1937, "atom smashers" were being installed in commercial electrical laboratories. A Westinghouse laboratory furnished some of the first pure uranium to be utilized in atomic energy productions; scientists from many of the electrical companies collaborated in the atomic bomb project during the last war, beyond its use for military purposes; electrical and other industrial scientists are engaged today in developing atomic energy for peacetime purposes. Yet the present military needs have not been forgotten. General electric announced in 1951 a research project upon an atomic-powered aircraft engine. Westinghouse engineers, in cooperation with others, have already succeeded in putting the atom to work to drive submarines. Some such vessels are already under construction.

Research proceeds upon the problem of extremes of heat and cold, as the need continues for electrical systems for airplane engines which can withstand such extremes. A new material, developed in industrial electrical laboratories, enables airplane generators to function in stratopheric flights. By a bewildering research process electrical scientists have even succeeded in rearranging the atomic structure of iron, to effect some of the economies mentioned previously.

The electrical industry continues to serve the medical and surgical professions with many new types of equipment. Improved infra-red lamps serve to ease muscular pain; ultraviolet lamps destroy germs by radiation. An x-ray camera with a photographic rate of a millionth of a second can catch fleeting actions inside human beings just as surely as it detects the internal action of rapidly-moving machinery. New x-ray image amplifiers enable physicians to gain clearer pictures of living tissue.

In what may be regarded specifically as the radio-electric field, marvel upon marvel has come into production from the laboratories of RCA, Westinghouse, General Electric and others. Much of this research has been concentrated upon the development of short-wave radio, as for instance the frequency-modulated (FM) or staticless radio, first developed by Major Edwin H. Armstrong in 1935. It is possible that the future will see FM replace the traditional amplitude-modulated or AM radio, and will thus work a revolution in radio broadcasting.

Of great importance in modern radio communications is the radio relay, which makes use of the quite recent discovery that high frequency radio waves may be directionally beamed from one point to another. First tried out extensively overseas under war conditions, after a brief experimental period from 1936-1941, the radio relay proved so valuable in securing quick telephonic and telegraphic contact with distant points that its use during the postwar period became almost a foregone conclusion. Immediately after the war, five of the great communications companies sought and obtained authority from the Federal Communications Commission to begin experimental installations. These relay systems operate in the ultra-high or super-high frequency portions of the radio spectrum, varying from 2,000 to 7,000 megacycles (millions of cycles). Their efficiency is greater than ordinary radio wave propagation, as the signals are beamed directionally, not dispersed in all directions; a very small amount of power will suffice to carry signals long distances; furthermore the higher frequencies offer greater width in the spectrum than ordinary frequency limits. It has been suggested that in time the remaining wire circuits for overland long-distance communication may be on the way out, to be ultimately replaced by the radio relay system.

Television has undergone extensive and successful development since the end of the war. Equipment has been improved and has become cheaper in price, through the introduction of production-line methods in its manufacture. Network transmission of televised programs has arrived, through introduction of coaxial cable into the television circuit; still greater broadcasting distance may be obtained by the addition of radio relay to the existing circuits. "Stratovision" has been experimented with, by which a plane, or a number of planes, can transmit television and FM radio programs from the stratosphere. Westinghouse and Glenn L. Martin engineers, who have been working upon this scheme, have estimated that a chain of aircraft, each plane cruising over a fixed spot 30,000 feet above the earth, could simultaneously transmit five FM and four television programs to listeners or viewers on the ground; that with fourteen planes suitably equipped they could furnish television service to three-quarters of the population of the United States. In this plan, the planes would act as the antennae of the stations on the ground where the programs originate. Experimentation in color television has been proceeding very fast in electrical laboratories. During the year 1951 the Supreme Court upheld the approval by the FCC of the field-sequential system as standard for broadcasting in color, although in mid-1953 the project was re-opened to all manufacturers. Several companies, working through the National Television System Committee, have pooled their laboratory and field experience in evolving a color system which can be received on black and white television sets. Color television is about to be offered to a wide buying public.

This is a little, but only a little, of what has gone on in the field of electricity and electronics during the past century or so. That the future will see much more progress along the lines which we have been considering, is beyond argument. It is no part of the purpose of a brief chapter to peer into the vast future of electronics to find what may come of industrial value. Yet the temptation is strong. We would hope for a situation in which transportation equipment and its machinery will be controlled and safeguarded by electronic means; in which world-wide communications will be not only possible, but reasonable in price. We would hope that commercial aviation will be made safer by electronic devices still to be invented; that somehow the catastrophic hazards created by bad weather and poor visibility may be completely overcome. Electronics has come to the aid of the pilot of the multi-engine passenger plane in many ways, yet terrible accidents continue to occur. Complete electronic control of the plane may come when equipment for this purpose has developed to the point where its performance is completely reliable. Electronic computers are solving in record time some of the most complicated mathematical and engineering problems for us.

Most impressive of all, probably, are the enormous possibilities of atomic energy. The first great atomic energy project, the production of the 1945 bombs, would have been impossible to complete without the aid of electronic devices. The first large-scale source of uranium 235, the explosive material of the bomb, was a great electrical and electronic apparatus called the calatron. For an entire year this calatron separation plant was the only source of this vital material. During every step of the process the effects and changes were recorded by delicate electronic instruments; at the Bikini tests, after the War, the effects of the great atomic bursts were observed by equipment which was either radio-controlled or operated photoelectrically from the light of the explosion. Limitless possibilities for peacetime use are suggested by the enormous energy which may be produced, transmitted and controlled by the future work upon the atom. It is not being neglected by the great electrical companies, which have set up divisions in their organizations to carry on atomic research; staffed with scientific personnel trained in the newer aspects of nuclear physics, they will work out industrial applications of atomic energy which will astonish us.

CHAPTER SEVEN

MODERN MINING AND METALLURGICAL INDUSTRY

One of the strongest motives for commercial ventures and colonization schemes during the seventeenth century was certainly the hope that valuable minerals, especially precious metals, would be found in such places as the English colonies in North America. Although the more ambitious of these speculations proved to be vain, mining of a crude sort became a lesser, though a promising phase of the early American economy. In the first chapter we have noted the extraction of bog iron in early colonial days, and the development of some resources of higher grade ore in the northeastern colonies before the Revolution. The mining of coal began about 1800. Just before the mid-nineteenth century gold was discovered in California, followed by the great Nevada silver strike shortly after; the Pennsylvania oil fields came into production two years before the Civil War began.

Almost simultaneously with these valuable discoveries, the presence of vast iron deposits in the Superior district was disclosed. The mining of lead had been begun by the French in the previous century, and as early as 1820 the Missouri region had a number of lead mines in operation. Modest quantities of zinc were being found in New Jersey by 1850 and the mining of copper had begun in Michigan even before that; a few small copper mines had been worked earlier in the eastern states. Mercury was first extracted in California in 1850.

Other minerals, though in many cases already discovered, became important commercially only as settlement advanced. For a long time lumber and imported brick were the chief building materials, but by the middle of the eighteenth century the quarrying of stone for this purpose had become relatively common. Salt mines and salt wells were of early origin; large deposits were discovered in central New York, in Michigan, in western Virginia, and elsewhere. Phosphate rock was not quarried until after the Civil War. The deeper and more extensive veins of minerals could not be unearthed until improved methods of extraction were evolved, nor could they serve a commercial purpose until improved processing methods and better transportation developed. The early discovery of a vein of ore or the presence of other useful minerals did not necessarily assure their immediate use in our industrial system, since finds of this sort were often in such inaccessible places as to make their extraction impracticable if not impossible. The crude and wasteful methods of early mining soon depleted the supply of minerals which could readily be taken from the ground.

The separation of the mineral from its source is of course only one stage of making it valuable to industry; the product must be broken up, brought to the surface, cleaned and purified. Transportation must be available near the source of supply, and very soon in the process the product enters the domain of manufacturing. Indeed there is no very clear distinction between the mining and manufacturing processes of metals and non-metallic minerals. For our purposes, it may be wise to first discuss the general technological methods which have been applied to mining in the nineteenth and early twentieth centuries.

The Technology of Mining

Early American mining was largely a surface process carried on by individuals. On such a scale, it was inevitable that only the richer and most accessible veins of coal and metal could be dug. During the past one hundred years, however, the industry has changed completely into a large-scale affair, in which great quantities of low grade mineral ores have been successfully unearthed and processed by mechanized methods. The revolution which has changed all industry has made large mining operations by heavily capitalized companies profitable. Markets for more and more mineral products have been opened up by the swift growth of the chemical, electrical, petroleum and steel industries, which began to use great quantities of iron, coal, copper, lead, zinc, limestone, aluminum, gypsum, phosphates and alkaline products. Extraordinary demands upon the extractive industries, as for instance those in wartime, have stepped up production enormously and have called for speed in the mining of metallic and non-metallic minerals.

The mining industry is unique in that it deals with resources which are large in quantity but limited in supply. Exhaustion of mineral veins has occurred, and still does occur, a fact which dictates a constant shift of mining operations from place to place and makes necessary a large capital structure in order to finance changes from exhausted areas to those which give promise of profitable operation. The spectre of ultimate exhaustion of some of the most necessary minerals has been frequently present, especially in wartime. Therefore much emphasis is constantly given to the discovery of new mines and to developing substitutes through chemical and metallurgical research. During the last war, one mineral after another came into short supply, dictating its placement on the critical list, and the launching of projects of conservation and substitution.

To some extent, the backwardness of the mining and metallurgical arts in early days can be attributed to the fact that much good mineral ore could be easily secured. Mining in the early days was often combined with farming; ordinary tools sufficed to scrape or dig the surface ore, and simple washing and heating was often the only refining needed. If water flooded the small mine, it was usually abandoned, since the limited demand for the product did not justify the installation of pumps to dry the working level, or the sinking of shafts to reach the deeper veins. In the old Mississippi Valley lead mines, crude methods persisted as late as 1880. A local storekeeper or farmer would stake the workers to tools and the men would work the ground after a summer of farm labor; after the raising of the ore to the surface, it could be crushed and cleaned by hand or by horsepower.

Some improvements in the mining of metals were under way before 1880, particularly in the extraction of precious metals. The placer mining of the prospectors was being replaced by hydraulic methods, in which large streams of water were used to sluice down gravel banks into trenches where the gold or silver might be recovered. Capital was necessary to furnish the hydraulic equipment and to build the necessary ditches and canals; at this point the mining company began to replace the individual. In such an area as Virginia City, Nevada, where such an immense amount of silver became suddenly available in one place, no individual could hope to reap the full benefits without the aid of extensive funds and of organized manpower. An important innovation, in the Comstock Lode of Nevada, was the development of what was termed "square-set timbering." Heavy timbers in rectangular arrangements were placed in sections from which the ore had been removed; the spaces between the timbers were then filled with waste rock. This permitted the digging into weak walls of ore which could not be handled by the method of temporary supports. This system was adopted later at Butte, Montana, to provide access to large and deep veins of copper. Meanwhile the Comstock area at Virginia City became in fact a "school of mining" in which the newest methods of underground operations and the latest European machinery were utilized.

With the great industrial build-up in the years just before 1900, mining began to enlarge the scope of its operations. For one thing, it brought the railroad to the mine pit and opened up a wide demand for mineral raw materials. Anthracite coal mining, which had an early development, went through much the same evolution as the mining of metals, but much earlier. As in the case of metals, unskilled individuals with crude tools could provide enough coal to satisfy the markets of the early days. Power pumps were unnecessary prior to the 1830's, for most of the coal was mined above the water level. By about 1850, however, demand had increased to such an extent that it was necessary to exploit the lower levels of coal mines; steam power came in at this point, to pump the lower tunnels dry and to operate the coal breakers which prepared the anthracite product for the market. As greater efficiency in operation became necessary, the small operators fell aside, and large coal corporations were organized, possessed of capital enough to dig very deep perpendicular shafts and to take the fullest advantage of veins in high slopes.

The anthracite part of the industry advanced more rapidly than its bituminous counterpart, but both were well-developed by the end of the century, as was also the mining of metals. Deep shafts were now possible and underground tunneling with roofed supports was effectively accomplished; steam power provided for the hoisting of ore and coal to the surface, dried out the lower levels, and blew fresh air into the underground working positions. Meanwhile the actual separation of the raw material from the "working face" was still the job of the miner, who had become a skilled craftsman. Compressed air drills came into use about 1900, and dynamite began to replace the less efficient black powder blast.

The intensive mining of rich ore veins, the usual procedure in the late nineteenth century, gave way eventually to extensive methods, highly mechanized, especially when and where the rich veins gave out, and the need arose to produce a greater tonnage of ore in order to continue at a profit. Steam shovels were introduced to speed up the rate of extraction in the great Mesabi range of the Lake Superior iron district, since this area was readily adaptable to open-cut mining. In Bingham, Utah, where the first open-cut mining of copper was attempted, steam shovel mining was not the only key to the situation; here a concentration method was needed, to convert the ore from a low grade to a higher grade product; this was called "beneficiation": the breaking and concentration of the ore at the mine site before shipment. The change-over from selective (or intensive) to non-selective mining has not been complete, since in some cases mines may still be worked profitably by the older method of underground mining, where the cut and fill system, with much square-set timbering, is still practicable.

New methods have been introduced into underground metal mining. One of these, called "shrinkage stopping," calls for the sinking of two shafts, on either side of an ore vein, down to a haulage level. Along this lower level, cars haul away the ore which comes down into them from chutes which reach upward to the main body of ore. Atop this main ore body the miners blast and drill away the metallic ore, which falls down the chutes. As the cutting away continues, the miners gradually build up a floor of broken ore. While this method takes advantage of gravity in loading and therefore saves power costs, it can be used only in mines where the walls of ore are strong.

In "sublevel caving," another of the newer methods of underground mine operation, a number of horizontal channels are driven into the area where the ore is present, one above the other. The ore is then mined from the top downward by "caving," or by cave-ins; as a part of the channel caves, ore drops down chutes to a haulage level, as in the previous method. This system has been in use in the underground mines of the Lake Superior district. A more advanced method is termed "block caving." A large mass of ore of perhaps several hundred feet in height is undercut except for supporting pillars; these are later blasted out, so that the entire mass of ore and rock caves in for a short distance, thus breaking up the ore; the resultant product is then drawn off into mine cars as it falls through the chutes to the haulage level. As the removal of other "blocks" continues, the rock continues to cave until all the ore is removed. This method works best with soft or "fractured" ores and resembles open-pit mining in that all the ore of the area is eventually removed.

These developments, which have contributed to greater production of metallic minerals, signify that "mass" mining has almost wholly succeeded the selective extraction of rich ore veins. They also point up the fact that the success of a mining venture depends no longer upon the skill or resourcefulness of the individual miner, who was once a skilled craftsman. It is now the engineer who designs the operation in general, with full consideration to the extent and nature of the ore deposits to be worked; likewise it depends for success upon the mechanical efficiency which is next applied to the removal and beneficiation of the ore. The miner, meanwhile, has become a machine operator, with his part in the process rigidly defined.

Mechanization of metallic mining proceeded rapidly after the invention of the steam drill in 1899. This clumsy but helpful device was first used in driving the Hoosac railroad tunnel in Massachusetts. A Census Report on Mines and Quarries states that by 1902 three-quarters of the gold and silver and eighty-five per cent of the copper mined underground came from mines which used power drills. Even by 1880 mechanical coal cutters had begun to replace the miners' picks; both the coal cutter and the power drill have undergone marked improvement since those pioneer days. In coal mines, the percussive puncher of the old days has been replaced by the so-called "chain-breast" machine which has an endless chain of removable steel bits which revolve about the outer edge of a cutter bar or plate about four feet in width; this cutter bar is fed automatically against the body of the coal.

In the mining of metals, the ordinary drill, in which the piston moved backward and forward with the drill bit, has been discarded for a hammer-type device, in which the drill stays in the aperture, while the piston delivers hammer-like blows against it. This enables a miner to drill upwards more easily than before; by means of hollow steel drills, he can force a mixture of air and water through the drills to keep the hole clean; detachable bits supplied the miner for a day's work assure continued operation. Following the introduction of the improved drill into copper mining by 1909, its use became general in the extraction of metallic ores. Compressed air has become the most satisfactory power element for drilling.

Improvement in blast explosives has helped the mining industry. Dynamite proved to be more effective and much safer to use than the old black powder or than the newer nitroglycerine; dynamite came into common use at about the same time as mechanical drilling. In the past few years a variety of blasting materials has been added which give any desired type of fragmentation. More important than drilling and blasting in breaking up ore veins are the improved types of mining already referred to, in which gravity is utilized, especially in the copper and iron mining industries. Block caving and sublevel caving make unnecessary much of the drilling and blasting formerly needed; great savings in labor cost have resulted also, costs which were incurred formerly in maintaining a large force merely to break up the ore. Other equipment, such as power ore scrapers and a variety of mechanical loaders, have also saved time and labor. Mechanical loading seems to have been most highly developed in the mining of coal. Haulage by power means within the mine came to be general only in the present century, succeeding hand and animal traction. Steam hoists came earlier, and were already in extensive use by 1902. About the time of the first World War, storage battery locomotives were introduced into mines, and mechanical haulage has spread quite rapidly since that time, though as late as the 1940's hand and animal haulage was still common in the Tri-State lead and zinc district, and also in many small mines and in newly-made drifts.

Technology has made important changes in breaking, loading and in transporting minerals, but in large degree these improvements would have been impossible without better drainage, lighting and other auxiliary services. While ventilation by artificial means was not considered necessary in the early metallic mines, it was soon observed that miners who labored in high underground heat had a low rate of efficiency. In the Comstock Lode operations, temperatures sometimes reached 130 degrees Fahrenheit; even with air pipes, forcing fresh blasts from powerful pumps into the mines, the men could work only a few minutes in each hour. Improved ventilation systems have reduced this serious factor of heat exhaustion in the present century; similarly, improved mine pumps have increased the depths at which it is possible to operate.

Of very great importance, as an aid to efficient mine labor and as a protection against explosions from mine damp, has been the increased use of electricity in the mines, not merely for lighting up the various levels and shafts, but also in the perfected electric portable miner's lamp. The newer methods of mining already described eliminate in large measure the time-consuming labor of constructing square-set timber sets. For such timbering as is still used, carpenter shops at the surface, using machine methods, decrease the time needed. Preservatives applied to the lumber lengthen the life of the sets placed underground.

Probably the most efficient type of mining, and the one which applies equally well to stone quarrying, coal mining and the extraction of metallic ores, is the exploiting of an open pit. Of course mining from the top down is not always possible, but it has become increasingly so with the improvement of power machinery and of methods of concentrating the minerals at the mine site. Surface mining eliminates at once all the difficulties of driving shafts and tunnels and of providing supports for undercuts. So too are eliminated the expensive methods of ventilating, of pumping out the lower levels, and of safeguarding against underground explosions. On the other hand, enormous amounts of waste materials, or "overburden" have to be dug. Therefore, to be profitable, open-pit operations must be very large mechanized processes. Proceeding to greater popularity with the evolution of the power shovel which provided both backward and forward propulsion, open pit mining, notably in the vast operations of the Mesabi iron range, has taken great strides forward. In hauling away the ore, the railroad lines constructed close to the mine pits supplemented the advantages of mechanical extraction; indeed the development of the Superior iron area awaited the construction of these lines before coming into full production, as we have seen in a previous chapter.

The power shovel, first used extensively on railroad projects, and limited in its arc of operation, gave way in the present century to the full revolving, caterpillar traction electric shovel. This single improvement alone accounted for more than a one hundred per cent increase in daily production in certain western copper mines, between 1923 and 1934. The increased use of mechanical drills was marked between 1900 and the end of World War I, especially in the copper and iron industries. With the change to standard gauge railroad tracks and the increased size of locomotives and cars, the introduction of the mechanical car shifter and the large motor truck, great increases in the rate of shipping mine products away from the mine were realized.

Technological changes in mining have involved the preparation of ore or non-metallic minerals for the market at the locale of the mine itself. While many minerals are shipped to the smelter or the consumer as they come from the mine, a large proportion of them undergo concentration or separation before shipment in order to make them more valuable. Coal is sized and cleaned in a surface plant, metallic ores are ground, concentrated and sometimes separated, before they leave the mine site. A very low grade ore, for example, may be concentrated into a higher percentage product; complex ores, such as silver mixed with lead and zinc, may be separated into separate silver, lead and zinc ores and command regular simple ore prices, whereas as mixed products composed of all three, they would sell for considerably less per ton.

This practice was employed first in the mining of anthracite coal, in which sizing at the mine began as early as 1830. A modern hard coal breaker screens the coal and crushes it into six or eight sizes; it includes picking tables for washing the coal. Hydraulic means are used for cleaning small-sized coal; jigs are employed which make use of the fact that rock and other impurities are heavier than coal or water, so in the process the former settle to the bottom and the good product remains on top. To a lesser extent, these methods have been applied also to the bituminous industry.

Iron ore is frequently crushed and concentrated by the removal of waste matter at the mine site, though most iron ore is shipped directly without such treatment. By the use of heat or by washing, many impurities in ore may be removed. In Minnesota and Alabama much iron ore is found mixed with fine sand particles; by a jigging process, using gravity, coarse particles of rock may be removed from some ores. In eastern iron ores, found in New York, New Jersey and Pennsylvania, magnetic characteristics make their concentration possible by the use of horseshoe magnets; hematite ores, which are non-magnetic, may be changed to magnetite ores and similar means used for their concentration.

Ore dressing of the non-ferrous metals near the mine is in many ways more necessary than in the case of coal or iron, for these are often found with a very low degree of metal content; some copper ores, for example, assay as little as two per cent. Therefore there has been, in the non-ferrous field, great emphasis on ore dressing techniques; the separation of waste from the metal results in large freight savings when the smeltery is some distance away. Ore of this type is first broken into smaller pieces by various methods; this is especially advantageous because a better percentage of metal may naturally be obtained from pieces which may be more nearly pure, when separated from others which may be nearly all made up of impurities. This same principle applies when there is a need to sort one metallic ore from another. The smaller the ore pieces, the greater chances of a high grade, single metal product. A series of grinding mills, each in succession effecting a finer ground mineral, will produce, with the aid of chemical reagents, the best separation possible for the market.

In the case of gold and silver-bearing ores, mechanized treatment began with the stamping mill, in which ore was placed upon a die, and broken by causing a heavy cylinder to fall upon it, the ore and the die being covered with water. This mill, at first a wooden affair operated by water power, evolved into a steel machine powered by steam or electricity. After stamping, the gold ore was washed over tables covered with mercury; the mercury would catch the gold particles and hold them in alloy. This method would work only with ores which had gold in them, and could not be used in the separation of complex ores from waste matter, or of each ore from the other. In such cases the ores had to wait for the smelter, or undergo the action of some chemical process. The best of the chemical processes, in treating silver and gold ores, is the cyanide method; very finely ground particles of gold and silver in the stamping product can be amalgamated with sodium or potassium cyanide. Gravity concentration devices have also been used from early days for ore concentration. For example, ore particles may be carried over several tables on a stream of water; by a process of shaking the tables, up, down, and sideways, the heavier metallic particles sink lower than the lighter weight waste materials; the water flows away, leaving the metal particles against the table ridges. By cyaniding and by shaking them in this fashion, ores which formerly could not be used can now be treated economically. In 1880 only gold ores which could yield $100 to $200 per ton were bothered with; with modern methods, ores as low in yield as $40 per ton can be handled.

With the copper, lead and zinc ores, methods of concentration and separation made slower progress than in the case of the gold and silver ores. Copper ore, in particular, is mixed so intimately with large amounts of waste that especially fine grinding must be used to reduce it to a size ready to be treated. A series of jaw and gyratory crushers, plus the aid of ball or tube mills, are needed to reduce the particles to a few thousandths of an inch in diameter. For some years methods similar to those used for the precious metals were applied to these baser metallic ores. But recently there has been greater use of the "flotation" process of concentration. By adding a minute quantity of oil to a pulp of water and ore, the ore is caused to float above the waste material when air is introduced into a flotation cell by a blower at the bottom of the cell; the mineral particles attach themselves to the air bubbles formed and may be recovered at the surface. Flotation makes for a higher recovery percentage of ore and produces a cleaner concentrate. This process has been further refined by introducing chemical reagents which aid in recovering the desired minerals, and which prevent those not wanted from floating and becoming mixed with the others. This "selective flotation" helps to produce a lead concentrate nearly free from zinc and a zinc concentrate fairly free from lead; thus the process is of immense importance in the recovery of byproducts of the original mining process.

Recent Developments in Mining and Metallurgy

Looking backward over the history of our mineral industries, one is impressed not only with the continued importance of the older minerals, as coal, iron, oil and copper, but also with the great changes which are always taking place in this dynamic phase of the American economy. The most obvious fact in the entire story is the vast increase in mineral wealth which has paralleled the expansion of the nation from its situation along the Atlantic seaboard to its present continental proportions. For a considerable portion of our history, this physical expansion, with the continued discoveries of minerals, sufficed to meet the demands of a growing population and the needs of mechanized industry. With the coming of the present century, however, and the intensification of our industrialism, mere geographical expansion is no longer an assurance of the adequacy of what once was an ample storehouse of valuable minerals. The growth of transportation, the advances in electrical power and communications, the increase in construction projects of all kinds (to say nothing of the unforeseen requirements of two world wars), have placed pressure upon domestic mineral resources which calls for a new planning technique.

During the past half-century, and particularly in the past thirty years, the ultimate exhaustion of many of our most necessary minerals has loomed as a distinct possibility. There has been instituted therefore, by industry and by governmental agencies, an extensive conservation effort. The aim has been not merely to explore all possible ways of saving our national mineral stockpile, but also to enlarge the available reserves by bringing into operation more and more mineral ranges and fields, not only in the United States, but in foreign countries as well. This activity, which seeks to assure us of enough of these vital raw materials to keep our industries moving at a high level, may well be the most important economic project of the next fifty years.

The iron and steel industry has become well aware of the need for continual effort in developing new resources. Thus far, sufficient iron ore of a high quality, as well as the coal and limestone necessary for its smelting, has been available. Large reserves still exist in the Lake Superior area, and substantial amounts are still to be found in the Appalachian region and the far western states. To date, close to three billion tons of iron ore, so fine in quality that it could be shipped directly to the steel mills, have been extracted from the ranges lying almost wholly within the states of Michigan, Wisconsin and Minnesota. To a large extent, it has been this vast storehouse of ore which has fed our giant steel industry during the first half of this century and which has enabled the nation to meet the very large production demands of two world wars. Yet when one realizes that the last conflict cost us half a billion tons of iron ore, it is clear that a policy of conservation and expansion must be furthered to assure future supplies. To a considerable extent, superior technical methods and the development of new alloys have taken a part of the pressure from the available reserves. But new sources must be found, for the rate of their attrition is alarming. The United States Steel Corporation alone plans to develop, in the next twenty years, sources which will yield twenty-five million tons annually, from areas which are not now being tapped. In this way one company can increase its future production by half a billion tons, but it will take two decades to do so.

The new sources of steel, which U.S. Steel and other companies will seek to exploit, consist first of all of the billions of tons of low grade ore still available in the United States, and secondly, of ores which can be found in foreign countries. One of these low grade ores is taconite, a hard rock which contains about twenty-five to thirty per cent of iron. Research in utilizing taconite ores is being intensified. One subsidiary of a large steel company, the Oliver Iron Mining Company, has constructed a plant with an annual capacity of half a million tons of taconite concentrate, of sixty per cent iron content. This is the beginning of a very large program for concentrating that type of ore. Steel companies are now beginning to utilize the low-grade ores of the Superior district; they are also beginning to pay more attention to similar ores in New York, New Jersey and Pennsylvania. The ore of the Alabama district, while of lower grade than that of the Superior ranges, continues to hold its own because of its close proximity to the steel manufacturing area of that state.

Steel companies have found some promising ore deposits outside continental United States. Some of these are in Liberia, West Africa. Another field has been found in the wilderness of Quebec Province, where, it is estimated, are some 400,000,000 gross tons of high grade "direct shipping" open-pit ore. Several hundred miles of railroad, it is true, will have to be constructed to get this ore to the St. Lawrence. A new region in Venezuela, possibly as rich as the Quebec findings, has been discovered and is being developed by the Orinoco Mining Company, a subsidiary of U. S. Steel. Ore from Venezuela will be conveyed directly by a fleet or ore boats to the steel plants along the Delaware. For the necessary accessories to steel making, limestone and coking coal, mines and quarries already in operation have been greatly enlarged, and new fields are being placed in operation. The newest of these coal mines, the Robena mine near Pittsburgh, is completely mechanized.

The mining of copper, one of the oldest known metals, has

had its share of the ups and downs so characteristic of mining in general. Originating as the successor of small prospectors and small mining concerns, who sold out their claims when their money ran out, the large copper mining companies of the United States have developed across the years through alternate boom and depression. The first great boom in copper followed the rise of the electrical industry, which has utilized about half of all the copper mined in the United States. World War I stimulated copper as it did all industry, but between the two wars copper was in overproduction, with resultant depression in price and cutting of sales. With the abrupt change from surplus to shortage in the virgin copper field about 1940, a new boom began, leading to new production levels in World War II.

The copper companies, during the slow-up of the 1920's, reacted generally to the bad market conditions by integrating their operations from the mine to the fabrication of finished products. In this way the Phelps Dodge Corporation and the Anaconda Copper Mining Company sought to control outlets and guarantee markets. Kennecott Copper Corporation acquired the Chase Brass Company in 1929 and the American Electrical Works in 1935; the latter became the Kennecott Wire and Cable Company. Anaconda developed a profitable market for copper wire through its wire and cable company, and Revere Copper and Brass (then known as the Taunton-New Bedford Copper Company) joined with five other copper and brass firms during 1928-1929 to assume its present form. The greatest success enjoyed by these enlarged and integrated firms was in the brass and wire fields.

At the end of World War II, a different sort of problem confronted the copper mining companies. The profits from war contracts had been large, and available reserves, even after taxes, remained substantial. In some cases such reserves amounted to a large proportion of the firms' total assets. To allow such reserves to rest, uninvested except in banks and government bonds, would subject them to the inflationery trends of the times. Yet the prospect of putting large sums to work, with available copper reserves declining, was a puzzle. The Kennecott Corporation revised its charter in 1945, permitting it to extend its interests beyond copper to all natural resources; it set aside in its annual budgets half a million dollars, and later larger sums, to carry on explorations to replace worked-out fields, not only in the copper industry, but also in mineral fuels and in other metals. In the petroleum field, Kennecott made an agreement with the Continental Oil Company, to prospect and drill for them. In South Africa, the Corporation joined forces with the Anglo-Consolidated Investment Co., and after a series of explorations, acquired a half interest in two gold mines in Orange Free State, which are expected to come into production in 1954. The most successful of the Kennecott explorations was the discovery in Quebec of the richest titanium mine in the world, which in collaboration with the New Jersey Zinc Company, it has already begun to develop. More will be said later of titanium and other light metals.

In some ways, Kennecott is more fortunate than Anaconda in the distribution of its available copper resources. Only sixteen per cent of Anaconda's copper production comes from the United States, while some three-fourths of Kennecott's is mined here. A large part of Kennecott copper ore comes from its four western mines, of which one, that at Bingham, Utah, is the largest open-pit copper mine in the world. A subsidiary of Kennecott, the Braden Mining Company, operates the El Teniente copper mine at Sewell, Chile, which is the largest underground copper mine in the world, but its productive totals run second to Anaconda's enormous open-pit at Chuquicamata, also in Chile.

Copper has been mined in the area of the El Teniente mine for centuries, but the area was first opened up in a big way by William Braden, an American engineer, in 1904; the Guggenheims acquired it in 1909 and it came into the hands of the Kennecott Corporation six years later. Since 1904 the El Teniente property has produced seven billion pounds of copper and at the present rate of extraction will probably last only for another generation. The Braden Mining Company's properties occupy 300 square miles of land, extending high up into the Chilean Andes. From a terminal called Rancagua a forty-three mile railway winds to an altitude of 7,000 feet into the mine opening, then circles through tunnels around an extinct volcano. Up the mountain slope are placed the living quarters of the population of Sewell, a mining town of 16,500 men, women and children, entirely dependent upon the extraction of copper ore from the mine, which goes on at the rate of 28,000 tons a day.

Kennecott does its own milling and a part of its smelting, but the larger part is performed for it by the American Smelting and Refining Company. Charles R. Cox, president of Kennecott, has led in the decentralization of the various properties of the concern, rendering them administratively independent. He has also increased the budgetary allowance for the exploration of possible new mining properties. This project goes on under a chief explorer with the aid of thirty geologists and twenty apprentices. To facilitate this vast scheme, teams are organized to cover various sections of the United States, in east, center and west; teams are also at work in Canada, South Africa and South America. The shortage of trained men for this work is acute, for few trained geologists are willing to go so far afield. These prospectors, quite different from the older type of aimless, free-lance workers, are looking for big, not small, deposits of copper, lead, zinc, titanium, petroleum, aluminum and uranium.

It is by these means that large mineral resources, extensive enough to justify expensive long-term development, may be uncovered. However the explorations may turn out (and they have already turned up important finds of gold and titanium), the policy is progressive and in line with the needs of the present period. In any case, Kennecott has a number of alternatives to choose, in its plans for the future. It can expand its Chase Brass subsidiary; it may extend its mining of low grade copper ores; it could add to its production of copper wire; it could go into aluminum production, as the Defense Production Administration has urged mining corporations to do; or it may expand its smaller oil and gold projects. Thus has a typical large mining company adjusted its activities to the changed conditions of the age.

* * *

Among the so-called major metals, in addition to iron and copper, the United States has been for most of its history a large producer of lead and zinc, but has almost no native tin. Imports of the latter metal are very large, since tin in bars, pigs or in granular form is utilized in many industrial processes. With imports from Malaya, the largest source of the metal, tin enters the manufacture of bronze, solder, tubing, foil, coating and plate. Tin is also a factor in the making of printer's type.

Lead is one of the oldest metals known to man and is one of the most useful. It is heavy, durable, and has characteristics which make it an important factor in the electrical, automobile, shipbuilding and munitions industries. Welders and plumbers use it, and for many years its red and white pigments enjoyed a unique position as raw materials for paint. Recently the value of lead pigments has declined due to competition with other metallic pigments, as zinc oxide, lithopone, and titanium compounds. World production in lead is confined in large part to four major countries: the United States, Canada, Mexico and Australia, which between them furnish about three-quarters of the world supply. In the United States, the Missouri region is the leading area of production; five of the largest mines, including the largest lead mine in the nation, are operated by the St. Joseph Lead Company. Other large operators of lead mines are the National Lead Company, Phelps Dodge, Anaconda Copper and Kennecott Copper. Other operators include the "smelting and refining group": the American, the United States, and the Federal Smelting and Refining Companies. The fact that lead ore is frequently found in combination with zinc, silver or copper accounts for the interest of general metal companies in lead mining.

Zinc has frequently been classed with lead, and spoken of in similar connection, often in the same terms. The two metals are alike in that they are found and mined in the same locality, even in the same ore masses. Both are important sources of paint pigments. Zinc also serves industry as a protection for galvanized materials through its resistance to rust and corrosion; it gives to copper added strength and hardness in the manufacture of brass products. The zinc market also includes manufacturers of roofing materials, electrical products, rubber tires and chemicals. The New Jersey Zinc Company, St. Joseph Lead, Phelps Dodge, Anaconda Copper and the smelting and refining group are the leading operators of zinc mines. The areas richest in zinc are: the tri-state area of Missouri, Kansas and Oklahoma; the Tennessee-Virginia district; St. Lawrence County, New York; Sussex County, New Jersey; and the Rocky Mountain mining states, particularly Idaho, the largest of zinc producing states.

Not only have larger quantities of the older and better-known metals become available with the passing years; new metals have been discovered, some of them suitable to replace or supplement the others; in some cases they have qualities which the others do not possess. The most valuable characteristic of metals has always been their strength, a quality which has usually been associated with great weight. It was very important, therefore, when research in ore reduction disclosed metals which possess considerable structural strength, but which have also the advantage of lightness.

* * *

Sir Humphry Davy suspected that the earth called "alumina" might be the oxide of a metal. Some years later H. C. Oersted, a Danish chemist, proved this to be the case when he isolated a small bit of aluminum. Other scientists, including Friedrich Wohler in Germany, and Henri St. Claire Deville in France, followed up Oersted's pioneer effort. Deville was able to produce a solid bar of silvery aluminum for exhibition at the Paris Exposition of 1855. Aluminum made a sensation when it was first seen and handled; attractive in appearance, its weight was only a third of that if iron. It resisted corrosion, would conduct heat and had reflective qualities. While not as strong as iron, it might be strong enough for many purposes. Furthermore, it was prevalent everywhere; one-twelfth of the earth's crust, it was said, was made up of aluminum.

Grave problems, however, confronted scientists who saw in this fascinating new metal the answer to metallurgical difficulties. Its extraction and reduction were slow and expensive operations. Early investigators had used heat and chemical agents, principally potassium and sodium, to produce small amounts of the metal. But even Deville, the most successful of the early aluminum researchers, could not reduce the cost below eighty-five francs, or about $17 per pound, which precluded commercial use. In the 1880's a young chemist of Oberlin, Ohio, succeeded in reducing aluminum oxide to metallic aluminum by passing an electric current through a mixture of the oxide with cryolite, a sodium aluminum fluoride. At about the same time as Charles Hall's experiment, a Frenchman named Paul Heroult performed a similar electrolytic break-down of the metal. The two similar processes became basic in the manufacture of aluminum.

The great significance of the two discoveries was that the expense of manufacture was cut far below its former excessive cost. Both men, Hall in the United States, and Heroult in France, had some difficulty in persuading capitalists to back their ventures. At length Swiss and French investors came to Heroult's aid. In the United States, Hall found a group of men in Pittsburgh, under the lead of Captain Alfred E. Hunt, who were willing to attempt the development of aluminum. In 1888 the Pittsburgh Reduction Company was formed; in 1907 it became the Aluminum Company of America, one of the great American industrial corporations.

The basic geological and metallurgical data about aluminum are quite well known. The ore from which it is prepared is called bauxite, perhaps from the fact that large amounts of it were discovered early in the vicinity of Les Baux in southern France. When the Pittsburgh company began operations, it started with aluminum oxide, or "alumina," a white powder prepared chemically from the ore. Since the Pennsylvania Salt Company had large amounts of aluminum oxide available, the new aluminum manufacturers purchased their supplies largely of them, instead of going into ore smelting, or entering the mining field.

The oldest and largest source of bauxite ore in the United States is in several southern states. The first discoveries were made in Georgia in 1883; the most valuable deposits are in two counties of Arkansas, which were found to have bauxite in 1887. After about fifteen years of processing aluminum from aluminum oxide, the Pittsburgh Reduction Company entered the mining field. In 1904 the General Bauxite Company, a subsidiary of the General Chemical Company, came into the hands of PRC; in 1909 the Republic Mining and Manufacturing Company, which had operated the Georgia and Arkansas bauxite fields, was also acquired. These fields, however, long ago became insufficient for the needs of the large aluminum companies; Alcoa, as the Aluminum Company of America is popularly designated, secures its principal supplies of bauxite from Surinam, 2500 miles from the United States. Kaiser Aluminum and Chemical Company and the Reynolds Metal Company, competitors of Alcoa, secure large supplies of the ore from Jamaica, British West Indies. Aluminum Limited, the largest Canadian Company, obtains ore from the Los Islands, off French Guinea in West Africa. Other large bauxite fields are being prepared for use in the East Indies, Africa and South America.

By 1900 the Pittsburgh Reduction Company, Alcoa to be, began to experience difficulty in obtaining an adequate supply of alumina and so was led to enter the mining and refining fields. At first it refined bauxite by a precipitation process with carbon dioxide. After 1911, however, a better method became available to all aluminum refiners through the expiration of the exclusive Bayer patent, which had run since 1894. This process included "digesting" the ore under pressure with hot caustic soda liquor to form sodium aluminate, which is soluble and can be passed through the silica, iron oxide and other impurities. The dissolved alumina is then precipitated from the filtered liquor as alumina trihydrate. After washing and calcining, it becomes aluminum oxide, ready for the electrolytic process of making it into metallic aluminum. The two processes, Bayer's, for treating the ore, and the Hall method of reducing the oxide to the metal, are today the most common methods of aluminum preparation.

In the great aluminum plants, successful manufacture is dependent upon the quality of bauxite used; while aluminum can doubtless be made from almost any alumina-bearing clay, to use all such clays would be impracticable, since most of them have too much silica and too little of the oxide. Alumina has an annoying affinity for soda and silica in the digesting process. Minerals like kaolin contain so much silica that their alumina content cannot be economically extracted. The richer ores are very scarce in the United States, and hence aluminum companies must continually import bauxite to supplement the inadequate supply of native ore. During the last war, twenty-six ore ships of the Alcoa fleet were torpedoed by submarines between Surinam and the United States, a fact which calls attention to the need for safeguarding that particular sea lane, as well as others, in times of national emergency.

Few of the aluminum pioneers, whether scientists or capitalists, had very clear ideas about the markets which could be opened for the finished metal. Some uses of bauxite and alumina were known, even before the metal was made in any quantity. Abrasive manufacturers used bauxite; salt refiners used alumina. The early price of the metal, two dollars per pound in 1889 and seventy-five cents in 1893, was still too high for any large scale demand. Small quantities were used early by novelty makers, in surgical instruments, and as a deoxidizing agent in making steel. An interesting tale is told about the way the Pittsburgh Reduction Company got into the aluminum utensil business. In the early years of the company, Mr. A. V. Davis, one of the founders of the firm and later chairman of the Alcoa board, visited Ely Griswold, a utensil maker, to show him a teakettle which had been molded of aluminum. The object of Davis' visit was to induce Mr. Griswold to go into aluminum ware manufacturing and thus become a large buyer of aluminum. Griswold was greatly impressed with the kettle, but instead of deciding to make such ware, he gave Davis an immediate order for 2,000 of the kettles which he said he would try to sell. So there was nothing for PRC to do but add a fabricating unit to its New Kensington plant and make the kettles.

With the turn of the century, many new uses of aluminum appeared. The metal interested the makers of fruit and preserve jars, and orders from such companies opened up the very large field of covers and seals for all kinds of bottles and containers. A significant purchase of aluminum from the Pittsburgh firm was made in 1903 by two brothers named Wright, as a light-weight material for the airplane engine which they were trying to build. Just before the advent of the motor car, the company perfected an aluminum alloy hard enough to use for horseshoes; even though the automobile soon opened up a much larger market than the horseshoe business would ever have afforded, it is a fact that the race horses of today still wear the strong but light aluminum alloy shoes.

Near the beginning of the century, two very efficient salesmen of PRC convinced large power companies on the West Coast of the superior virtues of aluminum wire. With orders for a substantial quantity of wire, but with no suitable mills in existence, the company had to build its own wire mill. The results, with single and multiple strand wire, were not satisfactory and it became necessary to evolve a steel-reinforced aluminum cable. Some two million miles of this type are in use today on high-tension transmission lines.

In the last generation aluminum has come into its own. A list of its uses is so long and so well-known as to be unnecessary to enumerate. It has become well-nigh indispensable in the airplane industry for engine parts, propellers, frames, outer coverings and parts for landing gear; it plays a vital part in the making of railroad cars, automobiles, and commercial buildings; it has become a factor in the electrical power and communications industry and enters into the manufacturing of a number of household appliances; it serves the chemical industry and has helped to revolutionize the packaging of food and drugs.

A unique feature of the aluminum business is its great dependence upon electric power. It takes ten kilowatts of electric current to manufacture one pound of aluminum. In 1943, the peak year of production of this metal, the industry consumed twenty-two billion kilowatts of current, making it the largest single user of electric power in the United States. The Pittsburgh Reduction Company was an early customer of the Niagara Falls Power Company; many years later, as Alcoa, it purchased billions of kilowatts from TVA. It has been compelled, as have its competitors, to seek greater and greater power resources. The growth of the aluminum business has thus been closely tied to the improvements in hydro-electric and turbo-electric power machinery. To produce a billion and a half pounds of aluminum annually, which is Alcoa's present rate, some fifteen billion kilowatt hours of electricity are needed.

It will be recalled that in the Hall electrolytic process, large amounts of cryolite were needed, and still are, as a solvent material. This chemical compound is hard to obtain, as its main source is Greenland, from which only a few shiploads a year can be secured. It was therefore necessary to supplement the supply of natural cryolite by a synthetic product, made from fluorspar. Aluminum makers had also to pioneer in building rolling mills to run aluminum ingots into usable form, as the steel and brass companies, which had the only available facilities, did not wish to undertake the job. They also had to develop the process of extrusion, or of pressing aluminum into desired shapes; it was quite similar in the making of tubing, forgings and screw machine products. Between the process of making an aluminum teakettle, which the public had to be taught to desire, and the present high development of the industry, lie sixty years of research, adjustment, expansion and effective salesmanship.

* * *

Magnesium is another light metal, the usefulness of which, as a structural material, has become important only recently. During World War I it was used as an ingredient for flares and explosives. At the close of the war, the principal manufacturer of magnesium in the United States was American Magnesium Corporation. Through the latter's need for capital, it was willing to allow Alcoa to acquire a controlling interest in its firm, and Alcoa acquired such interest between 1919 and 1924. The metal was manufactured by an electrolytic process at Alcoa's magnesium plant at Niagara Falls. Beginning with the oxide of the metal, it was reduced by use of a molten salt bath of mixed fluorides. Meanwhile the Dow Chemical Company began to produce magnesium from magnesium chloride, which had been a useless byproduct of other chemical processes. Since the Dow method was cheaper than Alcoa's fluoride process, the latter stopped its metal production and began to purchase its materials from Dow. American Magnesium, the Alcoa subsidiary, continued to operate thereafter as a manufacturer of mill products with metal furnished by the Dow firm.

Germany had been a very large producer of magnesium, especially before World War II. She possessed dolomite and other magnesium-bearing ores in large quantity, but for her mobilization needs she lacked copper and had to import bauxite for aluminum production. Faced with such a situation, she turned more and more to magnesium production for aircraft and other war material, even for products which on this side of the ocean can be made of other metals of lower cost and better performance. While Germany was making great strides in magnesium manufacture, metal manufacturers in the United States were slow to enter the field. Dow Chemical was the only producer in the United States in 1939, at the outset of the second World War, and American Magnesium was the largest fabricator of products made from it. The Bureau of Aeronautics was just beginning to urge manufacturers to consider its use for "unstressed" parts. Yet as late as the end of 1940, American Magnesium was still losing money. To overcome its lack of control of magnesium processes, the parent firm, Alcoa, had earlier entered into an agreement with

I. G. Farbenindustrie by which a new corporation was set up, called the Magnesium Development Corporation, of which each company, Alcoa and I. G. Farben, would own half the stock, and would share between them the patents formerly controlled by I. G. Farben. After a complicated series of transactions between the two above and Dow Chemical, which attracted the unfavorable attention of the U. S. Department of Justice, a new war came and the demands for magnesium rose at once to high levels; in a single year, 1943, American production reached a total of 370,000,000 pounds. In the postwar years, the uses of magnesium show a large increase over the earlier years; large amounts are required for the printing industry in photo-engraving, much for aircraft manufacture, household appliances, yard tools and textile machinery.

* * *

One of the newest prospective competitors of steel, copper and aluminum is titanium. The United States Bureau of Mines pioneered in pilot plant experiments in titanium manufacture, under the direction of Dr. Kroll, a Luxembourg chemist; favorable results were announced in 1948. While first efforts to produce the metal have thus been quite recent, the pigments obtainable from titanium oxide have competed with lead and zinc pigments since the 1920's. Since 1940 a few small mining properties, among them a du Pont mine in Florida, a National Lead project in New York State, and a mine in Virginia operated by American Cyanamid, have been the principal producers of ilmenite ore, which contains large proportions of iron and titanium, and of similar ores from which titanium may be obtained. The reduction of ore to metal on a commercial basis was apparently begun by the du Pont Company in pilot plant operations at Newport, Delaware, in 1948.

Unquestionably the most important developments in the titanium field have come about as a result of the exploration activities of the Kennecott Copper Company toward the end of World War II, which have been already referred to. Near Havre St. Pierre, in the Allard Lake region of Quebec Province, Canada, claims were staked out in 1942; as a result of observations conducted jointly by teams of prospectors of Kennecott and the New Jersey Zinc Company, these claims were purchased by the two firms during 1944-1946. Preliminary investigations showed that in a region of some 1500 square miles, overlaid with swamp and glacial drift, lay a body of 125,-000,000 tons of ilmenite ore, a figure which might reach double that amount when fully worked. To tap this ore body, the two companies formed the Quebec Iron and Titanium Corporation. The Lac Tio mine, principal working face of this immense mass of ore, is undoubtedly the largest ilmenite ore mine in the world.

The arrangement between the two companies is an extremely advantageous one. Kennecott has gained a promising use for its reserve capital, and New Jersey Zinc stands to gain a large share in the profitable pigment resources which will become available, pigments which already compete seriously with those of zinc, which are so large a share of NJZ's commercial business. In addition, the zinc people have evolved an effective process of smelting titanium, an almost unsmeltable metal, which thus becomes available to the new joint enterprise. Kennecott likewise gains a share in the valuable pigment business, which is still the most profitable use of titanium, and will also gain large amounts of the metal-bearing ore to meet future demands for the metal itself. While the cost of titanium production is still very high and knowledge of its usefulness as an alloy is still incomplete, it is likely that the metal will find a high place among aluminum alloys and stainless steel products. The weight-strength ratio of the metal is said to be greater than those of steel and aluminum; its resistance to rust and salt-water corrosion is high. Its possibilities are viewed with great interest by Navy and Air Force personnel, always on the watch for metals of lightness and durability.

Our brief survey of some of the principal metallic resources has included both the older metals, the traditionally vital iron, copper, lead and zinc, and some of the lighter metals which have become important in more recent years. Space forbids extended discussion of the many metals which serve the steel industry, as vanadium, cobalt, tungsten, chromium and molybdenum. Feldspar serves the ceramic industry; fluorspar and cryolite are vital to the manufacture of aluminum metal; uranium, of which more will be said elsewhere, finds its chief importance in the development of atomic energy. A serious feature of our mineral economy is the lack, so far as is known today, of vital metals which occupy an important position in American metallurgical processes. Our domestic supplies of manganese are much too small to supply the demand of the steel industry, principal user of such ores. The Soviet Union, which shipped to the United States a third of our total import requirements of manganese up to 1948, cut this item of trade to token shipments thereafter. This made it necessary for manufacturers of steel and of electrical batteries to look elsewhere; with the aid of an inter-departmental government committee, headed by the Director of the Bureau of Mines, new sources of the metal were located in the Gold Coast, India and the Union of South Africa which will go far toward remedying this serious scarcity.

It cannot be urged too strongly that there must be no letting down in the search for new mineral sources and study of the most efficient uses of those which we have. Even in the case of metals of which we still have vast deposits, as of iron and copper, our brief analysis has shown that we have used up dangerously large proportions of what seemed at one time to be inexhaustible reserves. The many-sided conservation and expansion program which has been described earlier is the only way to assure the continued operation of our industrial economy at a level which will maintain the present standard of living. Strong supports of the planning which must take place in this field are the continued advance of metallurgical technology and the fostering by all peaceful means of the vital foreign trade which is a necessary feeder of our mineral raw material supply.

CHAPTER EIGHT

THE NON-METALLIC MINERALS

No one can deny that continental United States is one of the world's richest storehouses of valuable minerals. Its large area and the infinite variety of its land mass, taken in connection with the geological history of its physical structure, have made this true. Nature has laid down in the depths of the nation hundreds of different kinds of minerals, many of which serve important purposes in American industry. The metals, mined and processed in the most effective ways known to modern industrial science, have been basic to our industrial growth. But the metals form only one facet of the total industrial picture.

Deep in the physical structure of the land are countless non-metallic minerals which furnish the energy to run the industrial machinery and which also act as raw materials for hundreds of manufacturing processes. Without them, the great veins of mineral ore would be worthless. There are the hydrocarbons—gaseous, oleaginous and solid; there are basic chemicals imbedded in rock, dissolved in liquids, or lying in an almost pure state under the surface. Almost any stretch of coastal water, any range of rocky hillside, even the most ordinary farm or field, may conceal valuable mineral wealth. To a large extent, the richest raw material of the industrial age is hidden from view, and may only be found by unearthing what lies below the top cover of the land, or by delving beneath the swamps, rivers, lakes or coastal waters. The ocean itself, scientists have stated, is probably the richest storehouse of valuable minerals and chemicals.

Petroleum and the other Fuels

The great mineral fuels—coal, natural gas and petroleum—have long played an important role in our industrial economy. During the nineteenth century the basic industrial operations depended upon our massive resources of anthracite and bituminous coal. We have noted that coal, in that period, supplied heat for homes, offices and factories; provided the best energy source for steam power in manufacturing and transportation; and became the largest single item of commercial freight in the formative period of our railroad transportation system. As one views the growth of industrialism in the late nineteenth century, it is not too much to say that ours was a "coal" economy. The growing steel industry, the railroads, and manufacturing in general all depended exclusively upon it. The accessibility of coal deposits close to other mineral resources was an important factor in the development of nineteenth century American industry.

Before the century came to a close, the picture began to change. While coal still furnished a large proportion of the heat and power needed to keep American industry humming, it no longer had the unique position of a former day. During the last generation of the century, the energy base of American industry had been broadened by the advance of mining technology, while in the first fifty years of the present century we have seen added the vast resources of natural and synthetic gases, petroleum, hydro-electric power and atomic energy. Not all of these energy sources are completely independent of one another, we have found; in some cases, they represent conversion, either by nature or by man, of one form of energy into another. In the case of petroleum and of natural gas, however, we have an immense addition to the available energy resources.

It is a far cry indeed from Colonel Drake's shallow oil well in Titusville, back in 1859, to the gigantic worldwide industry which has emerged from such humble beginnings. For many years petroleum was used largely for illuminating purposes, plus minor use as a lubricant, and even as a medicine. Even during the vast build-up of Rockefeller's Standard Oil Company of Ohio, a concern which, with its various affiliates was able to gain control of seventy-five per cent of the refining and ninety per cent of the pipelines in a few short years, there was still a comparatively limited use for the product. The future of the oil business, it seemed in the 1880's, lay in the development of oil technology and wider and more diversified markets. In those early days, before the wide use of the internal combustion engine or the extensive use of electric lighting, there was possible a worldwide demand for kerosene; lubricants were needed more and more in factories; vasoline, greases, and sulphur were obtained from petroleum, but the gasoline was thrown away as useless.

The coming of the automobile provided the strongest stimulus to the extracting and refining of petroleum since its original discovery, yet despite the pleasure which the sudden demand for the more volatile byproducts of petroleum gave the producers, the new development involved the oil companies in a dilemma. The newer petroleum fields of the mid-west and Gulf areas which became the chief source of "crude" after 1900, did not yield as high a gasoline-bearing product as the lighter Pennsylvania type. Therefore the producers were forced either to accumulate large stocks of unsaleable byproducts, after the gasoline was removed, or find a method of increasing the ratio of gasoline to the amount of crude oil which was refined. This latter alternative became possible through the "cracking" process.

"Cracking" petroleum involves the breaking of large molecules into smaller ones by means of heat, a process also termed "pyrolysis" by organic chemists. While such a molecular change is to some extent an unpredictable process, nevertheless research was successful in reaching almost uniform results when the same cracking stock was used successively under identical conditions of temperature, pressure and time. Though numerous experiments had been made during the nineteenth century in combined cracking-distillation processes, no practical use was made of any of them until about 1910. At that time the demand for motor gasoline had begun to exceed available supplies and there was a powerful demand for securing more gasoline from crudes than the ordinary distillation methods could provide. This older process seldom produced as much as a twenty per cent yield of gasoline, whereas by cracking and re-cracking it has been found possible to obtain yields of fifty to sixty per cent. Cracking also produces better gasoline, as well as more per volume of crude oil, for in the new product are more unsaturated hydrocarbons than in the old "straight-run" gasoline; these newer fuels, together with anti-knock compounds such as tetraethyl lead, have brought in an era of high compression motors of greater efficiency than those of a generation before. This series of developments was naturally of the highest importance to the petroleum industry. It helped to solve the problem of adjusting petroleum refining to new demands, and effectively added to our potential reserves of gasoline, since most if not all of our untouched oil reserves may be processed in the same way.

The volume of gasoline has also been increased by the conversion of natural gas into "natural" or "casing-head" gasoline. While much of the liquified production of natural gas wells is too volatile to be used as a constituent of gasoline, it can usually be processed into suitable fuel by mixture with heavier gasoline produced from crude oil. As high as ten per cent of our annual gasoline production has been furnished by use of natural gas.

The petroleum industry has thus resorted to a number of measures to increase the available gasoline supply, which, since the developments in fast land, water and air transportation, has assumed high importance in these fields. Of all these measures, it is probable that the cracking process, developed by several companies at about the same time, has been the most important conservation move. The Standard Oil Company of Indiana had its Burton process; the Texas Company the Holmes-Manley process, and there were several others. Along with improvements in refining, improved methods of extracting petroleum from the ground have been devised, and an intensive search for new oil fields has gone on. For not only has transportation placed extreme demands upon the oil supply, but also the widespread use of oil as a heating fuel and the increased demands for lubricants in our expanded manufacturing industries have added to the pressure.

* * *

The extraction of oil from the ground depends upon certain physical conditions. First of all, it is found only in sedimentary rock layers, and where these are not extensive, oil cannot be extracted with profit. The position of petroleum in strata of sand or porous rock poses a tough problem, for the oil adheres closely to these strata; the oil film stays close to the sand grains, whether near the surface or thousands of feet below. The only part which is available for pumping operations, usually, is that which remains atop the sand beds after the sand particles below are thoroughly saturated. Thus, under ordinary conditions, a vast proportion of the oil stays in the ground and cannot be brought to the surface by merely pumping at these oil beds. Various means have been tried to force larger proportions of the stubborn substance to the well head. Natural gas under high pressure is frequently forced into the sand stratum; in some cases water is forced into deep wells under pressures of 750 to 1000 pounds per square inch. Solutions of hydrochloric acid are pumped into wells in which a part of the underground structure is made up of limestone; the fact that a part of the rock or sand structure is soluble in the acid can lead to greater porosity near the bottom of the well; into these new porous spaces oil may flow and be pumped out. None of these methods has been found to be wholly efficacious. Other proposals include the mining of the oil sands to recover the petroleum content, a scheme with many obvious difficulties. In the first place, there is the danger of inflammable gases; the petroleum lies usually at too great a depth to make mining operations easy; the rock structures are fragile and to brace the shafts and tunnels is very difficult. When we realize that oil wells sometimes penetrate to depths of nearly three miles, and that even greater depths than this are under consideration, it can be readily seen that, at present, mining the oil sands is impracticable.

The problem of getting all the oil, or even a large part of it out of the ground thus presents one of the most difficult technical problems of the age. Since such a large part is almost bound to remain in the ground, and since, of the amount extracted, a considerable part is lost by fire, seepage, or evaporation, it has been estimated that perhaps less than twenty-five per cent of the underground oil reaches the pipeline. Waste in utilization also lessens the amount of energy which is available to perform useful work.

Other large resources in oil exist which are not at present tapped because pumping petroleum from the ground is still much cheaper than production from oil-bearing shales, for instance, would be. Some of the chemical and engineering problems connected with the recovery of oil from shale have been studied in pilot plant operations by the United States

Bureau of Mines, but no methods have as yet been devised which are economically feasible. Synthetic motor oils have been produced from coal by distillation and hydrogenation processes. There may be a possibility of production of enough grain alcohol to save a material amount of our petroleum resources by using the alcohol as a blending agent with gasoline. The grain requirements for this are huge, however, and it is likely that even limited amounts of such blended fuel would add to the total cost of motor fuel per gallon. Long range planning for conserving our oil reserves still leaves much to be done, and the danger of their ultimate exhaustion is no idle fear.

Much of the foregoing refers to future possibilities not applicable to the present situation. There is no shortage of petroleum for predictable needs in the next few years, large as our present requirements are. The increase in petroleum production during the past half-century calls forth figures of astronomical size. World consumption at the opening of the century was 160,000,000 barrels a year; by 1949 this had increased to 3,600,000,000 barrels, an increase of 2000 per cent; of this latter amount, North America consumed about two-thirds. The United States has for many years used more oil than the rest of the world put together, and consumed more than two billion barrels in 1949. On the last day of December, 1949, the Bureau of Mines estimated that our "proved" resources of liquid hydrocarbons, including crude oil reserves and natural gas liquids, amounted to more than twenty-eight billion barrels, a supply sufficient for about thirteen years, at the present rate of consumption. However, it is also estimated by oil men that there exist some 1,500,000 square miles of territory in the United States, or about one half of its total area, in which oil may yet be found; a large amount of this huge stretch of land has never been explored with the possibility in mind of finding oil. In addition, oil technology has developed to such an extent that deeper wells are being driven all the time, and depths to 25,000 feet are considered possible. Furthermore, many important discoveries have been made of oil fields under water along the continental shelf; extensive extraction of this "off-shore" oil has been going on for some time in the Gulf of Mexico and along the Pacific Coast.

To a considerable extent, the discovery and development of our natural gas resources have paralleled the advance of petroleum as an item of mineral production and as an industrial raw material. Gas is often found in association with petroleum, and its importance in the extraction of oil from the ground and its place in the production of hydrocarbon products have been mentioned. The proved recoverable reserves of natural gas in the United States at the end of 1949 were placed at the impressive figure of over 180,000,000,000,000 cubic feet. In 1948 more than seven trillion cubic feet were produced, of which about two-thirds came from the gas wells, one-third from oil wells. Of this total production, about five trillion cubic feet were marketed, one trillion reserved for further refining, and the rest wasted or lost. The large increase in consumption of natural gas is accounted for by greater use in electric power plants, petroleum refineries, portland cement plants and other industrial establishments. A number of cities have converted from mixed gas usage to natural gas. The introduction of combination gas-oil burners in the appliance industry and the improvement of the gas turbine have both had some influence on natural gas consumption, and may have even more in the near future.

Thousands of miles of pipelines carry natural gas great distances from its source to its area of consumption. The interstate shipment of gas, largely by the pipeline method, has been widely extended in recent years. The increase in interstate shipments was twenty-five per cent during 1948, and the total of new pipelines authorized by the Federal Power Commission in 1949 was 7,500 miles. Probably the most arresting alleged fact about our natural gas reserves is that their total represents an energy asset about one-third greater than all our proved reserves of crude oil. Since this vast amount of gaseous energy is partially available for conversion to liquid fuel, its importance in the general energy picture is evident.

Vast as are the still-untouched, though not unlimited, supplies of energy in the form of oil and natural gas in this country, they still do not equal the energy represented by our proved reserves of coal. So far as known world reserves of minerals are concerned, the United States is in the best position in coal, with a total of forty-seven per cent, nearly half, of the world supply. The estimated coal reserves of the world, according to figures compiled a few years ago by the Bureau of Mines, total more than eight trillion tons; of this figure, according to the same estimate, North America possesses reserves of more than five and one-half trillion tons. Since United States production of all types of coal did not reach 500 million tons by 1949, it would seem that hemispheric, and United States, coal reserves will be sufficient for many, many years to come.

Much speculation has gone on concerning the competition between coal and petroleum, and between the latter fuel and gas. While such statistical comparisons may be meaningful to individual mining corporations in the promotion of their special products, the view which is more significant to most Americans is that these storehouses of fuel are supplementary. The tremendous reserves of coal, and the smaller though very considerable amounts of energy in liquid and gaseous form add up to a very full storehouse of fuels for all predictable use for quite a long time. While a bituminous operator may worry over the sharp decline in soft coal consumption, technology and business efficiency can probably solve his problem; he may seek new markets among the chemical manufacturers, or among steel and electrical industrialists. An important new market for coal products lies in the coal-chemical field, especially for ammonia, crude light oils, and tar products.

Valuable byproducts of the coke process, which in itself provides a huge market for the bituminous operators, include benzol, toluol and xylol, derivatives of light oil. Ammonia and tar derivatives include creosote oil, tar acid oil, phenol, and pitch-of-tar.

The coal industry, modernized into the coal products and coal-chemical industries, and holding in addition a considerable portion of its traditional markets, seems in little danger of being superseded by the other fuels. There is probably an insufficient supply of the other fuels to meet the increasing demands of the steel, electrical, chemical and transportation industries, to name but a few. Viewing the fuel situation comprehensively, we should begin properly with coal as the basic energy fuel, since it is universally used and is in such plentiful supply, and we should still consider petroleum and natural gas as alternative or supplementary fuels. It may even be that ultimately we will be forced back into complete dependence upon coal with the near exhaustion of the recoverable supplies of the others. We can still function industrially, even then, with the help of chemical science, by processing coal into whatever form we wish.

The Non-Metallic Minerals, other than Fuels

Metals and fuels are the most valuable and necessary of our mineral resources, yet industrial requirements are so complex that a wide variety of minerals usually classed as non-metallic are vital in the utilizing of our metals and fuels in the most effective way. Just to name the more important ones would establish the truth of this statement. In some cases these minerals are extracted only, by firms which specialize in that type of mining operation; in other cases, as for instance in the asbestos and gypsum industries, the mining of the raw material is integrated closely with the fabrication and marketing of products manufactured therefrom; in the case of another group of minerals which are closely related to chemical industry, the mining of the materials is usually carried on by the chemical companies themselves.

A large group of non-metallics include the various mineral pigments which have been mentioned in connection with their related metals; zinc oxide, lead oxide and titanium oxide are used extensively in the manufacture of paints and lacquers. Sulphur is one of the most important of all non-metallic minerals, furnishing as it does the basic element for sulphuric acid, highly useful in the chemical and petroleum industries and in various metallic fabricating businesses.

The halogens, which include chlorine, iodine, bromine and fluorine, have considerable industrial importance. Chlorine is most important, as the base for hydrochloric acid, which is an important industrial aid. Chlorine itself, which is prepared by an electrolytic process from common salt or brine, has uses in the paper and textile industries, as a bleaching agent; water purification requires large amounts. Combinations of chlorine, as calcium chloride, furnish road dressing, are used in refrigerating brines, in the dust-proofing of coal and as a dehumidifier for storage areas and basements.

Bromine, an element highly noxious to human flesh, has many important uses. The photographic industry uses a good deal in preparing emulsions; it serves the pharmaceutical industry; gasoline refineries take a large supply for the making of ethyl fluid. The production of bromine in the United States in 1949 exceeded 88 million pounds. The source is largely from salt brines carrying bromine in the form of sodium bromide or in some other compound. Until the early 1930's, the chief area of supply was in the vicinity of the Thumb district in Michigan; since then a very large source of supply has been the ocean; Dow Chemical Company pioneered in this field. One very large plant, at Wilmington, North Carolina, was developed by this concern and has taken bromine from the Atlantic at the rate of 50,000 pounds a day.

Iodine is much less important than the two halogens just mentioned. Used chiefly as a medicinal, it is also useful with potassium (as potassium iodide) as a photographic emulsion and as a component of animal feeds. The principal producers of iodine are the Dow Chemical Company and Deepwater Chemical Company, who recover a large proportion of this mineral from waste oilfield brines in California. In Japan and in some parts of Europe, iodine is obtained, though with considerable difficulty, from kelp, a kind of seaweed. Before American chemical industry undertook its domestic production, most of the world's supply came from the Chilean saltpeter firms, who controlled as well the world price, which remained at four dollars per pounds for many years; since the entry of competition into the iodine market, the situation has changed and the price of crude iodine dropped as low as a dollar a pound; it is now about $1.50.

Fluorine, a highly corrosive gas, has no particular importance in itself. Processed into hydrofluoric acid, it has a use in the etching of glass. The source of fluorine is fluorspar, or calcium fluoride, which turns up in lead and zinc ores frequently. Crystalline fluorspar is used in the optical industry, as a fixing material; it may also serve the glass manufacturing industry as an ingredient which lowers the melting point of the glass mixture. The United States is the principal producer of fluorspar; the principal areas of production are in Kentucky and Illinois and to some extent in the southwestern states.

Asbestos is a name applied to a class of minerals, which, though mined in rocky formations, have a textile-like quality. These minerals can be spun and woven into threads and clothlike material; the best-known characteristic of asbestos is its fire-proof quality and in various forms finds its place in electrical insulation, brake-lining, steam-line insulation, in roofing and other building materials and to some extent, in corrosion-resistant chemical ware. Consumption of asbestos is rather large in the United States, having exceeded half a million tons in 1953, but the domestic reserves of the mineral group are less than a tenth of that, making it necessary to import large supplies from Canada, East Africa and other locations. The largest domestic producer is the Vermont Asbestos Mines Division of the Ruberoid Company.

Graphite has a number of necessary uses, but the American supply is far from adequate. We use this mineral in "lead" pencils, lubricants, paints, stove polish, dry batteries and on brushes for electric generators. It is interesting to note that graphite and diamonds are chemically identical but have different crystalline structure. Besides being found in mineral form in many parts of the world, graphite may be manufactured from coke or anthracite coal by subjecting them to very high temperatures in an electric furnace. American graphite is of inferior quality; better grades are found in Czechoslovakia, Germany and on the islands of Ceylon and Madagascar.

The fertilizer group, as it may be called, constitutes an exceedingly important series of non-metallic minerals. Of these, the most necessary are nitrogen, phosphate, potassium and calcium. In the case of nitrogen, the atmosphere provides an inexhaustible supply and has been available to chemical companies since the Haber process for making synthetic ammonia from the nitrogen of the air was perfected in 1912. This process revolutionized the production of nitrogenous products for fertilizer material and for explosives.

The forcing of nitrogen into some useful chemical combination intrigued scientists for many generations; various methods were tried as research went on independently in several different countries; the efforts were stimulated by the possibility that the saltpeter beds of Chile, the world's chief supply, might be depleted. While there are differences between the various fixation processes, they all follow the Haber method, perfected in Germany just before World War I. This discovery was incidentally of much value to Germany after 1914, when the Allied blockade effectively shut off imports of Chilean nitrates, as well as other strategic materials, into that embattled nation. The United States is the largest producer and importer of nitrogen and nitrogenous products; nearly five million tons of nitrogen compounds were produced here in 1953, principally ammonium nitrate, anhydrous ammonia, and ammonium sulphate. The greatest source of these is the air, the next largest amount comes from coal, as a coke-oven product, and a very small amount from Chilean nitrates.

In the production of phosphates the United States is peculiarly fortunate in that its output of phosphate rock is larger than that of any country, and indeed has amounted to more than half of the world supply since 1949. The largest source of mined phosphate rock is in Florida, where about two-thirds of the national supply is found. A large proportion of the extraction is performed by chemical companies such as American Cyanamid and Monsanto; the second largest producing state is Tennessee, in which a sizable amount is extracted by the Tennessee Valley Authority.

Potassium is one of the most abundant elements in the earth's crust, but a large share of it is locked up in the insoluble igneous rocks, as feldspar. When these rocks are eventually broken down, the potassium compounds become soluble and are carried down with water, often into the sea. Yet many large deposits are saved by finding their way into land-locked bodies of water, such as the Dead Sea in Asia and Searles Lake in California. During the first World War potash was in serious shortage, but since that time the situation has been improved by the development of large sites like the Searles Lake region, where enough has been found to satisfy our national needs for many years; more recently much larger deposits have been found in Texas and New Mexico, in many cases by oil drillers and in some cases by Bureau of Mines explorers. On the Pacific coast the only producer is the American Potash and Chemical Corporation; in the southwest, the deposits are being worked by the United States Potash

Company, the Potash Company of America, and the International Minerals and Chemical Corporation. Dow Chemical Company produces some potassium chloride from its brine wells in Michigan; there are a number of others. American production of potassium is at present more than sufficient for our own needs.

Calcium serves an important need as a plant food and as a mineral soil conditioner. In its role as an aid to plant growth, extraordinary amounts of this mineral are needed; tons of lime are required. It is fortunate, therefore, that calcium in the form of limestone or hydrated lime is a common material throughout the United States (and in the entire world), thanks to the concentrating activity of early marine organisms, which have been the chief agency in building up nature's vast supply. Besides its large use in agriculture, lime serves the building supplies industry, the drug industry, the chemical manufacturer, and aids in water purification. The United States produced over six million tons in 1953.

A very large group of non-metallic minerals includes those which are basic to our construction industry and which are vital in the building and maintenance of our roads and other public conveniences. Here are found the minerals such as cement, gypsum, clay, sand, gravel and the various building stones. While the very hard and most desirable of our building stone may soon be depleted (the marbles, granites and limestones), there is certainly an immense supply of the moderately strong and useful igneous and sedimentary rocks. A wide variety of products useful to the public come from the stone quarries, the sand and gravel pits, the gypsum mines and the clay beds of the United States. Clays are utilized by the pottery and china industries, the paper manufacturers, the makers of paint and cement products, the rubber companies, and the brick and tile industry. Mineral cement is the basic element of concrete; the tonnage of concrete, vital to our construction industry, far exceeds in weight our total production of steel. Our cement production far surpassed all domestic requirements in 1953, amounting to more than 250,000,000 barrels of 376 pounds each.

Sand and gravel, produced in 1949 in the quantity of more than 300,000,000 short tons, serves the building industry, provides paving material, and is used as railroad ballast. Considerable amounts of high-silica sand are used in glass manufacture, and suitable types for molding, grinding and polishing purposes. Stone is customarily considered in two classes: dimension stone, cut into definite sizes for structural use; and crushed or broken stone, which has a wide variety of uses; much is used along railroad road beds, in the manufacture of concrete, for furnace flux in smelting, and as lime, to spread over arable land. About 1,750,000 short tons of dimension stone was sold or used in the United States in 1949, while the sale and use of broken stone reached the enormous total of nearly 300,000,000 short tons in the same year.

One of the most interesting of these highly useful materials is gypsum. Its largest producer in the United States is the United States Gypsum Company, which in 1952 completed the first half-century of its corporate existence. Formed in 1902 to supply the American public with plaster, plaster-of-Paris and wall plaster, it has grown in the past fifty years to tremendous size, with some forty-six plant locations, a fleet of ocean-going ships, some 61,000 acres of gypsum-bearing mineral lands, and over 80,000 acres of pulp wood timberlands. Sales of its products rose in 1952 to a value of nearly $200,-000,000.

An amazing thing about this material is that it has been known to man and used for thousands of years. The Chinese were familiar with it and the Egyptians used it in the building of the Great Pyramid of Cheops some three thousand years B.C. Gypsum is a form of rock, usually white, but often found in shades of pink, yellow, brown and occasionally black. When heated and ground to a powder, it readily may be formed into a plaster. Its plastic quality and surface beauty attracted artists and architects for many hundreds of years. Great artists of the Renaissance period used this serviceable material to construct fine panelling on walls, ceilings and doors. Even more serviceable in the modern period, gypsum serves a wide variety of purposes in the building industry, in blocks, sheets, laths, and as material for roofing tile, and in plaster which may be applied directly over concrete. In recent years gypsum manufacturers have moved into the cement field and into the manufacture of insulation materials and paint. An interesting use of gypsum is in the construction of patterns and mock-ups for the aircraft and automotive industries.

Producers and processors of non-metallic minerals are of such a bewildering variety that space forbids anything like a complete analysis of their operations. The reason for this is that the minerals, once secured, find their way into almost every kind of industry. To some companies, the use of minerals is a minor part of a larger set of operations; the item happens to fit their needs and that is all. Other firms are specifically set up to mine and process and sell a certain group of non-metallic minerals. This is true of the gypsum companies, which seem inclined to complete integration of their activities, from mine to customer. To a certain extent it is also true of the glass companies, though they tend to specialization in certain commercial types of glass. Pittsburgh Plate Glass Company and Libbey-Owens-Ford Glass Company are the chief producers of window and plate glass; Owen-Illinois Glass Company is the leader in the glass container field; Corning Glass Works leads the field in the manufacture of electric light bulbs and has a strong line of signal and optical glass. The two greatest cement companies are the Universal Atlas Cement Company and the Lehigh Portland Cement Company. The largest producer of asbestos products is the Johns-Manville Corporation.

The producers of dimension and crushed stone, clays and other types of very common and useful building material are so many in number as to preclude special mention.

It is abundantly clear that in the field of non-metallic minerals the United States is especially fortunate. The extent and variety of our underground mineral structure provides the nation with almost every one of them in sufficient quantity for all foreseeable purposes. We are short of graphite, mica and asbestos but are very rich in supplies of stone, clay, cement, gypsum, potash, phosphates, nitrogen, the halogens, sulphur and the metallic pigments.

CHAPTER NINE

RUBBER AND RUBBER PRODUCTS

Rubber is in many ways a unique material. Its traditional source is a plant, the *Hevea Braziliensis;* rubber production began, therefore, as a type of specialized agriculture. The peculiarities of the plant have, so far, restricted its large-scale cultivation to three principal areas: South and Central America, tropical Africa, and the Indo-Malay region. Therefore it has been for centuries an item of international trade, since its tropical habitat has been, and to a large extent still is, nonindustrial. The limited regional sources of the plant, even after the introduction of plantation cultivation, has made the production of natural rubber an object of commercial rivalry; individuals and nations have sought to create a world monopoly by control of the sources of supply.

Rubber became a commodity of primary industrial importance only with the development of machine industry. Its unusual qualities when refined and compounded with other materials gave it an industrial importance possessed by no other substance: it is highly elastic; when compounded with sulphur and other materials and subjected to heat and pressure, rubber may be soft and pliable or hard and tough; it is waterproof, may be made highly resistant to abrasive wear, is an electrical insulator, and resists the action of most gases and chemicals. Several of these qualities were utilized in early rubber processing, but the most important factor in the history of industrial rubber has been the development of fast transportation.

The rubber tire, upon which a large proportion of all moving units roll, is only one of the uses of this valuable material; it has literally hundreds of others. During the past half-century, in which rubber has evolved into a major industrial raw material, its peculiar distribution in the natural state, taken in connection with an international emergency which threatened for a time to tear down the economies of the free world, stimulated one of the most difficult projects in chemical research in the history of industry.

The Early History of Rubber

In the days of the early Spanish explorers, it was noticed that the natives of Central and South America made use in a primitive way of the hardened juice of a native tree. As early as 1536, Gonzalo Fernandes D'Oviedo y Valdes wrote about a game played by South American natives which they called "batos"; in playing the game they tossed a ball made of strips of sticky gum. It is often claimed that Columbus, during one of his voyages, noted the elastic nature of the ball used by Santo Domingan natives. Other Spanish travellers noted that this material came from the "milk" of a tree, and some of their countrymen are reputed to have attempted the waterproofing of their garments and the constructing of crude overshoes of the gum, in imitation of the native, but with little success.

Europeans found no practical use for this gummy substance for two centuries, and the occasional chunks of rubber which found their way to the Old World ended up in museums as curiosities. In the early eighteenth century, however, two Frenchmen, La Condamine and Fresneau, were sent on a geographical expedition by Louis XV's government. After some five years of exploration, La Condamine sent back from the Amazon some rolls of the substance which he called "caoutchouc" or "weeping tree." He reported that the natives used it for waterproofing cloth, made shoes and even molded unbreakable bottles of it. His colleague Fresneau studied native production methods, and suggested a number of uses which might be made of the gum, but he concluded that any processing of caoutchouc would have to take place where the trees grew because the juice hardened so quickly. Later in the century, scientists in France and England began attempts to dissolve the rubber lumps in order that some useful items might be made from them. Various oils, ether, and turpentine were used and seemed to be effective. Meanwhile the chemist Priestley obtained a piece of it, found that it rubbed out pencil marks and presented small fragments to his friends. Thus caoutchouc became "rubber" to the English-speaking world. By 1791 another Englishman, Samuel Peal, patented a process of waterproofing fabrics by treating them with a solution of rubber dissolved in turpentine. A factory was set up in England by Thomas Hancock in 1820, and three years later Charles MacIntosh established another in Glasgow, where he began the manufacture of the waterproof garments to which he gave his name. His fellow rubber industrialist Hancock is often credited with the invention of two key devices, fundamental to all rubber manufacture: the "masticator," for softening crude rubber, and the "calender," to coat fabrics with the softened gum. He also pioneered in the use of heat and pressure to mold rubber into useful articles.

Early in the nineteenth century interest in rubber began in the United States with the importation of some articles from South America and others from England. In 1832 John Haskins and Edward M. Chaffee founded the Roxbury India Rubber

Company in the Massachusetts town of that name. This firm, the first of its kind in the United States, was a large one for its time, having started with a capital of $400,000. Its principal item of production was waterproof garments, though it dealt in a rather wide variety of imported rubber goods. Coats and shoes of rubber were extremely unsatisfactory in those early days; they hardened and became brittle in cold weather, and in summer they became sticky and smelled badly. It became a common pre-occupation of scientists and "tinkerers" to search for some way to offset these undesirable features. The greatest success in making rubber resistant to changes of the weather followed the mixing of various amounts of sulphur with the gum. Experimenters in both England and Germany had made improved rubber compounds before Charles Goodyear's famous discovery, but it remained for the American to invent the first practical process of "vulcanizing," or heat-treating rubber, in 1839. Over in England, Hancock, who may have seen the results of Goodyear's first efforts, obtained a patent for a process similar to the latter's, in 1843. The next year Goodyear was granted his U. S. patent on applying high temperatures to a compound of sulphur and rubber.

The vulcanizing processes of Goodyear, Hancock and others really started the industry. Rubber factories sprang up in the New England states and New Jersey, as well as overseas, to make waterproofed clothing, rubbers, overshoes, toys, bumpers for railway cars. Nelson Goodyear, an inventor like his brother, discovered that a harder rubber could be obtained by putting more sulphur in the compound and continuing the heat treatment for a longer period. Before the Civil War, the United States rubber industry had grown to be a $5,000,000 business and employed some 10,000 workers. The pioneer manufacturers labored to improve the original vulcanizing process by adding other materials to the rubber-sulphur mixture. Each rubber maker had his secret compound, the formula for which was guarded carefully and locked up at night in the company safe. In general, the new elements added were usually "accelerators," often lime, magnesia or white lead, to decrease the time needed for vulcanization.

While much attention was being given to a better product and a faster process, new uses were constantly being found for the finished material. Tires for carriages, as well as for bicycles, were made of hardened rubber at first, but were succeeded by the pneumatic type, developed by John B. Dunlop, a Scottish veterinary, in 1889; the invention of the automobile and the airplane brought in limitless variations of the tire business, as thousands and later millions of passenger cars, trucks and airplanes required larger and stronger casings and treads. An entire world of industrial rubber goods was developed, in belting, hose, tank lining, cushions, flooring materials, gaskets for machinery, rings for glass containers and a hundred-and-one other items in constant daily use.

Modern Rubber Technology and Production

Before the invention of the automobile, important changes were already taking place in the production of natural or crude rubber. The principal supply, until well into the twentieth century, about ninety per cent of the world supply in fact, came from the wild jungles of the Amazon in tropical Brazil, and most of the remainder from tropical Africa. Its production by native gatherers was a slow and painful process; much of the Amazon rubber had to be shipped 2,000 miles by river to the seacoast before its shipment abroad. In Africa King Leopold of Belgium built a profitable rubber business, as a part of his notorious enterprises in the Congo. In England, in the 1870's, a project was taking form to introduce rubber culture into the tropical British colonies and in that way to break the monopoly enjoyed by Brazil. Sir Henry Wickham, a distinguished explorer, was commissioned to obtain from Brazil sufficient seeds of the Hevea plant to enable the British authorities to carry on a large transplanting experiment. Sir Henry obtained 70,000 carefully selected seeds, which were taken to England and planted in Kew Botanical Gardens near London. From these seedlings, plantations were started, first in Ceylon, and later in British Malaya and in India. After many years of disappointments, these plantations began to produce successfully and started to furnish a larger and larger part of the world's supply. Meanwhile the Dutch introduced the plant into their East Indies and the French began successful culture in Indo-China. American rubber companies, from 1910 on, began acquiring tropical plantation lands. The total amount now produced on the plantations of the Far East far exceeds the production in all other areas; some 8,000,000 acres of rubber plantations were in operation at the beginning of World War II.

In 1870, the year in which Dr. Benjamin Franklin Goodrich established the first rubber factory west of the Alleghanies at Akron, Ohio, the United States imported about 4,000 tons of crude jungle rubber from South America; by 1900 this total had risen to about 27,000 tons. In an average year at the present time this nation alone consumes about a million tons of rubber; the proportion of natural rubber to this total varies from year to year. Nearly all of the world's total production of natural rubber, amounting to more than a million and-a-half tons in 1941, comes from the plantations. Most of the trees which produce this enormous amount are descended from Wickham's trees. Instead of the slow and arduous gathering of jungle rubber from widely-separated trees, natural rubber is now taken from scientifically-managed "orchards" of Hevea trees which begin to yield the milky latex after five years of growth and may continue to yield from eight to ten pounds per year for twenty-five years or more.

The great rubber companies own and operate their large tropical plantations and maintain processing plants nearby, where by chemical and mechanical means the original juice is concentrated and as liquid latex is shipped to the United States and the other rubber manufacturing countries. There it is processed into foam rubber products, elastic yarns, adhesive tape, carpet adhesives, rubber toys, tire cord solutions, wire insulation and rubber gloves. A part of the tree product is converted into solid rubber by mixture with acids, and after the water is squeezed from the slabs, they are dried, smoked and pressed into bales. This "smoked" sheet rubber, after shipment to the United States, becomes available for the manufacture of tires and tubes, footwear, hose, golf balls and hot water bottles. At the plantation end of the industry, laboratory methods continue, to improve the young rubber stock; bud-grafting is carried on in order to fight a dangerous leaf disease by grafting onto the rubber tree a bud from a tree which resists the disease; through soil experiments and fertilization the yield per acre has been constantly increased.

Along with this effort to improve and expand natural rubber production in the great Malayan, Netherlands Indies and Ceylon plantations, has gone a wide experimental project to produce plantation rubber in the Philippines, Panama, Costa Rica and Liberia, among many places, by American rubber companies. At the outset of World War I, Great Britain controlled about one-half of the world's acreage of plantation rubber, the Dutch about forty per cent, and Americans, French and Japanese the remainder. The advancement of natural rubber production is conditioned by the fact that, thus far, good rubber production is possible only in tropical lands with about 100 inches of annual rainfall and with average temperatures of 70 to 90 degrees. Such conditions are found mostly in areas about ten degrees north and south of the Equator. It may be that guayule, a rubber-bearing plant grown in Mexico and southwestern United States for forty years will eventually be improved. The United States Department of Agriculture has carried on experiments with the goldenrod, which contains a gummy fluid in its leaves; the kok-sagyz or Russian dandelion for its root content; and the crypostegia, which has rubber-like fluid in both stems and leaves.

Into this unsatisfactory production situation, so far as the United States, the Netherlands, Great Britain and France were concerned, came the impact of World War II. So long as Japan and the United States were neutrals in the war, some hope could be entertained that the vast raw materials of southwest Asia might still be utilized to help maintain the industrial economies of the nations which drew upon them so heavily. This hope was rudely dashed by the advance of Japan into some of these areas in 1940, and was completely demolished by the shattering attack upon the colonial possessions of the Netherlands, Great Britain and the United States in December, 1941.

The fact that many vital raw materials, among them iron, copper, coal and petroleum, were in quite plentiful supply in the United States, but that rubber was not, created a highly critical industrial situation. Even had the United States not been attacked, and hence obliged to defend herself, the cutting off of the major source of raw rubber would have been a serious problem to the United States, for she had already moved into a position close to belligerency by the legislation of March, 1941; she had engaged to lease and lend materials of many kinds to those nations under attack by the Axis powers. But after December of the same year the United States was in a much worse predicament, because she was obliged to assume an immediate war status; at the same time she felt it necessary to continue Lend Lease aid to Britain, Russia and her other new allies.

It should be realized that rubber was not merely a raw material which could be used or dispensed with at our convenience. All industrial production had come to depend upon rubber and rubber-like materials. Without them or substitutes for them, it would be difficult if not impossible to gear our industries to wartime levels; airplanes, tanks, motor trucks, field artillery—in fact all heavy equipment which was motorized or pulled into action by motorized units—needed rubber. Moreover, our powerful enemies, Germany and Japan, were in a far better state than we, so far as rubber was concerned. Germany had for a number of years produced large quantities of synthetic rubber, while Japan, if she were allowed to establish herself permanently in the Southwest Pacific, would thereby come into possession of the world's largest sources of natural rubber, and could deny access to it by the United States, the world's largest importer and manufacturer of this vital substance.

Thus arose an industrial and military dilemma of serious kind. Industry, were it to supply the material for a huge army and expand the shipments of desperately needed military supplies to our allies, could hardly wait for the re-conquest of the Pacific, since industry and our armed forces could not move into full-scale activity without the very raw materials which had been in large part cut off by Japan's swift conquests. The problem was an urgent one; science and industry would have to provide the answer, and quickly. The Baruch Committee stated in 1942:

> Of all critical and strategic materials, rubber is the one which presents the greatest threat to the safety of the nation and the success of the Allied cause . . . if we fail to secure quickly a large new rubber supply our war effort and our domestic economy both will collapse.

Mr. Baruch was not exaggerating the danger of the situation.

The Search for Chemical Rubber

Let us trace briefly the efforts of scientists to analyze the chemical nature of rubber and to duplicate it by chemical means. For nearly a century this effort had gone on, with the result that the basic constituent of it was found to be a substance which was termed isoprene. Methyl isoprene, chemically derived from potatoes, grain or other starchy foodstuffs, was for some years the nearest chemical approach to synthetic hard rubber.

During the first World War Germany had been confronted by a serious rubber shortage; in the latter days of that struggle her motor transport was running on worn-out, patched-up casings while her good rubber was being carefully saved for gas masks, surgeon's gloves, airplane tires and storage batteries. In this critical situation, her transportation authorities resorted to methyl rubber for tires, but it proved a very poor product.

After the war the United States also was faced with a rubber shortage at the very time her motor car industry was expanding; this was partly occasioned by the limitation by Great Britain of her Far East plantation production. Congress appropriated $500,000 to aid in surveying possible rubber-producing areas in the Philippines and Latin America. Thomas A. Edison joined experts of the Department of Agriculture in experimenting on cactus, goldenrod and other plants for possible supplies of natural rubber. In 1925 chemists of the du Pont Company began a search for a chemical substitute. Although Great Britain repealed the Stevenson Act in 1928 and therefore alleviated the world shortage somewhat, the efforts to find new sources and to produce a satisfactory synthetic continued.

As the 1920's advanced and the German economy collapsed, that nation dropped for a time its research on synthetic rubber, but with the coming to power of the Nazi regime in 1933, scientists were again put to work, for with the probability of war as a means of Germany's recovery of her former position, another rubber shortage in the Reich would be disastrous. Two types of synthetics were soon produced by German rubber scientists: Buna-N, of low quality but useful for such purposes as gasoline hose; and Buna-S, a higher grade product, more elastic and workable, and which came to be used extensively for tire treads. These efforts naturally interested American manufacturers, in both the rubber and petroleum industries. Efforts were made to secure American rights for Buna-N, as the prospect of having to make tires of synthetic seemed rather remote at that time. The Germans were quite willing to allow American industrial representatives to examine their new rubber, but would give out no information as to how it was made. It had been possible, however, for the Standard Oil Company of New Jersey to purchase from I. G. Farben in 1927 the rights to certain German patents and processes which related to oil. Through a number of clues received in these visits to Buna plants, plus its own long experience, the Goodyear Company was able by 1937 to produce experimentally a synthetic which it called "Chemigum" and by 1939 to manufacture this rubber, which resembled the German Buna, and to try it out in auto tires.

Meanwhile the du Pont chemists had been continuing their experiments. They may also have received a hint from German chemists, who had begun to utilize acetylene as a raw material for their synthetic, in place of the grain and potatoes formerly used. Food was scarce in Germany, and acetylene was easy to prepare there from coal, limestone and water, which were in plentiful supply. The chemical process as the Germans carried it on was a long and difficult one, however; first it was necessary to produce acetone from acetylene, then to process the acetone into methyl isoprene. The objective of the du Pont chemists was to eliminate the first steps in the process and to produce a synthetic directly from acetylene, without going through the acetone stage. At first their efforts failed, for in order to secure a rubbery material this way, it was necessary to treat diacetylene with natural latex.

Still far from success, the chemical director of du Pont's dye-stuff department, Dr. E. K. Bolton, came into contact with Dr. Julius Nieuwland of the University of Notre Dame, at an American Chemical Society meeting in Rochester, New York. Dr. Nieuwland had no interest in rubber or in the experiments going on to produce it synthetically, but he was an outstanding authority on acetylene gas. At the meeting in question, the Notre Dame chemist had read a paper in which he described his experiments in treating acetylene with various copper catalysts; in one reaction he had produced a yellow oil which came to be known as di-vinyl-acetylene. The possibility of working with this yellow oil occurred to Dr. Bolton and the other du Pont men who had listened to the report, and after a consultation with Dr. Nieuwland, a period of joint experimentation followed. It was discovered finally that if the gas which was given off in the experiment which produced the yellow oil (which did not prove to have any synthetic rubber possibilities) were combined with hydrogen chloride, a thin, clear liquid resulted. This was a new material, structurally like isoprene but possessing superior properties. This new substance, which was named chloroprene, when treated, polymerized into a material like rubber quite rapidly. Experimentally it seemed as good as rubber, in some ways better.

The discovery of this synthetic was announced by the du Pont Company at the American Chemical Society meeting at Akron, Ohio, in 1931; it received the commercial name of neoprene and a factory was already under construction for its manufacture when the announcement was made. The time was not propitious, for the depression was becoming worse and natural rubber was selling for five cents a pound in New York City. Yet in 1932 neoprene came on the market, at $1.05 per pound, a figure which was materially reduced when the du Ponts got it into full production.

The reasons why neoprene sold, to rubber processers, to the tune of half a million pounds during the depression years of 1931-1935, included the fact it was made of raw materials which were of local origin and not dependent upon the foreign supply. It resisted action by oils, gases and gasoline better than natural rubber, and was moreover less affected by heat and hence less apt to crack in direct sunlight. It could also be mixed with the same materials as natural rubber, processed by the same machinery, and could be vulcanized in the same manner. Its quick success permitted substantial price reductions and by 1939 it had an even wider industrial use.

Meanwhile several of the rubber and petroleum companies had been making progress on the problem. The Firestone Company put its chemical research division to work, studied samples of various German and Italian synthetics in 1936; two years later, after large amounts of the German Buna-S had been imported to this country, it made tires of the synthetic and gave them road tests. The next year a Firestone pilot plant also produced Buna-N. By mid-1940 Goodyear prepared to expand its chemigum manufacture to 2,000 tons per year. However, the commercial manufacture of Buna-S rubber had been patented by I. G. Farben and the American rights were held jointly by Standard Oil of New Jersey and the Farben combine; toward the end of the year 1939, Standard bought sole rights for this country, and offered licenses to produce Buna-S to American rubber companies. An agreement was concluded with Firestone in February, 1940 and that company began production of Buna-S synthetic in April.

The Standard Oil Development Company, which had been a research division of the New Jersey Company, and a separate corporation after 1922, began a strong interest in synthetic rubber development about 1930. The Germans had produced butadiene, the main ingredient of their synthetic rubbers, from coal and lignite. Standard and other firms interested in synthetics began the development of butadiene from petroleum. This proved a sound idea, for we possessed large petroleum resources, and the gasoline cracking process left a large butadiene residue which was sometimes as large as two-and-one-half times, by volume, the total amount refined. Here was a large available supply of an important ingredient. Another principal item of the synthetic process which the oil and rubber companies were perfecting was styrene, a product of ethylene, which was made from petroleum and benzine (a coal tar product); styrene had already been developed commercially for other uses. Standard followed up its work in the butadiene field by developing butyl, another rubber-like petroleum product, which later became a preferred material for inner tubes for tires. It is less elastic than rubber, but only one-tenth as permeable by air.

To summarize synthetic development by the end of 1940, American industrial research had produced a number of synthetics, some of them closely resembling German Buna, others entirely new. It was significant that the use of petroleum butadiene, instead of the more expensive and less satisfactory product prepared by the Germans, had been proved to be practicable. It was highly important that industrial research had proceeded so far, at the emergence of the great crisis in our foreign relations.

By August, 1940, the United States Government became interested in the rubber situation. The Rubber Reserve Corporation, which had been set up two months before under the Reconstruction Finance Corporation, called into conference representatives of the large American rubber companies, and gave strong encouragement to the building of new plants for rubber manufacture, to increase the national stockpile. In March, 1941, four plants, each of 10,000 tons annual capacity were planned; when Singapore fell to the Japanese, this program was increased to 400,000 tons annually, and after the attack on Pearl Harbor, to double that amount.

Three measures were to be taken in order to place our rubber supply in line with the wartime demands which came immediately after December 7, 1941. First of all, there was the building up of the stockpile of natural rubber, which had already been increased by the strong efforts of Rubber Reserve, assisted by the Goodyear, Goodrich, Fisk, Firestone and U. S. Rubber firms to the respectable total of 600,000 tons by the end of the year 1941. A second project included the effort to expand the production of wild rubber in areas free from Axis invasion, and to make these sources available to Allied industry. An agreement was completed with seventeen Latin American nations at Washington on March 2, 1942, to make their wild rubber resources available to the Allied cause; the efforts put forth to implement this agreement were moderately successful, although the wild rubber obtained was not large in amount; its procurement was a very costly process. The third measure taken by the government was naturally the organization of the synthetic rubber possibilities on a large scale.

This latter project involved extensive cooperation between government and the rubber industry, and other industries which could furnish technical and manufacturing assistance. A committee of the top men in industry, headed by Mr. B. M. Baruch was set up by the Rubber Reserve Corporation; a study was made of the status of synthetic rubber research in an effort to find the most practical product for quick manufacture. A composite of several types was decided upon, called GRS, or Government Reserve, Styrene, to distinguish it from various kinds which did not use styrene. These included GRA, the oil-resistant Buna N; GRM, which was du Pont's neoprene; and GRI, which was Standard Oil's butyl. The nation's large chemical companies, including Monsanto, Dow, Koppers, and Union Carbon and Carbide, were given the job of producing styrene; oil companies such as Standard, Shell, Sinclair, Cities Service, Phillips and Humble were to produce the butadiene. At the end of the process the four largest rubber companies were to operate two or more copolymerization plants each, and two more corporations were organized among the smaller rubber companies, also to produce synthetic rubber. In the summer of 1942 the entire program was put into the capable hands of William Jeffers of the Union Pacific Railroad, as Rubber Director. By 1943 the program got well under way, and by 1944 the huge figure of 800,000 tons, set by the

Baruch Committee, was reached and passed. A critical situation had been well-met, and rubber, natural and synthetic, had been geared to total war.

Big Rubber

There are many characteristics of the rubber industry which are similar to all other large-scale American industries. It has been highly competitive and was at first composed of hundreds of smaller firms; as the years passed many of these firms gave up completely, or combined with others to form larger units. In time a few very large companies, through more ambitious policies and display of greater imagination, or by making more successful adjustments to the needs of modern life, outstripped the others and a condition of oligopoly resulted. This was much the way the other great industries had developed. Thus rubber, natural and synthetic, followed an industrial pattern already familiar.

"Big Rubber" has followed another practice which is quite common among industrial companies. It is usually an integrated industry. A large rubber organization commonly has its own rubber plantations in the Far East; in the early days of the business it usually got its start by manufacturing footwear and waterproof clothing; with the coming of the automobile it invariably went into the tire business, since seventy per cent of all rubber goes in that direction. With the advent of synthetic rubber, an important chemical phase was added to all rubber production. The importance of textile fibers as an ingredient of tires dictated the entry of rubber companies into the textile field; since their arrival as operators of cotton plants they have taken part in the development of new chemical textiles. Many recent technical innovations have made rubber industrialists factors in the insulation business, suppliers of drug sundries, and of plastics, building material and even insecticides. That has been the conventional picture in large rubber industrial growth: integration from plantation and chemical plant to the manufacture and distribution of a wide variety of rubber products to the consumer, by a single organization. To attain such a position, the larger companies have expanded horizontally across industry, and have become heavily involved in the chemical, petroleum, textile, clothing and footwear specialties.

Rubber also became "big rubber" because of the strategic nature of the materials with which it works and because of the enormous demand which peace and war have created for its products. The difficulty and complexity of securing adequate raw material at such a distance, added to the technical obstacles of building a more adequate supply of synthetic materials, have made it an exciting and highly competitive business. The necessities for quick expansion, and for acquiring control of chemical and textile plants to perfect newer processes, have placed pressures upon smaller companies which have added to the casualties among them; this trend has also placed a higher premium upon the daring experimenter than seems to be the case in other industries. No business except certain types of mining has experienced such variations of "boom and bust" as the rubber industry. Only the strongest and best-managed companies have succeeded in the grueling contest which has gone on in the century and more since Charles Goodyear's famous experiments.

The United States Rubber Company was organized in 1892 as a successor of several companies which had already been functioning for a generation. Its footwear plant at Naugatuck, Connecticut, had at one time borne the name of "Goodyear's Metallic Rubber Shoe Company" because Charles Goodyear himself had at one time been a director; another plant in New Haven, L. Candee and Company, which also entered the U. S. Rubber combination, was an original licensee under the Goodyear vulcanizing patents; so was the Naugatuck plant.

U. S. Rubber started out as a kind of central purchasing

agency for a number of eastern rubber footwear companies; later these companies lost their identities, as U. S. Rubber became, and still remains, the largest maker of rubber footwear. Its transition from this specialty into the field of waterproof clothing, foam rubber and coated fabrics was a natural one, and it now operates seven plants devoted to these products. While some of its subsidiaries had been making solid rubber tires before the days of the automobile, the success of the pneumatic tire on power-driven vehicles first induced U. S. Rubber to take a major interest in the tire manufacturing field. In 1905 it acquired the Rubber Goods Manufacturing Company, which controlled the Hartford Rubber Works, and also tire and tube factories in Detroit and Indianapolis. At the present time its manufacture of tires and tubes ranks with that of the four largest companies in the United States and is the highest single item on its production list.

From these beginnings U. S. Rubber branched out into mechanical rubber goods, drug sundries, lastex yarn and other specialties. It acquired natural rubber plantations in Sumatra in 1910, making it the first of the American rubber companies to do so; it added to this acreage later and expanded further in Sumatra and Malaya. Here the Company pioneered in experiments to improve the quality of plantation rubber and in increasing the yield per acre, which it nearly doubled by 1941. In 1917 the firm began acquiring textile companies of which it has nine at the present time; it operates two synthetic rubber plants. By 1947 there were more than 66,000 employees, including some 7,700 on the plantations, and the Company sold in that year over $580,000,000 worth of goods.

The Goodrich Company stems from the enterprise and ambition of a young Civil War surgeon, Dr. Benjamin Franklin Goodrich, who after the War found that he had a livelier interest in business than in the practice of medicine. His connection with the rubber industry began when he and his partner in the real estate business decided to trade about $10,000 worth of property for stock in the Hudson River Rubber Company, a company in not too prosperous a condition. Finding prospects not very rosy in two New York State towns where the factory functioned briefly, Goodrich decided to try his luck farther west, and boarded a westbound train with the object of finding a town where his rubber company could find opportunity for expansion. It is said that a stranger aboard the train sang the praises of Akron, Ohio so effectively that Goodrich decided to look over that place, at that time a town of some 10,000 population. With the encouragement of the citizens whom he met, and who later advanced him considerable sums to aid him in locating his factory there, Goodrich, Tew and Company, later the B. F. Goodrich Company, launched its career in Akron on the last day of December in 1870.

It was a momentous decision for both Goodrich and Akron as well, for within a generation this little Ohio canal town was well on its way to becoming the rubber capital of the world. The new firm began its business by making cotton-covered rubber fire hose, a product which was very successful and which is still on its manufacturing list. This was followed up by bicycle tires which were greatly improved by the use of a fabric in them which did not have cross threads; the firm developed a pre-stretched rubber transmission belting; when the automobile came in, Goodrich was engaged by Alexander Winton, a Cleveland bicycle maker, to make tires for his pioneer horseless carriage. In 1895 the Goodrich people pioneered in establishing a regular laboratory in which to test raw materials and to work out better rubber compounds.

B. F. Goodrich, as well as many other rubber manufacturers, benefited markedly from a process for reclaiming old rubber, an objective upon which many experimenters had worked, but most of them in vain. Arthur H. Marks was a young Harvard chemist who in 1898, at the age of twenty-four, became superintendent of the Diamond Rubber Company, an Akron competitor of Goodrich. This was in itself an event of some importance, for chemists were not then considered of much use in the industry except for some routine testing. Marks went to work and within a year had perfected a method by which scrap rubber was reclaimed. He heated ground rubber waste in a solution of dilute caustic soda for twenty hours at a temperature of about 360 degrees. This treatment destroyed the cotton fabric, dissolved the sulphur and some of the lead compounds in the old material, and plasticized the rubber. The method seemed to work with all kinds of scrap; in the case of old tires, the reclaimed rubber was equal in weight to about one-half of the scrap tires so treated.

It is hard to over-estimate the importance of the Marks process to the rubber industry; it added an element of conservation to the rubber supply; it resulted in enormous savings to rubber companies, which soon made use of the process, and decreased the price of raw rubber and of finished products. In view of the fact that crude was selling at better than a dollar a pound in 1899, when Marks' process was patented, the invention was very timely. It was a good thing for Marks also, as he profited enormously by it, and gained important recognition in 1912, when his Diamond Rubber Company joined Goodrich and into the merger came Marks as vice-president and general manager of the consolidation.

Marks also deserves credit for introducing another great chemist, his friend and classmate at college, George Oenschlager, into the Diamond Company and assigning to him the problem of doing a research job on organic substances which would speed up the process of curing rubber. The development of organic accelerators has been called a discovery second only to the vulcanizing process itself. The term "accelerator" is in a sense deceptive, for it calls attention to only one factor, that of speed. What the rubber chemists before and after Oenschlager really were looking for were chemicals which would not only improve the slow sulphur process, but would also improve the quality of the product. In the early days of rubber "chemistry," if it may be so-called, a wide variety of chemicals had been mixed with the crude rubber. Even Goodyear had used white lead in his first experiments, and had tried lime and magnesia as dryers of rubber surfaces. He and others found that these chemicals accelerated vulcanization. Later ammonia and various salts were used, and still later various chemists, both in Europe and America, used bromides and iodides of half-a-dozen metals, both as accelerators and to color or otherwise improve the appearance of the product.

Early in the twentieth century a chemist named C. O. Weber reported upon favorable results which he had obtained by treating rubber with various soapy substances; the fatty acids acted to soften up rubber mixtures and by a complicated chemical process aided the acceleration of vulcanization. However, Arthur Marks, of "reclaimer" fame, apparently preceded Weber in the use of fatty acids in rubber chemistry, while engaged in 1896-1897 by a Boston rubber firm which made rubber stamps and bicycle tires. Marks was also interested in those days in trying to make good rubber out of cheap Borneo jelatong, which cost only one-third as much as high-priced para stock. He made a fairly pure rubber out of jelatong but it did not compare in tensile strength with the more expensive product. It was Marks' fixed idea of making good rubber out of cheap gum that led him to engage George Oenschlager after his appointment as superintendent of the Diamond plant.

By 1905 Oenschlager proceeded with this task, and by a long and thorough period of research he accomplished it. He had first to determine if the chemical composition of cheap rubbers was the same as that of the best grades; this he found to be the case. He next embarked on a long study of the effects of adding various organic compounds to a mix of poor rubber, sulphur and zinc oxide. He tried compounds of all the common elements, and found most to be of no value; he had also to take care lest some which might be of value as accelerators would also act as carriers of oxygen and hence do harm to the resultant compound. In various experiments he cut the vulcanizing process down to a few minutes; but tests of the product created were unsuccessful. After almost endless trials he came up with one which seemed to be the answer. Trying aniline as one of the simplest, cheapest and most common of the organic bases, he got excellent results, so good that the Diamond factory by June 22, 1906 began to use it in inner tubes and in the carcasses of tires.

A difficulty immediately appeared: aniline is a poisonous liquid, so Oenschlager looked about for a less toxic product and soon hit upon an aniline derivative called thiocarbanilide. He found that a six per cent mixture of the latter raised the tensile strength of the compound to 3,000 pounds and reduced the curing time to ten minutes. By September of 1906 tires were being produced in the Diamond plant with the aniline derivative in the tread and aniline in the rubber which was used to coat the fabric. In tests the new ingredients worked well, and seemed to act as a preservative; later, in 1912, aniline in the mixture was replaced by another accelerator discovered by David Spence, an English chemist in the employ of the Diamond Rubber Company; this bore the impressive name of "para-amino-dimethylaniline." Efforts to patent these mixtures seemed impracticable, and the secrecy which was at first maintained regarding them was soon abandoned and they became common property, as workmen moved about in the Akron area, as employees of first one, and then another rubber factory.

An interesting postscript to the pioneer work of Marks, Oenschlager and Spence with the Diamond firm, and later as part of the enlarged Goodrich organization, was the fact that due to the earlier secrecy surrounding these experiments, Wolfgang and Ostwald, two German chemists, came close to duplicating them in 1908; four years later two others, Gottlob and Hoffman of the Bayer Company, completed research on a process similar to Oenschlager's aniline method and patented it. Much later, in 1926, the Grasselli Chemical Company, assignee of the Bayer patent, sued the National Aniline Company for infringement of its patent rights, but lost the case, the American court holding, in 1927, that the patent in question covered too many substances and was based upon insufficient experimentation. British chemical circles likewise began to claim, during the first World War, the honor of having first discovered a successful organic accelerator. At length Spence, who had perfected one of the important processes, published the facts regarding his and Oenschlager's pioneer work back in 1906-1909. The upshot of the ensuing controversy was the admission of their prior claims and the awarding to Oenschlager in 1933 of the important chemical distinction: the Perkin medal. He was the first rubber scientist to be so honored. In 1929 President Hoover's Committee on Recent Economic Changes credited rubber age resisters as saving motorists fifty million dollars each year. Rubber scientists have estimated that organic accelerators have been responsible for saving a much large sum in capital investment which would have been necessary had the enlarged volume of rubber manufacture been maintained without them.

The Goodrich Company has developed a tough adhesive rubber isomer which locks metal to rubber in the lining of tanks, tank cars, pipes, valves and fittings and serves to protect the metal surfaces from the corrosive action of acids. This is commercially known as vulcalock. Another useful compound known as koroseal finds its way into floor covering, electric insulation, garden hose, luggage and window draperies; the versatility of this rubber-like polyvinyl is such that it can be processed very hard or very soft and resists most influences which cause deterioration of the older rubbers. An important achievement of the 1930's, later very helpful to the United States Army Air Force, was the development by the Goodrich firm and others of de-icing "boots" which could be attached to airplane wings and tails. The Company's important participation in the synthetic rubber program of 1941-1945 has been mentioned. The firm organized its own chemical company in 1943 and opened a new research laboratory center at Brecksville, Ohio, in 1948. Among its postwar accomplishments the Goodrich tubeless, puncture-proof tire, developed in 1947, ranks high. In size Goodrich ranks with the four largest rubber companies, with its twenty-nine plants, some 500 stores, many sales offices in this country and abroad, and a rubber purchasing office in Singapore.

* * *

Another of the Akron "Big Three" is the Firestone Tire and Rubber Company, which in 1950 celebrated its first half-century of corporate history. To a large extent, the most interesting feature of the "Firestone Story" so ably related by Mr. Alfred Lief in his book, is that the great organization is largely the success story of Harvey S. Firestone, a business man of outstanding ability, courage and imagination. Product of an Ohio farm community, young Harvey began his business career as a salesman for the Columbus Buggy Company. When the company collapsed in 1896, Firestone boldly embarked in a business of his own, and in four years had built a profitable carriage tire business; he sold this business and another he had acquired two years before to Edward S. Kelly, who was intent upon organizing a huge rubber combine.

Firestone up to his arrival in Akron had bought his tires from rubber manufacturers, but now had the ambition to become a rubber manufacturer himself. With about $40,000 which he had secured from the sale of his business interests in Chicago, Mr. Firestone moved into the Akron arena and looked the ground over. For a time he acted as manager of tire sales for Whitman and Barnes, an Akron carriage tire firm. This was only temporary, however, since Firestone's possession of a patent for attaching tires to carriage wheels, and his inclinations towards manufacturing, led directly to an association with others who could contribute to these plans.

In Akron lived James A. Swinehart, patent owner of a "side-wire" device which would free carriage tires of the danger of being cut by wire and seemed likely to prevent their creeping on the rim; three other Akron men, Dr. Louis E. Sisler, W. D. Buckman and James Christy, Jr., had been assigned a share in the Swinehart patent. Firestone's arrival, with fresh capital and another patent to his credit, fitted in with the plans of all four of the men. The upshot was the formation of the Firestone Tire and Rubber Company in the summer of 1900 with capital stock of $40,000; Firestone put up half the cash investment and received half the stock. The other half was paid in by the others, who shared the rest of the stock between them. As vice-president and general manager, the main direction of this enterprise for approximately the next thirty years fell upon the shoulders of Harvey Firestone. To the new corporation the partners contributed their patents and Firestone also added an option which he had received on the Whitman and Barnes tire business.

The company took over the unexpired leases on Whitman and Barnes stores in three cities, New York, Chicago and Boston; it arranged with that firm to sell the tires which the older company would manufacture and rented small quarters on the Whitman and Barnes premises for office and shipping space. Firestone now plunged into a sales campaign with the energy and skill which had already been so marked a part of his earlier success. Letters were sent to owners of livery stables in the cities where the tire stores were stationed, other letters to physicians and other interested groups. New stores were opened in St. Louis, Philadelphia and other large centers. The firm began to attract the attention of baggage transfer firms, bus operators, breweries and other extensive users of horse-drawn vehicles and early automotive vehicles. The results were good; sales for the first year rose to $110,000; and to $150,000 the next year. The facilities of Whitman and Barnes were insufficient to meet the demand; it was clear that Firestone must make his own tires; but all of his money was already in the firm and his partners, enthusiastic at first, began to drift away.

Again it was the Firestone salesmanship which did the trick. He had to sell stock and obtain credit for the means to acquire and equip a factory. A banker friend had some old iron foundries on his hands; he agreed to transfer to the Firestone Company one of these buildings and a few vacant lots for future expansion, in return for a note which the firm would pay when it was able to do so. By hard persuasion Firestone next interested several new stockholders, among them the important Akron industrialist Will Christy, brother of his partner James. With the proceeds of his vigorous stock-selling program and the generous credit extended by the banker, machinery, much of it second-hand, was installed in the factory; from a goodly supply of skilled workmen in the town, a foreman and eleven helpers were picked, and with Firestone taking on the duties of superintendent of the plant, it was ready to open in December, 1902. In 1903 profits first began to appear and the Firestone Tire and Rubber Company began its long climb into the ranks of leading American industrial firms.

Space forbids narration of all the details of Firestone's rise from carriage-tire selling to its multifarious interests of the mid-century. However, a number of incidents which occurred along the way are of interest not only in the story of industrial rubber but also involve other industry as well. It is noteworthy that Firestone built a sizable business on hard rubber tires alone and that before expanding into the pneumatic tire field, it had developed a profitable business in supplying heavy vehicles, both horse-drawn and automotive, with a popular and high-grade product; operators of brewery electric delivery vehicles, city fire trucks and baggage transfer vehicles showed a decided preference for Firestone hard tires. At the St. Louis Exposition in 1904 the Company received a gold medal for its achievements in the hard rubber tire field. The same year it exhibited hard tires up to eight inches in width at the New York Automobile Show.

The progress of the motor car by this time, however, showed clearly the trend to pneumatics and Firestone determined to move into that field. But pneumatic "clincher" tire patents, based upon the bicycle, were controlled by an association of manufacturers which alloted quotas to its eight members, and to no others. So Firestone put his men to work to develop a tire which would avoid the clincher principle; this resulted, after a slow and costly period, in the development of the "straight-side" type of tire. Soon after putting this tire on the market, it became known that Henry Ford was planning to make a car to sell for $500; so to Ford went Firestone, with the result that he bagged orders for his new tires. This was the largest single tire order ever secured up to that time; in three installments in the spring of 1906, Firestone engaged to deliver 8,000 casings to Ford. This led to the liquidation of the clincher association, and the gradual emergence of the straight-side type as the tire of the future. Tire manufacturers went to work on rims, with the result that a universal rim was perfected which would accommodate wire-cable clinchers, standard clinchers and straight-side tires, whichever were desired. By the end of 1906 the Firestone Company was in the field with a complete line of pneumatic tires and its own universal rim for all kinds. Its sales for 1906 grossed more than a million dollars.

In the years which followed, the Company kept pace with tire and motor car requirements, with safety treads, demountable rims; by 1909, when threatened with serious losses due to the control of the rim supply by a single firm, it went into the rim manufacturing business. Increased capitalization and new factory expansions again were needed from time to time as the business kept pace with the development of the motor age. Pneumatic tires for trucks, cord tire construction, low pressure tires,—all arrived in the 1920's and with them the rubber shortage due largely to British restrictive legislation in the Stevenson Act. Harvey Firestone placed his influence behind the movement to repeal the law and just to make sure, his company began one of its most important policies, entry into the rubber planting business, in Liberia.

It was also at the end of the first World War that Firestone's expansion of manufacturing into foreign countries began, with a tire, tube, and industrial rubber products factory in Hamilton, Ontario. In the 1930's this type of expansion was extended to Brazil, India, South Africa, Switzerland, Spain and other countries. With the approach of a new war emergency, the Firestone Company took on a considerable share of the synthetic rubber project and converted many of its plants to even more specialized defense manufactures. By the end of the second World War Firestone sales had climbed to the enormous total of $681,000,000 yearly. From selling tires which had to be manufactured by someone else, the Firestone Company had progressed to a stage of world-wide interests and activities, with enormous African plantations, with factories and agencies in all the continents. It had reached out beyond the tire business, its first and principal interest, to the chemical, steel, plastics, aircraft, footwear and industrial rubber goods lines. At the end of the war it ranked second in the rubber industry and thirty-second among all American industrial companies, in total assets.

* * *

There remains one other monster concern in the rubber field. This is fittingly enough the Goodyear Tire and Rubber Company, a firm which has grown to the greatest size of any in the industry. Its name commemorates the most famous individual in American rubber history, but it had no connection with the life or work of Charles Goodyear, except as all rubber industry has profited from his pioneer invention. As in the case of Firestone, but to a lesser extent, the Goodyear Company owed its beginning and much of its early success to an individual. He was Frank Seiberling, who landed in Akron with nothing but debts in the spring of 1898. He belonged to a family which had accomplished many things in business; they had made mowers and reapers, cereal, strawboard; they had operated banks and street railways. But the panic of 1893 had forced them to liquidate most of their enterprises. It was in a "down" period of the family fortunes when Frank Seiberling acquired an old strawboard factory, with a small power plant and seven acres of land in Akron, for a price of $13,500, of which Seiberling borrowed $3,500 as a down payment and gave his note for the rest.

Though Seiberling had no money at all at the time and very hazy ideas about the future except that he wanted to manufacture rubber, he knew people who had money, and, like Firestone, he had the gift of salesmanship which served to induce some of them to put their money into a new rubber firm. It was a good time to begin, as the bicycle had put new life into the industry, and Akron already, by 1898, had four thriving rubber companies: Goodrich, Pfeiffer Brothers, Ohio C. Barber (otherwise known as the "Match King"), and Kelly-Springfield, which had moved to Akron from downstate. Seiberling got his company incorporated by August, 1898 and started with a capital stock of $100,000, of which $43,500 was paid in. Heavy stockholders of the company included two of Seiberling's brothers-in-law, Lucius Miles and Henry Manton, and a wealthy clay products manufacturer, David E. Hill.

The story of Goodyear's start was rather typical. The old plant had to be repaired; floors had rotted, windows were broken, most of the machinery had rusted. Two men were hired to get it ready. Second-hand mills were picked up to mix the rubber in, and calenders to spread it on fabric. Seiberling found a couple of old boilers in the scrap heap of the street car company he had formerly operated. Late in November, after obtaining a few orders for bicycle and carriage tires and hiring a few operatives, he turned on the steam and began business.

While Seiberling had solved some of his initial problems, he soon had others. For several years his raw rubber and other ingredients were shipped to him on a C.O.D. basis; if the money was lacking, the rubber could not be removed from the boxcars. Workmen had to be paid every Saturday night, or they would drift away to Goodrich or Kelly or Barber, where they would be more certain of being paid. For years it took ingenuity and energy to secure the cash to meet the Saturday night payroll. The new company had to win its way, against competition from better-established companies, in the same town. No one was clamoring for the Goodyear product.

Seiberling was wise in realizing, during these years of struggle, that for some time his job would be that of finding capital funds on which to operate the business; therefore it behooved him to seek out, and find, men who were especially capable in other lines, who would bring their skill and training into the business. He was very fortunate in this direction. Men he found not only made Goodyear a well-balanced organization in the early years, but also were able later to undertake the more responsible positions in policy-making and top level management.

One of these was P. W. Litchfield, a graduate of Massachusetts Institute of Technology and one of the first technically trained men to go into the rubber business. He had deliberately decided to make rubber manufacture his career, not because it was an outstanding industry, but because it was so backward, and therefore was a challenge to an ambitious engineer. Litchfield's career as an American industrialist and as head of the

Goodyear Company for many years, is so well-known as to need no extended treatment here. Another of Seiberling's early aids was G. M. Stadelman, a quiet, intellectual type of person who became the highly effective head of the Goodyear sales division for twenty years; his was the job of analyzing markets and planning sales campaigns. Frank Seiberling's brother Charles, his colleague in many enterprises, became an important and popular member of the Goodyear team, especially in the area of employee relations.

Frank Seiberling mapped out a production range which included bicycle tires, carriage tires, and the new field of automobile tires; he also added rubber bands, horseshoe pads, druggists' sundries and other items, in order to provide revenue while major items were being built into full scale production and sale. So into the whirl of the tire business went the firm, to engage in its share of the troubles of that very difficult and confused industry. For some time its operations were threatened by the control of bicycle tire patents by Hartford Rubber, carriage tire patents by its Akron competitor, Kelly-Springfield, and the favored clincher automobile tire by the clincher association, the same group which had caused Firestone some early headaches. The Goodyear Company obtained a license from the Hartford Rubber Company, owners of the Tillinghast bicycle tire patents, and had developed its bicycle tire department to the sizable daily production of 4,500 tires, when the patent owners stepped in and withdrew the license. Goodyear went to work and evolved a variation of the tire which they had been making under license. Although sued for infringement, the company went ahead producing 4,000 tires a day; the patent owners, fearing they would be unable to sustain their case in court, withdrew the suit.

A long litigation came in connection with the Grant patent on carriage tires, controlled by Kelly-Springfield. Having defied the latter's notification, late in 1898, that Goodyear could continue to manufacture only under specific license (which was refused them), an infringement suit resulted and a desist order was clamped down on Goodyear by the court, from which Goodyear appealed. While the final decision was pending, the company had to post a large bond and to deposit in escrow all profits made from continued manufacture of the tires. At length, near the end of 1902, decision was rendered on the appeal and it was in Goodyear's favor.

Meanwhile, much thought was being put upon the problem of making automobile tires. Under the dictum of the clincher association, controlling the making of such tires, the Goodyear Company was allowed but two per cent of the nation's auto tire business. As Firestone had done in a similar case, the Goodyear Company evolved a straight-side tire which bypassed the attempts of the association to tie up the auto tire industry. With its share of the tire business in all fields fairly well assured, the Goodyear Company began a long period of expansion and growth by 1908. The story of that growth resembles in many directions the upward climb of Firestone, Goodrich and United States Rubber.

Looking backward, in 1948, upon fifty years of successful rubber manufacture, Mr. Hugh Allen, in his able account, *The House of Goodyear,* reminds us that in that eventful period the public has bought more than nine billion dollars worth of Goodyear products. By 1950 the company had passed the half-billion mark in the making of pneumatic tires, more than any rubber company in the world. Its net sales of rubber and related products reached a total of $1,101,141,392 for the year 1951. It began the production of plantation rubber early, in 1916. The Goodyear production line is extensive: tires, from the very small ones on the fork-lift truck to the monstrous types which ease great bombers off the ground, tire rims, and many aviation products, in addition to tires, mechanical goods, shoe products, neolite for luggage and accessories, air-foam cushioning, pliofilm wrapping material, vinyl plastic flooring, chemigum rubbers for oil-resistant products, vinyl fabrics and others.

The Industry in Mid-Century

It is evident from quick examination of the rubber industry, that it has climbed to large size very quickly. Fifty years covers the period during which it has really come to rank with the nation's greatest. It is also true that without the appearance of automotive transportation it would probably be quite small today. The rise, in little more than a generation, of companies begun on a shoestring by enterprising individuals with almost no capital makes the story of rubber an engaging one. But there is another side to this. The nature of the business was such, with its cyclone-like changes, its competitive pressure, and the insane behavior of the international rubber market, that it became one of the most difficult of American industries. To keep abreast of developments, to secure an adequate raw material supply, and to weather the war of patents required a maximum of skill, imagination and alertness on the part of management; it required moreover the optimism of a Harvey Firestone, a B. F. Goodrich or a Frank Seiberling to see the vast possibilities which lay beyond the troubles which always beset rubber firms.

Mr. Allen declares that the Goodyear firm, with its nine billion dollars of business throughout the years, has not made a great deal of money; that the overall average profit would not greatly exceed four per cent, a figure which was considerably less in the recent war years; this was also true of many other companies which converted largely to defense industry. The volume was enormous, but the net profits were not large in proportion. Furthermore, declares Mr. Allen, more than half the Goodyear Company's earnings were continually plowed back into the business; it had to do this, since the business was highly competitive, and the firms which survived had to scrap old equipment very quickly, as new designs were introduced. Old mixing mills, early tire-building machines, pot heaters, all had to go as improved methods, new machinery, and changed tire designs came in with great rapidity. The vast chemical expansion which was called for in the development of synthetic rubber was very expensive; new plants had to be erected, research laboratories established, and technical personnel of many kinds engaged.

* * *

The rubber business, with its sales of some four billions of products in 1950, is not one of the very largest of American industries; food products, iron and steel goods, fuels, automobiles and chemicals are far ahead of it, in total sales. But its importance lies not in its size or lack of it, rather in its strategic nature. It is vital to the mobility of men, raw materials and finished products through this vast country. Regarded as a chemical compound, as it should be, rather than a product processed from a tropical gum, as it once was, rubber and its products constitute one of the most necessary of our industrial groups. Finally, its importance may very well lie in the future, for greater possibilities than have as yet been suspected may emerge as the vast horizons of chemical research enlarge indefinitely.

A great rubber company of the 1950's is an organization vastly different from the rather crude raincoat, footwear and tire factories of the past century. As we have seen, the companies of today have become integrated from plantation through laboratory and factory to customer, and have crossed over into much of the rest of modern industrial organization. At the center of the entire industry is laboratory research, since the conditions of modern life which call for immense quantities of rubber and rubber-like products are constantly changing.

Greater demands are being made for better and stronger types of rubber, and new uses for rubber compounds are continually being found. Scientists have been endeavoring to learn, since the last war, why certain catalysts and materials work better than others. There are still thousands of hydrocarbons which may be experimented with, and there are many materials which may have valuable uses in the rubber manufacturing process.

The old methods of biting and pulling rubber to test its toughness have given way to the most scientific testing. Electron microscopes enlarge a bit of rubber 100,000 times; the sizes of the rubber particles furnish clues as to the most effective use of the compound under observation. An ultraviolet spectograph can give a chemical analysis of a tiny bit of material; it can detect, it is claimed, the presence of one millionth of one per cent of impurity. The laboratory has instruments to test the effects of temperature and oxygen on different samples of rubber. An interesting phase of rubber laboratory work is the use of miniature latex separators, sheeting and mixing mills; the miniature machinery duplicates the actions of the huge machines in the rubber refineries and their products may then be tested under exact conditions of temperature and pressure.

Ideas do not all originate in the research department; many arise among the chemists, metallurgists, physicists and various types of operators in the engineering departments; here new uses for rubber are planned, new designs made for rubber parts to do special jobs in machinery and in household equipment. They may come up with a new compound to cut down the noise made by an electric refrigerator; or the project may be to find a type of rubber which will seal the doors of an airplane built to operate in the stratosphere. Most rubbers would not stand the drop in temperature, sometimes from 150 degrees above to 90 degrees below. A new product, silicone rubber, of remarkable heat and cold-resisting quality, fills the bill.

Many kinds of new uses for rubber are probably on the way.

Rubber springs will likely be one of them; they may eventually take the place of coiled steel in office chairs, in cars and in buses. Air springs of rubber, which look like a tire lying on its side, have been experimentally tested on railroad cars; they would absorb the bumps with a cushion of air. An entire new future of rubber lies in the mixture of synthetic rubber thread, which is impervious to cleaning fluids, in clothing materials. The "give" of the rubberized material would add to the fitting quality and comfort of garments, it is urged. Much progress is being made in the sealing in of heat and the sealing out of dirt and noise by rubber insulation material; then there are the rubber paints, rubber wall coverings and rubber floorings.

An ambitious research project underway is the attempt to make more products directly from latex, without going through all the stages of coagulation, drying, baling, cutting and kneading. Vulcanization of rubber is more and more being accomplished by electronic heat instead of steam, which cuts the time element from five hours to a few minutes. A very important part of all rubber research is the steady improvement in the quality of synthetic rubbers, for though both natural and synthetic have uses for which each is best suited, the progress which has already been made in the chemical rubber field has given the United States some control over the price and supply of rubber. It would probably not be possible again for an enemy power to create the situation which existed in 1941, for when and if rubber demands skyrocket in the future due to some similar international situation, we will not fear the cutting off of the rubber supply. If necessary, we can manufacture our rubber within our own borders.

CHAPTER TEN

PULP, PAPER AND PRINT

Merely to mention paper is to call to mind some of the most momentous steps in the advance of modern civilization. Hardly had man advanced from the primitive stage than he began to put down on stone his ideas and the record of his achievements. When one ponders upon the kind of world we would live in were it not for this ancient tendency of our ancestors, the importance of a medium for making such a record becomes clear. The materials which have been used throughout the ages have differed very widely, depending upon whether permanence of the record or ease in making the inscription was uppermost in the mind of the recorder. Some early peoples, as the Babylonians, had no stone and so they devised a plastic material, tablets of clay which they later baked. This writing material was perhaps the first plastic prepared by humans. Wooden slabs, leaves of plants, wax and the skin of animals all served their turns as recordable materials, as human ingenuity sought natural surfaces suitable for use. But of all the media to carry the writing or the print, and indeed much of the artistic production of the past, paper has been the most important.

True paper is distinguished from papyrus, as first prepared by the ancient Egyptians, and parchment, of Near Eastern origin. Papyrus is a natural writing material, made by slicing the stalk of a rushlike plant into thin sheets and then gumming them together; parchment is of course a product of animal skin, usually of a sheep. Paper, on the other hand, is an artificial material made from vegetable fibers, usually cotton rags, later from many other materials, but chiefly from the processed pulp of wood. Invented in China, probably, near the beginning of the second century A.D. by Ts'ai Lun, the process of making paper did not spread to the West until the ninth century and perhaps later, when those famous carriers of civilized ideas, the Arabs, introduced it into European culture. From Italy and Spain the method of making paper spread north to France, the Lowlands, Germany and later to England; by the late Middle Ages paper began to replace the more expensive parchment on the European continent; the existence of a relatively cheap and convenient material helped to advance knowledge when printing by movable type was introduced in the mid-fifteenth century.

Paper-making, like other industry, felt the impact of the mechanical revolution of the eighteenth century. About 1750 Dutch inventors produced a contrivance for tearing rags into fibers for paper manufacture, which has fittingly come to be called the "hollander." In 1797 Louis Robert, a French craftsman, invented a machine which would mat rag fibers into sheets; this was altered and improved a few years later by Henry and Sealy Fourdrinier in England. Up to the middle of the nineteenth century cotton rags were almost the only material used for paper fibers, but research was continually going on to find a more plentiful material. Ages before, Ts'ai Lun had tried the inner bark of the mulberry tree, hemp, fish nets, and finally, cotton rags. Some results of a mediocre kind were obtained in the early nineteenth century with straw. About 1840 a German inventor named Keller produced a wood grinding machine, with the idea that the fibers obtained could be pressed into paper; after some fourteen years the wood grinder was a success, and although the pulp obtained did not match rag pulp in strength or quality, it served to make a cheap, non-permanent paper when mixed with the older, better material.

After the middle of the century, progress in adapting wood to paper-making was quite rapid, as chemistry came to the aid of purely mechanical methods. In the United States an inventor named Tilghman began experimenting with the action of acids upon wood about 1867; he found that sulphurous acid would dissolve the woody or ligneous elements, and leave a residue of cellulose fibers. While he was unable to adapt his idea to a commercial process, his efforts were carried further by Ekman in Sweden, by Mitscherlich in Germany, and by Ritter and Kellner in Austria. These adaptations of the sulphite process, as it came to be called, were in operation by the 1880's. Other chemical processes were soon introduced; Dahl's sulphate method was developed in 1884, and a caustic soda process which had been tried earlier, was gradually perfected. As a result, the paper industry was revolutionized completely and underwent a vast expansion process at the end of the century. Raw material was plentiful in every country with an ample amount of pulpable wood.

Curiously enough, it was not from England that the art of paper manufacture spread to America, for the English, who used wool and leather for their clothing, and hence had no supplies of cotton rags, were the last Europeans to establish paper mills. Dutch and German settlers in the province of Pennsylvania were starting paper mills almost as early as any mills were started in England. William Bradford, an English printer of early Philadelphia, solved the difficulty of securing a sufficient supply of paper by forming a partnership with William Rittenhouse, an immigrant paper maker from Germany. The establishing of the Rittenhouse mill in Germantown,

Pennsylvania, in 1690, ushered in the American paper industry.

In the American colonies, paper was made in small mills, by a few workmen, usually not more than ten in a single mill, and it was of course largely a hand process. Vats of "half-stuff" or pulp suspended in baths of water, were prepared; into them were inserted molds; the water was removed from the molded pulp sheets by hand presses and later the sheets were dried by being attached to poles. By 1810 there were 179 of these mills, but destructive European competition in the next few years cut down their number to about one hundred. The only machine used in the early American mills was some variety of the "hollander," or rag-tearing machine, but in 1827 a Fourdrinier machine was brought to the United States and installed in a mill at Saugerties, New York, by Henry Barclay. By 1830 a factory to make these machines had been set up in Connecticut; by mid-century most American paper mills were partially mechanized. After the Civil War came the wood grinder and the chemical processes.

The Manufacturing Process

All paper of whatever kind is essentially a product manufactured of vegetable fibers by a process which is partly mechanical and partly chemical. The amount of fibrous material used in, or separated from, the pulp depends upon the kind of paper or paper product which is desired. In the making of newsprint, for example, the original fibrous material of the wood is entirely used; in most other papers only the cellulose fibers are utilized. It is possible, therefore, to use the mechanical process of grinding wood into pulp to produce newsprint, providing a small percentage of pulp made by the sulphite process is added, to give the paper added strength.

Three types of materials come into the process of paper manufacture. First there are the natural fibrous substances; cotton rags, jute, straw or woodpulp; next are the chemical or mineral materials which are customarily used to treat the fibers, or as finishing agents, including sulphur, lime, caustic soda, clay or rosin; finally considerable amounts of water, coal and power are needed in the process.

By the 1880's, when the mechanical and chemical processes began to be effective in converting wood to pulp, the American pulp and paper business was heavily concentrated in New England, close to available supplies of spruce timber and water power; by the 1890's the industry had expanded to the Great Lakes area. These two regions continued to be the chief centers of pulp and paper processing until the first World War. The spruce resources of New England and the Great Lakes region were rapidly depleted, however, and the Canadian spruce forests were called upon to replenish the diminishing supply. Dominion spruce came into the United States at a faster rate following removal of import duties in the Tariff of 1913; after the war, northern European countries also began to furnish considerable amounts of our pulpwood.

Meanwhile domestic producers bestirred themselves to adapt the pines of the South to the manufacture of wrapping paper and paperboard by the sulphate process, an adaptation of the earlier caustic soda treatment; pulping techniques were further adapted to the balsam firs of the western forests and led to a tremendous increase in pulp production in Oregon and Washington. These new sources of pulpwood cut down the dependence of American mills upon imports from abroad, though Canadian production has since increased, largely under the stimulus of American capital. In the twenty year period from 1909 to 1929, U. S. Census figures indicated an increase of almost 500 per cent in the value of American pulp and paper products. Since 1929 the paper industry has moved into a very high position; by 1950 it stood twelfth in volume of sales among American industries, the total for paper and allied products for that year exceeding six-and-a-half billion dollars.

It is quite clear that this massive growth has given a great stimulus to forest operations, though it has also led, as has been noted, to an unfortunate "mining" of many of our oldest forest resources. It is significant that this immense recent demand has been superimposed upon an industry which has always been of great importance to the furniture, fixture and building industries. The pulp and paper business is basic to another industrial giant, the printing and publishing industry, of which more will be said later.

Spruce has long been the most desirable wood for pulp; balsam fir, western hemlock and southern pine have come into use as spruce has become too scarce and expensive, and as processes have been perfected to utilize the other types. In considering the manufacture of paper, it is possible to regard it as one process, from forest to final product, since the trend in the paper business is toward integration; such a move has already taken place among the larger companies. What gives variety to the subject is the existence of varying methods in preparing wood for paper-making, and the immense variety of purposes to which the final product may be put.

At the forest end of the process, the cutting and shipping (or floating) of pulpwood for eventual paper manufacture is simply a specialized form of logging, but early in the game it must be decided what kind of pulp and hence what kind of paper or paper product is desired. If mechanical pulp and hence newsprint are wanted, the logs are barked and cut up into blocks which are later ground up into mechanical wood pulp. Before being fed into the paper machine, it is mixed with unbleached sulphite pulp which may have come from the same forest, but which, as a chipped wood product, has gone through a digester and has been treated with steam and acid. The two pulp products, the mechanical part from the grinder forming about eighty per cent of the mass, and the sulphite product from the digester the remainder, are mixed together and enter the paper machine to emerge as newsprint.

A problem which has long influenced the production of mechanical pulp, and hence newsprint, is that hardwoods are not practicable for mechanical pulping due to the enormous amount of power needed. On the other hand, mechanical pulping is a cheaper process than chemical pulping, due to the great saving in equipment and materials, always a factor in chemical processes. Yet such a saving is difficult to achieve, for spruce, the ideal wood for the mechanical method, is difficult to obtain.

Modern pulp manufacture has become, therefore, largely a chemical process, or rather, a series of chemical processes. In any of the principal chemical pulp processes, the wood has first to be chipped before being fed into the digesters; beyond this point, the methods vary considerably. In the sulphite process, the wood is cooked in an acid solution, often bisulphite of lime, under high temperature and pressure; after some eight or ten hours of this, nearly half of the wood by weight is dissolved, leaving a high cellulose residue. This sulphite pulp has many uses; it may be bleached and mixed with rags to be made into bond, writing or ledger paper; mixed with soda pulp or waste paper, it can be processed into book paper; unbleached it may go into a mixture with mechanical pulp or newsprint, as we have seen; by itself it may become wrapping paper, bags, or container stock. Sulphite pulp is best if made from non-resinous woods, as the process does not eliminate resinous elements.

The soda process is similar to the sulphite one except that the cooking agent is caustic soda in place of an acid solution; many woods, including the short-fibered hardwoods such as aspen, gum and poplar are workable by this method. Soda pulp serves, as has been mentioned, as a filler for book papers. A modification of the soda method is the sulphate process and is so named because sodium sulphate, or salt cake, is introduced into the cooking solution in place of part of the soda. This chemical addition makes the pulping more rapid and uniform; it produces pulp of great strength, which is what gives both the process and the product the name of "kraft" from the German word for "strength." The sulphate process is adapted to resinous and pitchy woods, such as are found in our southern forests; it has contributed greatly to the container and wrapping paper divisions of the industry, for it gives those items much-needed strength.

Pulp of whatever kind, prepared by one or another of these methods, frequently is bleached, especially if a high grade paper is required, such as fine book stock. This is done while the wet pulp is in motion, by an oxidizing action, using hypochlorites of soda or calcium, or chlorine. Before the pulp enters the paper machine, whether it is bleached or unbleached, it is passed into beaters which roll it about over a bed plate to rub and bruise the fibers into a soft, gelatinous condition. This is intended to form the pulp into a sheet of uniform texture and to "felt" the fibers properly. In the beater, also, a number of ingredients may be mixed with the pulp: clay, talc and lime sulphate will make the pulp opaque and smooth; rosinsize may be added to secure a surface which will take ink and resist penetration by water. Another machine, called a "Jordan" is frequently used to attain the same effect as a beater in shorter time. In some cases both kinds of machines are used.

Paper machines which take this pulp and change it into the product desired are of two types: the Fourdrinier and the cylinder machine. The former is really a massive series of machines "about a city block in length." The same principle evolved by its inventor Robert in 1797 is still used; a bronze wire belt of about seventy meshes to the inch carries the wet pulp continuously; on this wire belt the fibers are formed into a continuous sheet as the water drains off through the meshes. The loosely felted web passes on between heavy rollers which press out more water, and finally through forty or fifty steam-heated cylinders. At the dry end, paper emerges at the rate of ten to fifteen miles per hour, sometimes as wide as 300 inches. The cylinder machine differs from the Fourdrinier type mostly at the wet end. Instead of a wire sheet-forming device, it has at least one large cylinder covered with wire cloth, partly submerged in a vat containing the "slurry" or wet pulp mixture. As the cylinder revolves, the water is drained off the wire cloth as the fibers cling to it. Some of these machines have several cylinders, in order that several types of stock may be picked up and laminated together. For example, one cylinder may provide a glossy white outer layer, a second an inner section of low grade stuff made of waste paper, while a third may add another outer layer like the first. All are sandwiched together and rolled through the machine, to emerge as fancy coated cardboard.

Some papers are delivered to users as they emerge from the Fourdrinier or cylinder machines, while others may undergo further treatment, according to need, by super-calenders or by baths in coating materials. Wax, mixtures of clay and casein, glue and starch are some of the surfacing materials used. It is indeed an impressive feat in mechanization, as the soggy mess which not long before, as pulp logs, was headed for the barking drums, emerges from the great paper machine as an enormous roll of smooth, finished paper. Some of these rolls weigh as much as 2400 pounds and contain several miles of paper.

Some Characteristics of the Paper Industry

This industry is often classified as one of the "chemical process" activities along with petroleum, rubber and man-made textiles. Another way of looking at it is as an intermediate or service industry, since a large proportion of its product is destined for other industries rather than for an immediate consumer. Its newsprint is exclusively for the newspaper publishing industry; book papers are made for the publishers of books and similar publications; paperboard and bagging serves the container industry, with a wide variety of customers among makers of shoes, clothing, foods, drugs and most all retail merchants. The thin papers and tissues are important to the above list, as protective or decorative material, and have also a wide use as towelling, for sanitary papers and others; the finer qualities of finished papers serve the stationery and gift card trade; very heavy grades are used in the building industry. The uses of paper products are almost limitless.

The paperboard section of the industry is very large; probably a half of all paper products are of this type. Production of paperboard has doubled every decade since 1899, the highest rate of increase of any paper product. This increase has come about partly because of the increasing cost of wooden boxes and containers and the lighter weight of their paperboard substitutes, which gives shippers a considerable saving in shipping costs; the technology of making paper boxing has advanced to such a degree that very large items may now be shipped in this material. The source of board has been waste paper usually, which in turn has depended largely on the collections of social agencies and junk dealers. An exception to this is the development in recent years of southern "kraft" paperboard, which is manufactured largely from pure pulp, as mentioned already.

There are many companies which make paperboard, and most of them are relatively small. The largest company in the field, as indeed in nearly every part of the paper industry, is the International Paper Company, formed in 1898 by an integration of some thirty-four paper mills in the United States and Canada. Other companies which produce large amounts of this material include Container Corporation of America, Robert Gair Company and Fibreboard Products, which is a subsidiary of Crown Zellerbach Corporation. The paperboard field has an element of stability in that it serves industries which ship steadily: drug firms, clothing manufacturers, soft drink companies and so on.

* * *

The makers of newsprint also have a stable market, since in the two hundred and fifty years of American newspaper publication the demand has grown to immense proportions, especially after the appearance of the cheap newspaper supported by extensive advertising, and with the appearance in the present century of the great newspaper chains. By 1940, due largely to the activities of Hearst, Scripps-Howard, Patterson-McCormick and the lesser newspaper chains, the daily circulation of newspapers reached thirty-five million, with twenty-four million on Sundays. Added to the newsprint demands of the daily and Sunday papers were those of some 5,000 weekly papers and about 2,000 monthly and quarterly magazines. Ten years later, by 1950, the combined daily circulation of English language newspapers in the United States had risen to 53,829,072, the Sunday papers to 46,582,342. Weekly papers and magazines showed similar increases.

It would have been impossible to meet such a demand, which calls for the manufacture of more than five million tons of newsprint, according to the last survey by the American Paper and Pulp Association, without the technical developments in wood pulping and the mechanical improvements represented by the latest Fourdrinier and cylinder paper machines; likewise had it not been for the improvements in making paper from wood pulp, no such production of cheap paper would be possible. Some twenty-five companies close to large sources of wood pulp, in Maine, New York, Wisconsin, Minnesota, Washington and Oregon supply the largest proportion of newsprint. Again the International Paper Company leads the field in this area of paper manufacture.

Newsprint is usually sold under long term contracts, which

makes for stability, a necessary feature of the business, since it takes about two years to organize and construct such a mill. During the period just before the depression of 1929 there was too great an expansion of the newsprint business, which caused a serious drop in production as the depression commenced. Competition from southern kraft paper makers led to the investment of much northern capital in the kraft newsprint business.

During a number of years after the first World War there was evident a steady drop in the prices obtained for newsprint which may have indicated excess manufacturing capacity and perhaps inflated capital structure in the newsprint business. This brings to mind a characteristic of the paper industry which was of great aid to paper makers in the depressed period. It is quite possible for a company making newsprint to enter another field of paper making; it may go in for wrapping paper or book papers, with minor mechanical adjustments in its process, since paper machinery and paper processes are fairly uniform whatever the product. A study made by the American Paper and Pulp Association of fifty-six newsprint mills between 1910 and 1934 noted that the mills made 733,022 tons of newsprint and 159,783 tons of other grades in 1910, but that by 1934, while they were making about the same amount of newsprint as before, they had increased their production of other grades to 714,205 tons. It is clear from such a study that newsprint companies did not hesitate to convert to other types of paper when the newsprint business was depressed, a practice which was of great value to the industry. It is fortunate that such flexibility exists. Of course it is true that the very large companies, like International Paper and Crown Zellerbach, have never confined their activities to only a few paper products; such a policy would be extremely unwise.

It is also characteristic of the paper and paper products industry that we find no high degree of integration in it, certainly to no such extent as has occurred in the chemical, steel and rubber industries. There were attempts to control certain areas of the industry, but they have never been very successful. Impressive positions in certain lines of paper manufacture were achieved by American Strawboard Company in 1889, by the International Paper Company in the months following its organization in 1898, by the United Box Board and Paper Company about 1902, and by the United Paper Board Company at the time of our entry into the first World War. Of these companies none continued to dominate any one field of paper making, though for a time International Paper processed about sixty per cent of the newsprint consumed in the nation. The rapid growth of the industry, with the flexibility alluded to, has prevented any striking monopoly within the industry.

In the 1951 report of the Federal Trade Commission on Interlocking Directorates, only two paper companies, International and Crown Zellerbach, are listed as among the one hundred largest manufacturing corporations in the United States. Yet fifty-nine paper companies are included as among the largest one thousand. Substantial but not exceptional size of companies has come to be the rule in paper manufacturing. There is also considerable diversity and specialization. There are bag makers, as Union Bag and Paper Company; paper coaters and glazers, such as the Kalamazoo Vegetable Parchment Company; wallpaper manufacturers, as United Wallpaper, Incorporated; makers of fine writing paper, as Hammermill Paper Company; and of specialties featured by the U. S. Playing Cards Company, the U. S. Envelope Company, the Gaylord Container Corporation, the Dixie Cup Company, and others.

While no complete integration is true of the paper industry, it is natural that the largest firms own their own timberlands, power plants, water supplies and filter plants, logging transportation facilities, and pulp and paper mills. International Paper Company, largest in the industry, produces annually approximately three million tons of paper, paperboard and pulp. Its forests, pulp and paper mills, and converting operations are scattered throughout northeastern United States, southern Canada, and in the South, from Arkansas and Louisiana to South Carolina and Florida. More than half of its production is in the kraft paper and board field; about a quarter of its annual product is newsprint; pulp for sale and miscellaneous types of paper account for the remainder. Crown Zellerbach, second largest in the industry, is the dominating paper and paper products company on the West Coast. St. Regis Paper Company, the third largest firm, manufactures paper, pulp, paper bags, plastics, bag-making machinery and bag-filling machinery. Kimberly-Clark, next in rank, produces book papers, rotogravure papers, wadding (both absorbent and building insulation), and hanging papers. Below the so-called "Big Four," the industry shows the diversity and specialization already noted.

It is hard to believe that today's average American uses some 340 pounds of paper products annually, about six times as much as did his grandfather in 1898, according to figures recently published by the International Paper Company. This average man reads about sixty-six pounds of newspapers, consumes commodities shipped or delivered to him in 125 pounds of paperboard containers and boxes, and in thirty pounds of wrapping paper and paper bags. Another 119 pounds is needed for his books, letter paper, railroad and theatre tickets, milk and food containers, telephone directories, catalogues, tissues. He or she may even have paper rugs and window curtains in the home, and ride in an automobile with seat covers made of twisted paper strands. This average person also wears, as does his family, rayon fabrics and uses many other forms of rayon in his household, all dependent upon the manufacturer of wood pulp, material which also goes into the manufacture of compacts, pencils, fountain pens, kitchen equipment, radio cabinets, and the cellophane which keeps cigarettes, chewing gum, candy bars, food, and a hundred other everyday articles clean and fresh. The multitudinous uses of the 26,000,000 tons of pulp and paper products annually consumed in the United States constitute an index of the high standard of living of the nation.

Printing and Publishing

It is difficult to imagine any feature of our civilization more important than the advance of printing and publication of books, newspapers and other printed materials. In its very nature, printing is a fine art, which has descended to us through the centuries as one of the most ingenious and most precious of heritages from the past. All learning and much of modern business and technology are dependent upon the ready accessibility of printed products. The preparations of these materials, based upon the techniques of paper-making, printing and all the arts related to publishing, constitute in themselves a large category of American industry.

To relate in detail any substantial part of the history of printing and publishing is quite beyond the scope of this volume. Nor is there time to speculate upon the enormous social and political influence of the printed word, whether in newspaper, book or periodical, upon the public mind. This section will merely call to mind some of the recent developments in the printing and publishing industries as a part of our great industrial system, in contributing to the advance of American industrial science.

Printing and publishing, as we have seen, are tied directly to the great technical advances made in paper manufacture, both of newsprint and book paper. A mechanical revolution also took place in printing itself, whereby the old lever and screw presses which had served most printers from Gutenberg to Ben Franklin were replaced by cylinder machines first developed by Friedrich Koenig and initially installed by the London Times in 1814. The cylinders rolled over the type, which was fixed in a flat bed, to make the impression; later of course the type itself was mounted on cylinders, the number of feeders increased and the capacity of such a press became five times as large as that of the early ones.

From this point it was possible to create the stereotype plate. A page of type, after being set and locked in the form, was used to mold a page-size paper matrix, or mat, from which a cast and any number of duplicates could be made. Also there appeared the continuous web of paper instead of the single sheets, and cutting and folding machines which reduced hand labor to a minimum.

It was necessary, especially in the production of the daily newspaper, to reduce the time of typesetting; machines were designed at first to pick the type out of the case and arrange it, but this system was soon replaced by the linotype machine of Otto Mergenthaler, in which the characters are actually cast in molten metal by an operator who plays a keyboard not unlike a typewriter. The first linotype machines of Mergenthaler were set up in the New York Tribune printing plant in 1886. These machines multiplied the speed of typesetting by many times and made possible the present voluminous newspaper. Inks were greatly improved from the usual linseed oil and lampblack of the early printers; rapid drying products, obtained by adding to the black material rosin dissolved in mineral oil, appeared. Color printing has been made possible on a wide scale by the use of coal tar dyes in the inks. Special inks have been produced for the rotogravure process so that the pictures or letters to be printed are engraved in the printing surface instead of being raised above it.

Mechanical improvements were not alone responsible for the development of the newspaper publishing business into the colossus we know today. With the increase of education and the massive increase of business activity, a news-hungry public developed a taste for immediate information on world events, domestic politics, business and financial changes, local events, literary and entertainment features. To satisfy this enormous demand required the work of scores of inventors such as Koenig, Hoe and Mergenthaler, of a considerable part of the chemical industry, of great editors like Bennett, Greeley and Raymond, and of great business promoters like Hearst, Pulitzer and Roy Howard. It also called for the organization of a huge advertising industry in order to stimulate revenues to pay for the immense costs of printing a multi-page newspaper to sell at a reasonable price.

It took more than this. Great news services like the Associated Press, the United Press, the International News Service and the world-wide services operated by many of the large papers themselves appeared. The technical aid of the radio, the teletypewriter and all sorts of automatic and electrical devices were mobilized to gather, edit, process, print and distribute the mass of information which flows endlessly from the presses of the great newspapers of the 1950's. Instances multiply of the constant improvement in the gathering and distribution of news by electronic means. The Associated Press announced on October 13, 1953 the opening of an around-the-clock teletypesetter service for its members along the Atlantic seaboard. The system carries continuous news transmission from the New York office of the AP to its northeastern members by teletype and perforated tape. At the newspaper offices the tape can be fed into line-casting machines to speed the production of type. The old newspaper, printed on the old-fashioned presses, personally conducted by the "great man" (who always added his daily wisdom on the editorial page), and carried about town by the hustling little newsboys, is a thing of the remote past. Instead we have a great corporation, often operating not one, but many papers, highly organized and streamlined, with thousands of employees, often operating its own news service in all parts of the world, and frequently owning its own pulpwood forests and paper mills. It has become in reality a large manufacturing establishment which uses enormous quantites of raw materials and conducts highly technical machine operations.

There are many financial ties between producers of newsprint and the great newspapers which are their principal customers. The large newspaper chain, usually called the McCormick-Patterson publications, which publishes the Chicago Tribune, the New York Daily News and the New York Sunday News, owns timber lands in Quebec and the stock of the Ontario Paper Company and the Quebec North Shore Company; these two paper companies naturally contract with the McCormick papers for their newsprint product. The New York Times also has a heavy investment in newsprint production; it owns nearly half the stock of the Spruce Falls Power and Paper Company in Ontario Province. The St. Paul Dispatch and Pioneer Press has an interest in the Blandin Paper Company and is under long term contract to buy paper from it. Conversely paper companies have sometimes held considerable stock in newspapers. International Paper at one time held part ownership in fourteen papers but dropped most of these interests about 1934, except its stock in one Chicago paper and in three smaller ones in the South.

The largest circulation of any paper in the United States is enjoyed by the New York Sunday News, with over four million; other papers with Sunday circulation over a million include the New York Times, the New York Sunday Mirror, the Chicago Tribune, the Philadelphia Inquirer, and the New York Journal and American. Only one of these papers has a daily circulation of more than a million; that is again the New York Daily News, which tops two million each weekday, but the New York Mirror is only a few thousand under a million. According to the Audit Bureau of Circulations, the authority for the above figures, there are some 136 newspapers in the United States which have a circulation of more than 100,000 copies a day, in either their daily or Sunday issues, or in the two combined. It is interesting to note that the largest newspapers tend to be morning issues.

* * *

"The earliest publisher," states R. L. Duffus in his stimulating little volume on books, "was anyone who possessed presses, type, ink and paper. Often his product was available only in a single bookshop, of which he himself might be the sole proprietor." Early authors often hired printers to process their manuscripts for them, and thus the authors themselves became their own publishers. Book publishing in this country is more than three hundred years old, since the *Whole Booke of Psalms,* the first American book, was published at Cambridge, Massachusetts in 1640. Such men as Stephen Daye, William Bradford and Benjamin Franklin were publishers as well as printers of books. In 1801 bookselling was launched as a regular retail business with the establishment of the American Company of Booksellers by Matthew Carey in Philadelphia. As late as 1820 about seventy per cent of all books published for American readers were by foreign authors, but this proportion declined until, by mid-century, some four-fifths were written by natives. Although British books were very popular in America throughout the nineteenth century, this fact did British authors and publishers little good, as we had no copyright arrangement protecting foreign authors until 1891. If an American publisher were scrupulous, he would arrange to pay a royalty to the foreign author or agent; if not, he simply did not bother.

Books were distributed in the early days by enterprising booksellers like Mr. Carey, with the keen rivalry, or sometimes assistance, of itinerant book agents who peddled lives of Washington or editions of *Pilgrim's Progress,* plus editions of the classics, about the rural areas. Even in the cruder days,

American taste for reading was lively and often directed to the serious side, as the huge sales of McGuffey Readers and Webster's Spelling Books testified. Some of the great publishing houses of today began their existence before the Civil War, among them Lippincott, John Wiley, Harper, Appleton (now Appleton-Century-Crofts), Scribner, Van Nostrand, Putnam, Little Brown, Dodd Mead, Houghton Mifflin, and E. P. Dutton. Others began their operations soon after the Civil War, as Ginn, American Book, Macmillan (after some years as American agent for the parent firm in England), Doubleday Page (now Doubleday) and many others.

Early publishers and booksellers organizations, started often to maintain list prices of books, were usually not permanent. The first permanent and successful trade organ of the industry was the *Publishers' Weekly,* started in 1872, which of course still exists. Later associations of publishers and booksellers were more lasting. The older publishers were often individuals of great talent and business ability, who placed an individual stamp upon their firms, many of which remained in family hands for several generations. This strong imprint of a single personality upon the publishing house has largely disappeared, as firms have become bigger and have had to adjust themselves to the rather impersonal trend of the times.

The book publisher follows what Stanley Unwin has called "an art, a craft and a business." He and his associates must be informed upon the background of their business, must have technical knowledge about printing, paper, binding, and the other arts connected with the physical making of books. He must moreover be informed about conditions in the book trade, and understand the law of contract and libel. He must know not only how to select publishable manuscripts and have them manufactured into books; he must also sell them. For every best-seller which he may have the good fortune to publish, he may very well publish a great many which lose money, or barely eke out their cost in sales. The average person, dazzled by the millions of sales which certain books achieve, often imagines publishing to be one of the most profitable of businesses; actually it is highly technical, difficult and hazardous. It is probably one of the easiest ways to lose money yet to be devised.

It was possible at one time to group book publishers into "trade" houses, or publishers of general books; text book publishers; and those who specialized in technical books, as medical or engineering. This classification is no longer strictly accurate. In a recent study of the book industry, by William Miller, we are able to learn many of the up-to-date facts about present-day book publishing, which has changed markedly over the years. For instance, Miller notes that "fiction rules the trade book world," however reluctant publishers are to admit that fact, and it is true that of the ten or twelve thousand titles annually published in the United States, fiction always leads by a wide margin. For example, in 1939 fiction titles totaled 1547, while juvenile books, its closest competitor, totaled but 949; in 1951 and for the first eleven months of 1952, fiction still led by an even wider margin. Since fiction sells in much larger editions than non-fiction usually, most trade publishers, to exist, must have a successful fiction list.

Most trade publishers, while they of necessity emphasize fiction in their operations, also include in their lists history, biography, religious books, poetry, science, travel, war and many other types, as the elements of a "well-balanced" list. "Sex" says Miller, "is a staple trade book subject and sells well in almost any literary form." The trade book, especially fiction, is the type most frequently promoted at book fairs, autograph parties and other special promotional occasions. A fortunate few of these trade books become the subject of the book gossip sections of the Sunday papers, of reviews in the book review section of the newspapers, or are featured by the book editors of the more literary of the weekly and monthly magazines. To some extent, the book publishing industry is under the spell of the "best-seller" urge and hopes and prays that the book being promoted may take hold of the public imagination, become a choice of one of the powerful book clubs and thus sell in many thousands.

The interest in those books which sell the best is a well-established one which dates from the publication in 1895 of a monthly best-seller list by the *Bookman;* this list was continued until 1933, when the *Bookman* went out of existence; since then this function has been performed by several of the large newspapers in their weekly best-seller lists, and also by the *Publishers' Weekly* and by *Bowker Book Guides.* A very interesting little study of the first fifty years of best-sellers (1895-1945) has been published by the Bowker Company, under the authorship of Alice Payne Hackett.

It is an interesting fact that the sale of non-fiction was so small prior to World War I, that the best-seller lists considered only fiction; this situation changed quickly when we entered the war, when for two years it became necessary to include three best-seller categories: fiction, war books, and non-fiction other than war books. After the war, the custom began of having two lists, non-fiction and fiction, which of course persists to this day.

For the half-century ending in 1945, fiction outranked non-fiction as a best-seller about three to one. Other types of books seemed to appeal at one time, not so much at other times. The early vogue of religious books came to an end about 1890, but became strong again in the late 1930's. Historical novels began to have great popularity in the first years of the century, and took another jump in the 1930's. Realism in fiction came along after World War I with Sinclair Lewis and others. Juvenile books continued to have a persistent attraction, the same titles often retaining their popularity for many years. Most publishers, says Miller, like to have a good juvenile list. These periodic tastes are punctuated at times by the sudden selling power of books on cooking, first aid, diet, personality, travel or biography, as they hit the public fancy.

Among recent best-sellers in the "hard-cover" field, the largest have been Margaret Mitchell's *Gone With the Wind,* with 3,625,000 sales; Dale Carnegie's *How to Win Friends and Influence People,* 2,750,000; Marion Hargrove's *See Here, Private Hargrove,* 2,500,000; Fannie Farmer's *Boston Cooking School Cook Book,* 2,500,000; and Betty Smith's *A Tree Grows in Brooklyn,* 2,000,000, according to Miss Hackett's compilations. The enormous sale of these and many other books is to be accounted for by several reasons: the importance of the larger book clubs, which by selecting a title can alone assure very large sales; the popular-priced reprint editions which often appear as soon as the original edition has assured the book's success. Older books, besides the *Bible,* which of course outranks all books, ancient and modern, in sales volume, have also sold in enormous numbers. But the measurement of their sales is difficult, since so many countries, publishers and editions are involved; no certain sales totals of British classics can be reached, for example, since so many of them were re-published here before the first copyright agreement with Great Britain was made in 1891; no record of these was kept prior to that date.

American publishers of trade books show evidence today of getting away from their traditional type of publishing and of entering any kind which shows promise of profitable results. The textbook publishers tend to enter the trade book field as well; for example Van Nostrand and John Wiley have added trade titles to their lists; the spread into both fields is illustrated by the widened activities of Harper, Harcourt Brace, Houghton Mifflin, Scribner, Holt and many others. Macmillan puts out books of almost every kind; its backlist of some 15,000 titles is more varied than that of any publisher.

An interesting type of publishing house is the university

press. Originally publishers of scholarly books almost exclusively, these organizations have recently come into competition with the trade book firms. Perhaps because of more aggressive management, or because of greater interest by faculty members in creative or unusual writing, perhaps to swell their rather sparse incomes, the academic presses have been increasingly taking over books which formerly would have been the responsibility of the commercial presses. Another reason why this is true may lie in the increased cost of book publication; since a trade book must now have sales of from six to ten thousand to make expenses, manuscripts which may seem to promise less than such a volume of sales have been in many cases rejected by the trade, and have been picked up by some university press or other, to whom sales of thousands seem better than sales of hundreds, to which they were formerly accustomed.

Whether such a departure is to be profitable to the university press group is doubtful, though it is true that they have a certain advantage not possessed by trade firms, in that they pay no rent. They may also reach a public not available to the larger publisher, in view of the prestige of the university which backs them, though this is rather doubtful; even if true, such a buying public would probably not be large. Mr. Datus Smith, Jr., of the Princeton University Press condemns the activity of some university presses, which seem to be searching for best-sellers, and claims that the financial well-being of a university press comes from its scholarly back-list, which sells steadily, though in small volume, rather than from a chance new title of great popularity. Yet it is true that the books of university presses sell more today than ever before, and that occasionally one of their titles takes hold with the public in a really impressive way, like the *Lincoln Reader* brought out by the Rutgers University Press, which reached very large sales.

Magazine publication partakes of the characteristics both of the book business and of the newspaper business. Like the newspaper, the rising costs of publication must be met by a very large advertising income; like the book business, the magazine's success depends often upon its choice of popular writers (and here competes seriously with the book trade), and the manner in which its publication policies suit the popular taste. Many of the magazines of a generation ago have disappeared: *Everybody's*, the *Century*, *Munsey's*, *McClure's* and the *Literary Digest* are no more; those which have survived have usually done so by striking changes in form or layout, and by improved promotion methods. Large corporate consolidation has helped many; Hearst Consolidated Publications, Incorporated; Time, Incorporated; the Curtis Publishing Company and Conde Nast Publications all publish a number of magazines each. Most of these organizations have been in existence for a considerable time, but Time, Incorporated is a comparative newcomer.

In 1922 Henry Luce and the late Briton Hadden founded *Time;* a few months after the crash of 1929 they began the publication of *Fortune*, and in 1936, a depression year, Luce began the issue of *Life*. It required several years to make a success of *Time*, but *Fortune* and *Life* were almost immediately successful. Along with the *Architectural Forum*, which Luce's organization acquired, the combined circulation of his four publications reached nearly eight million in 1947; five years later, *Life* alone was selling at a five million clip, being exceeded only by the *Reader's Digest* which claimed nine million sales for the United States and its possessions, in addition to various foreign editions. It is remarkable that publications like *Life*, *Fortune* and the *Reader's Digest* have grown so fast, registering great gains even in times of depressed business conditions.

According to the 1951 report of the Federal Trade Commission, twenty-six printing and publishing firms are among the one thousand largest American manufacturing firms. Of these, Hearst Consolidated Publications is among the largest one hundred; Curtis Publishing Company and Time, Incorporated are among the second hundred; Crowell-Collier Publishing Company is among the third hundred. The largest publisher of books, according to this report, is the McGraw-Hill Publishing Company, which also at that time ranked sixth in the printing and publishing industry; only three other book publishers appear among the largest one thousand firms: American Book Company, Macmillan, and Houghton Mifflin. Among the list are several large printing firms, among them the Cuneo Press, which prints and binds periodicals for several of the large publishers; and McCall Corporation, which in addition to printing and publishing magazines and patterns, also prints under contract magazines for several of the other publishers. The trend noted among newspaper publishers, of mantaining a financial interest in pulp and paper manufacture, is also true of a number of magazine and book publishers.

Paper and Print: Present and Future

Through all phases of the pulp, paper, printing and publishing industries runs a connecting thread: a common raw material has been put to work to provide a series of products which are of indispensable use to our civilization. From the great forests of America have come the materials which we process into numberless types of paper and paperboard through the miracle of paper chemistry and the evolution of paper machinery; a series of complicated machines has enabled printers to speed up and improve the production of every form of print and illustration; publishers have built large corporations to assure the publication of great newspapers and the publishing of every known type of book. Upon the continuance and extension of this series of industrial processes depends, to a large extent, the operation of our great educational system, the public awareness of world events, and the profitable employment of a large proportion of the skilled workers, scientists, and clerical personnel of the nation. Paper and print production involve a large part of the forestry industry. One paper company alone employs more than 30,000 persons; as noted, there are fifty-nine of them in the one thousand largest American manufacturing industries.

*   *   *

To a considerable extent, the cleanliness, satisfactory condition and the safe arrival of the merchandise sold and shipped depend upon the packages, containers, bags and cartons manufactured by the paper industry. Foods, drugs, books and magazines, household appliances, hardware and cutlery, tobacco, wine and liquors, soft drinks, scientific instruments and business machines, glass, china, furniture, textiles, shoes, sporting goods, automobile parts, ammunition, cosmetics and many others are only some of the products which the huge container industry is geared to protect, from their manufacture to their delivery. A sizable section of the paper and pulp industry supplies fiber for rayon textiles; another part supplies the building industry.

Not only are the great pulpwood forests of the United States needed to satisfy the huge demand for newsprint pulp; in addition a considerable part of Canadian pulpwood production is also utilized for that purpose. In the publishing of books and fine papers a tonnage of paper almost as great as that of newsprint is consumed annually. The per capita consumption of paper and paper products in the United States of 340 pounds per person each year, represents a total consumption by the United States of more than fifty billion pounds yearly.

These are some of the outstanding facts about the paper business and the publishing business, with which it is closely linked. What of the future prospects for greater totals, for more expansion? In order to attempt what can be only a guess in answer to this question, it is well to remind ourselves again of the interconnections between the entire group of industries with which this chapter is concerned. The availability of greater supplies of high grade pulpwood affects the production of paper; the conditions within the paper industry affect the production of newspapers and books; conditions within the building and the rayon textile industries affect the markets for building boards and building papers and for sulphite pulp for the rayon companies.

Basic to the preservation of our great pulpwood forests is an extremely careful and energetic conservation and reforestation policy carried on by the pulp and paper companies. The days of "mining" our forests have passed; indeed it would be corporate suicide to allow it to continue. The Southern Kraft Division of the International Paper Company exercises special care over its vast holdings of southern forests by keeping a forestry corps of one hundred constantly at work. International owns two million acres of forests along the Gulf and Atlantic coasts from North Carolina to Texas and up into Arkansas, which, it claims, are among the best managed in the United States. It plants some 3,000,000 seedlings each year. Its fire-fighting efforts are so effective that its record is several times better than that of unprotected lands in the South. The industry's new trees, which grow very fast in the South, are started in plantations and protected from disease and insects as well as fire. Furthermore, by marking and selecting individual trees for cutting, and by the use of power saws which cut close to the ground, the companies avoid waste. Mechanized handling of the pulp logs from forest to mill cuts damage and waste of materials. The goal of the leading companies is to utilize completely "forest growth," not the demolition and destruction of all forest resources. This should be the objective for the future.

Some companies maintain a highly interesting educational program; one contributes to the support of forest training camps to help farm folk to take better care of their own woodlands; another has for ten years distributed millions of seedlings to 4-H clubs as a part of a general reforestation program; in a single year this company, Container Corporation of America, distributed 2,000,000 of the little trees to the future farmers and woodmen. Another paper firm, Champion Paper and Fiber Company of Hamilton, Ohio, made nearly 2,000,000 replantings on their own 395,000 acres of forests in a single year. The only criticism of this reforestation plan is that it probably does not go far enough; it would be sensible to replace every tree cut with several, for only a fraction of the new trees will probably grow to maturity; it is better, however, to plant two or three million and give them special care than to plant ten million which the company might not be able to care for properly.

Guides for the future of the paper industry may also be discerned in the success which companies have had in developing new products. For instance, International, and doubtless other firms as well, have succeeded in utilizing hardwoods for the inner corrugation of container board; hitherto hardwood had not been considered usable for container board stock. Much success has resulted from rebuilding certain machines for higher speeds and by extensive conversion of certain types of machines to the making of products in greater demand. The widespread change-over of Southern forests to pulpwood manufacture is an excellent move, not merely for the South, but for the industry as well, for it gains thereby a huge new source of raw material in the thickly-wooded Southern states and also gains several years of growing time due to the warmer climate.

Another field of great promise to the paper manufacturer is its service to the newer industries opened up by the large chemical developments of the past few years. Paper companies have found chemists useful not only for devising the best ways to make paper and paper products, but also in developing new products from materials used in the paper process. International owns, jointly with American Cyanamid

Corporation, the Arizona Chemical Company. Here byproducts of paper-making are processed into pinene, which becomes synthetic camphor; also from the liquor remaining after pulpwood is cooked is made "talloil," useful in coating various materials, as a component of linoleum, and for the flotation process of ore refining. Other byproducts become paint thinners, rosin for soap making and other purposes, and fatty acids, which serve the manufacturers of soap. Research will probably continue to find many more uses of wood and woodpulp than merely for paper making. As John H. Hinman, president of International Paper, stated recently in an anniversary address, it is not past belief that in the years to come, the great forests of the United States and Canada, provided they are scientifically used, will serve not only as material for paper, but will also be processed into "quantities of foodstuffs, alcohol and chemical raw materials from parts of the wood which we are only beginning to use today." If, as ambitious paper manufacturers believe, wood can be processed so successfully that its container and board products can replace metal to a large extent, that is a fact of strong economic importance, for forests can be replaced, while metals cannot.

What about books today, which require our paper resources? Book production and sales have increased in 1952 over 1951. Serious books sold well. The *Revised Standard Version of the Holy Bible,* which did not appear until late October of 1952, quickly built up a sale of more than a million copies. Expensive books, such as the *Diary of George Templeton Strong,* at thirty-five dollars a set, sold beyond the publisher's expectations. A phenomenon of book publication, the paper-bound book selling for twenty-five to fifty cents a copy, has been an amazing success; estimates state that the total sales of such books may reach more than 300,000,000 copies for the year 1954. Freeman Lewis, of Pocketbooks, has stated that the paper-bound book business pays four million dollars annually to its authors, gives wholesalers a brokerage of ten millions more, retailers two millions profit, and represents an annual investment by the public of sixty-five million dollars.

The meaning of all this is that these are excellent years for business in general, and therefore good years for the book publisher. The most that one can deduce beyond such a simple fact, is that the book publishing business will gain as our economy grows in size; incomes will rise and a larger and larger book-buying public will continue to support it. Of course the matter is not quite as simple as that. Expenses can rise as well; if materials and labor costs continue to rise, as well as taxes, the rising totals will continue to cut sharply into the profits of manufacturing books. If, as many believe, we are approaching a plateau in taxation, the book publisher may be able to benefit from that fact, as will other business. One hopeful sign, for all who publish books, is the fact that the tremendous build-up of paper-bound book publication did not have the disastrous effect upon the general book business as at first feared by the conservative trade houses. So if book publishing can absorb the entry of so novel a development to the tune of hundreds of millions of items, it would seem to have a rather strong hold upon the public interest, which can carry it through lean years as well as good ones.

CHAPTER ELEVEN

FEEDING, CLEANING AND CLOTHING THE MILLIONS

Pasteur once wrote: "In our century science is the soul of the prosperity of nations and the living source of all progress. . . . What really leads us forward is a few scientific discoveries and their applications." These scientific discoveries and some of their almost innumerable applications make possible the raising of our standards of health and comfort, the liberation of man from the burdensome labor of the farm, and of woman from the endless drudgery of household tasks. Today, a woman can keep her family adequately fed and clothed, her home in order, and still have hours free to pursue some useful outside occupation:—in industry, business, education, or in other lines. Fifty years ago, all would have suffered had she not stayed on the job at home from dawn 'til dusk and later. The farmer may now cultivate twice as much land with far less labor and with increased production per acre. With our manpower needs so great, science has released more men and women to increase this national asset, at the same time permitting us to supply our fast growing population as well as that of less fortunate areas of the world.

What are the applications of scientific discovery which have wrought these changes? We find them in the food industries:—agriculture, dairying, meat packing, canning and preserving; in the textile and clothing industries; in the electrical industry and others, with all the near-miraculous facilities and products of the mid-twentieth century combining to so lighten the task of the home maker that she is able to develop her capabilities in the arts, science, business and industry, without impeding the vital activity of the home. With a riveter, a piece of chalk or a test tube in her hand, or perhaps a customer, a patient or a client before her, woman is seeing a challenge and is accepting it readily, knowing that science has freed her for newer duties. And the farmer, freed from hours of backbreaking toil, has gained time to experiment with newer machines and methods, to plan and study for better use of his acres, with his eye on the twin goals of better production and more security for himself and his nation.

Our scientific advances have gone beyond these first two phases, released manpower and increased production, and are working at the foundation of our national well-being, the land, to insure its conservation and continued productivity for food and other purposes, for the better nutrition of plants, animals and men. With our population increasing nearly 20,000,000 in the last ten years, and our industrial centers growing overnight, the American farmer has a rapidly growing home market for food products. He is also a supplier of industrial raw materials and produces both food and raw materials for the foreign markets.

Agriculture and the Sciences

Our first advances in scientific agriculture came with the advent of mechanical aids in the nineteenth century. Jethro Wood's invention of the iron plow, with standardized, interchangeable parts, in 1819, marks the true beginning of our drive toward mechanization. Other devices followed fast—the harvester, the reaper, the binder, the cultivator, the deep soil plow, and the seeder—beginning in the 1830's and continuing through the 1860's. In 1830, thirty-four agricultural machines were patented. The number steadily increased with the years until the Civil War, when it reached a peak, with 350 patents in 1861 and 503 in 1863. Then the number dropped sharply and averaged about 100 per year.

These aids released farmers from the cultivation of grains mainly, and permitted the development of new crops; they also assisted in opening vast areas of the middle West and the plains states. The invention and adoption of mechanical farming devices also expanded our machine industry.

With the coming of power driven machinery, which has now almost supplanted the horse and the mule, greater strides were possible. The earliest tractors, powered by steam, were too large, heavy and costly for any but the very large farms. But the internal combustion engine made possible the development of smaller, more practical machine tractors. At this particular time, it was World War I which provided great impetus for faster progress, needed to offset as far as possible the serious shortages in labor and animal power. It must be remembered that even as late as 1914, armies were still served to a large extent by animal power. Finally, the airplane found its place in American agriculture, when crop dusting by plane proved an effective means of pest control.

Electrification, as it was extended to the rural areas, carried the process of mechanization still further, permitting the development and use of electrical appliances on the farm, in the dairy, and in the home. In 1930, only nine per cent of the farms of the United States had central station power; by 1940, twenty-six per cent were using such power, and by 1951, more than eighty-seven per cent had high line electric service. Electricity now saws the farmer's wood, pumps his water, milks his cows, helps warm his chicks and piglets, and fakes daylight in his chicken coops. In the farmhouse it freezes his surplus produce, cooks his food, washes his clothes, and keeps him in touch with the outside world by radio, and even by television where the locality permits. It brings important crop and market news to his door, and helpful educational programs, too. TVA and other similar projects have revolutionized living for thousands of farm families.

Chemistry is another big factor in the story of feeding, cleaning and clothing our millions. It has permitted the machine-aided farmer to produce still more and better foods. It has enabled industry to convert surplus and new crops and by-products into many useful items: textiles, pharmaceuticals, plastic substitutes for wood, rubber and so on.

The large packing houses, soap manufacturers, and others have produced a wide variety of chemical by-products which are helpful to the farmer. The chemical companies, working in cooperation with the government agencies and farm groups, have developed superior fertilizers and pesticides to increase yield by soil nutrition and to prevent crop loss from insects. The fertilizer shortages of World War I, when our potash supply from Germany was cut off and our nitrogen supplies depleted, was fortunately not repeated in World War II, because of the activities of American chemical companies. Prior to the last war, farmers used inorganic chemicals mostly, and botanical products in pest control. By 1945 a great change was under way toward the use of synthetic organic pesticides which have proved more economical and effective. It is estimated that half a billion dollars is spent annually on chemicals to control pests.

The recent development of soil conditioners, with the promise of converting poor or useless land into fertile soil for the raising of crops, has opened up vast possibilities. Faster reclamation of waste lands and denuded forest areas, with consequent help in the controlling of watersheds, seem also to be within reach. Besides Monsanto's acclaimed krilium, products of other chemical concerns are now appearing on the market, such as loamium, based on a B. F. Goodrich soil conditioner, and other preparations promoted by American Cyanamid and the Celanese Corporation of America.

Working with the rare elements in the air, the chemist has produced a superior light refrigerant, solidified carbon dioxide. A product of the Air Reduction Company, dry ice first demonstrated the ideal refrigerating qualities which this solidified gas possesses in the transportation of perishable foods. It is light in weight, space saving and easy to handle, so particularly suited for air transportation. Perishable luxury crops may be picked in California one day, packed in dry ice, and flown to eastern markets for sale on the following day. Among the chief users of this form of refrigerant are meat packers, dairy companies and ice cream dealers.

The science of engineering, with its important role in irrigation, cannot be passed by. It has played, and still is playing, a significant role in our great Southwest, where the large irrigation projects have opened up huge tracts of formerly arid valleys, and turned them into richly productive fruit and vegetable gardens for the whole country. Engineers are controlling excessive moisture in our productive river valleys where flood waters can destroy not only the crops but the very soil in which they grow.

The chemist and the engineer, working together, may well provide the solution of the principal problem of agriculture: to provide a plentiful source of food permanently. The one will supply the proper elements for the enrichment of the soil, and the other will develop the necessary machinery to control the moisture necessary for growth, to bring water to the dry lands and restrain it in the watersheds.

Companies such as Armour, International Harvester, Pillsbury and Proctor & Gamble, to name but four as examples, have through their research laboratories given great help and service to the public. They have led the way to better equipment, more nutritious foods, superior fertilizers, more efficient cleansing agents, and more complete use of byproducts to promote better health and greater leisure.

Praise of the accomplishments of the many agencies of the United States Department of Agriculture, and the several agricultural experiment stations should be sung far and wide. Through the efforts of this corps of silently working scientists, we have blight-resistant grains and vegetables, earlier and later growing crops, new strains of beef and dairy cattle and hogs and poultry, better preserved foods, cleaner meat, improved vitamin-saving methods of food preparation in the home and in the factory—all these and many other benefits. The development of Burr-Leaming corn at the Connecticut Experiment Station in the 1920's, of Kandrell wheat at the Kansas Experiment Station, of meatier turkeys and leaner hogs at Beltsville, may be cited as examples. With the basic research in many cases undertaken by the Department, the results are passed on to private industry, there to be developed and benefits made available to us through the farmers, packers and dairymen.

The Department of Agriculture was established in 1862, and in the same year its present Bureau of Agricultural and Industrial Chemistry was organized, originally as the Division of Chemistry. Another important section, the Bureau of Animal Industry, came into being in 1883 as the Veterinary Bureau, changing to its present name in the following year. The passage of the Hatch Act in 1887, providing for the support of agricultural experiment stations in the land grant colleges of the several states, gave one of the greatest lifts possible to the progress of scientific research in this country. The first research station of this kind had been set up at Middletown, Connecticut in 1875. Two years after the passage of the Hatch Act, there were eighteen such stations. No one at the time could foresee the far-reaching effects of the discoveries which have been made in these stations, working in cooperation with the parent organization, the Department of Agriculture and its many bureaus, and with private organizations and interested farmers.

The Department added other agencies for vitally important research soon after the Hatch Act came into effect: the Division of Soils in 1894, the Bureau of Dairy Industry in 1895. More recently have been set up the Bureau of Human Nutrition and Home Economics and the Production and Marketing Administration, among others. All these agencies, working together, have given the American people incalculable benefits which have resulted from the labors of thousands of curious-minded and persevering scientists, working in all parts of this country and even in foreign lands, to keep the United States the enormously productive country which it is. Experimentation with oil crops has been of great value, especially during World War II, when the soy bean and peanut crops were greatly increased to provide additional food sources and commercial oils. The early work in selecting types of seeds and areas for their planting bore rich rewards in essential crops. Other oil crops, such as tung, are receiving careful consideration so that we may build up greater potential sources of supply.

At Beltsville, Maryland, the site of the Department of Agriculture's largest research station, the Agricultural Research Center, one can see experiments being carried on in all aspects of agriculture, and in problems of household and marketing science which are important for the general well-being of our population. Here, in cooperation with the outlying stations, improved farm stock has been developed; better poultry, including the popular new Beltsville turkey; newer methods of preserving and preparing food; of the care and construction of clothing. New machinery has been developed here, also packaging methods, to be passed on to the manufacturers for actual design and production. The Beltsville station is used by the various bureaus of the Agricultural Research Administration and by other governmental agencies which have use for their facilities. The Center covers about 11,000 acres, has over 500 laboratory buildings and barns, and large herds and flocks.

The discoveries of this network of research stations are made available in many helpful periodicals as well as in separate reports in pamphlet to book size. Some are highly technical in presentation, but a great many are written in clear, popular style for the use of the general public as an educational service. As examples we may cite the *Journal of Agricultural Chemistry,* which reports the work of that key bureau and its four regional offices; *Soil Conservation,* a well-illustrated journal containing many articles of general interest; and the new publication, *Agricultural Research,* which first appeared in 1953. Popular pamphlets on food preservation and preparation, on nutrition, on equipment and remodelling of the home, and others, are available free or at nominal charge.

In speaking of the research of the land grant colleges and the cooperating Department of Agriculture, Dean-Emeritus Mulford of the Missouri Experiment Station wrote, in the *Experiment Station Record* for December, 1940:

> It is not too much to claim that this national system of scientific research in the interest of agriculture is an educational phenomenon without parallel in the history of science. . . . The experiment station has revolutionized the attitudes of farm people toward the practical values of scientific research in the modern world. From an attitude of suspicion and even antagonism toward science, these agencies have won the enthusiastic support and approval of farmers, the most conservative element in our democracy. . . . They have popularized science, not by super-publicity methods but by utilizing science as an

instrument to solve the intricate and difficult problems of agriculture.

The Farms are Mechanized

The West and its farmers may claim some credit for instigating the early advances in agricultural machinery, because they realized that the problems they confronted in their sweeping plains, with relatively small manpower, were far different from those they had faced in the East with hand labor. The westward movement unquestionably stimulated evolution of important labor saving machines.

Until Cyrus McCormick made the first practical reaper at his father's farm near Steel's Tavern in Virginia, the scythe, the sickle and the newer cradle were the only aids available for harvesting the grain so necessary to feed man. With the coming of the reaper, one man was able to harvest as much as five men, working with cradles. With the possibilities of horse-drawn machinery once demonstrated by the reaper, other tools were soon adapted for use with horse power. McCormick, having made a start with his reaper, moved to Chicago in 1847, sensing a better market there. His company sold 1,500 reapers in 1849 and so started toward the larger success which led to the creation of the International Harvester Company, one of the great industrial firms of the world.

Other farm machinery enterprises soon thrust themselves into the fast-growing market. In Grand Detour, Illinois, John Deere, a blacksmith, began to manufacture steel-faced plows, and his business grew nationwide in a few years. Obed Hussey in 1833 put another reaper on the market, this soon to be followed by still other types. The first thresher appeared in 1834. Within a few years, improved harrows and plows were in use, forecasting vast improvement in the planting and cultivation of food crops.

Soon the farmer, with the aid of the horse, could cultivate

more land with far less labor and could secure greater profits. It has been estimated that in the ten years between 1830 and 1840, the harvesting time for an acre of wheat was cut from thirty-seven to eleven-and-a-half hours. In 1851, the McCormick reaper was shown at the Crystal Palace in London, and was given the highest award. Its importance to British farmers was hailed, and McCormick arranged for its manufacture by a British company. Before the Civil War, the reaper was being used on the European continent, and American agricultural machinery had opened a world-wide market.

Improvements on the harvester came fast. In 1852 a rake was added. The Marsh brothers of DeKalb, Illinois, perfected a machine which raised the grain to a platform so that it could be bound. McCormick adopted this idea for his reaper, and in 1872 marketed a wire binder, which practically eliminated hand work in harvesting. This was followed by a twine binder in 1879, the invention of John F. Appleby of Wisconsin. McCormick realized its worth and procured rights to the device.

Competition was fierce in the farm machinery business, especially with the impact of the Civil War and the concomitant manpower shortage. After the War, hundreds of companies entered the field. One of the largest of these was the Deering Harvester Company, founded by John Deering in 1869. With competition so keen, and companies mostly small, few had financial resources or the necessary organization to develop a large export market. The time for mergers had therefore arrived. In 1902 the McCormick and Deering companies, with three smaller organizations, combined to form the largest agricultural machinery company in the world, International Harvester, to develop implements and to tap the vast potentialities of the world market. This huge organization, keying its activities to the needs of the farmer, experimented with new lines and sought to include all kinds of farm equipment. With this in mind, the new Company turned to eastern markets, purchasing in 1903 the D. M. Osborne Company, which manufactured implements particularly suited to eastern farms. Gradually other concerns were absorbed until, at the present time, I. H. produces all kinds of tillage, seeding, fertilizing and harvesting implements.

While the American public has been, and still is, highly suspicious of big companies in industry, it must be admitted that without the greater financial and experimental resources made available by the massive size of such companies, many of our most important scientific advances would not have been made with the speed and success which characterize American industry. When International Harvester was formed, most of the plants of the component companies were centered around Chicago, where the new firm established its central offices. The results of greater resources were quickly apparent in the company's foreign trade, which doubled in five years, becoming particularly important in Europe, South America and Australia. Factories were established in several European countries by 1909.

Realizing the importance to both the company and the farmer of a diversified line of machinery, I. H. extended its line by 1912 to the making of corn binders, ensilage cutters, seeding machines, manure spreaders and harrows. In 1914 a light harvester-thresher was put on the market. The best features of harvesting machines made by the harvester companies taken over were kept, and gradually one line evolved, the McCormick-Deering, offering the best features of pioneer machines with benefit of further improvements. The experimentation of I. H. with tractors, to develop an all-purpose machine, was probably the most important of all its activities to date. Until the advent of the Farmall in 1922, tractor adoption had been slow. When the Farmall Works in Rock Island, Illinois was opened, general adoption of the tractor for farm labor was on its way. Furthermore, the success of the Farmall led other manufacturers to give more consideration to the needs of the small farmer. At the Louisville Works alone, the Company's newest tractor plant, about 6,500 are employed in the manufacture of 100,000 machines a year.

In Stockton, California, a large plant specializes in equipment for western farm needs, especially heavy-duty tillage machinery. At Memphis, a plant specializes in implements for the southern states, such as the cotton picker and some items for smaller farms. To promote service efficiency, I. H. has eleven large parts depots scattered over the United States and others abroad.

Foreign subsidiaries are expanding their plants and business as fast as possible. One of the new units is a truck plant in Australia which will produce 5,000 machines per year. In Mexico, at the Saltillo works, truck production will be doubled. In St. Dizier, France, a new plant for making Farmall tractors is meeting production schedules despite shortages of materials. In England and in Australia, diesel-powered tractors are being manufactured. Over 26,500 persons were employed by the foreign subsidiaries alone in 1951.

About 1900 the horse was definitely losing ground as a power source on the farm. Before the I. H. merger in 1902 the Deering Company was making fifty engines a day to mount on their machines. International Harvester Company made its first gas-powered tractor in 1905, and marketed some in the following year. There were still only a thousand in use on American farms in 1910. After World War I, in 1920, this number had grown to 246,000. Except for a temporary decline during the early depression years, the numbers increased steadily, and by 1952, after another great conflict, had reached the total of 4,170,000, an increase of 170 per cent since 1940. The small farm was not forgotten during those years; in 1945 the Farmall Cub gave the smallest farm a suitable tractor. An important added feature of the newer tractors was the introduction of the pneumatic tire in 1933, with its improved performance and more satisfactory operation.

Food must be transported. The old horse-drawn farm trucks

were time-consuming. Naturally, the next step was to extend mechanization to this need; after some experimentation, I. H. put the first 100 "auto-buggies" on the market. They resembled the old farm truck and found prompt acceptance. Improvements followed and automobile companies entered the field. After World War I, the number of automotive farm trucks had increased to 139,000, and by 1952 to 2,400,000. Tractors and trucks, gas-powered, have long ago proved their worth on the farm, the highway, in heavy construction work. Now industry is turning to the larger production of diesel-powered units for greater efficiency and economy in heavy work.

In the farm implement field, the re-designing of machines for use with tractor power began about 1930. A steady march toward greater efficiency with less labor has brought developments of great import, including the improved corn picker, of which there were 500,000 on our farms in 1951, an increase of 300 per cent in ten years; the one-man automatic hay baler which can turn out two to five bales per minute or five to six tons per hour; ensilage harvesters which cut, shred and load the corn into the trucks; and improved milkers. Two major labor savers are the sugar beet harvester, and the long-awaited cotton harvester, developed after decades of research. This machine saves up to sixty per cent of the cost of harvesting the crop, and does six to eight acres per day. Some models of corn planters, working faster and more accurately, permit twenty-four hour a day operation when weather conditions delay planting beyond the normal time. The planting of the 1943 Illinois crop was made possible in this way. Rains had postponed the planting until the very end of the season, and only tractors, working around the clock, could get the seed into the ground in time. Multiple-row sprayers permit quick salvage of crops which might otherwise be ruined by the sudden descent of pests. Touch control on tractors makes the adjusting of implements in the field almost effortless. These were introduced on Farmall models in 1948.

Research laboratories are at work constantly to promote greater efficiency and the anticipation of the farmer's every need. In 1951, I. H. alone employed 3,000 persons in such positions, including over 500 who hold degrees in engineering or other scientific fields. A large testing team is constantly on the job, checking the results of the researchers.

The combine, probably the greatest aid in our whole wheat production picture, was developed and put to use in its first form in the 1880's, using multiple horse power, sometimes as many as thirty horses, to move the heavy and cumbersome apparatus. By 1910, there were 1,000 of these in operation. This number increased to 4,000 by 1920. Improvements made it possible, by 1930, for one combine with two operators, to do the work of 175 men, reaping and threshing fifty acres per day. Mechanization of the combine, and development of smaller and lighter models, suitable for farms of smaller acreage, created a large demand. The necessity of heavy wheat production in World War II years still further popularized its use. In 1945, there were 374,000 combines on our farms. I. H.'s self-propelled machine, a fairly recent improvement, permits one man, using push button controls, to cut and thresh the grain in a thirty-five acre field in a single day, working across it in twelve-foot swaths.

As farms have become mechanized, land formerly needed to feed the draft animals has become available for the production of food and industrial crops. Between World War I and 1945, there was a decrease of 15,000,000 horses and mules on our farms. According to Johnson's *Changes in American Farming,* "the shift in mechanical power from 1918 to 1945 made available about 55,000,000 crop acres, or about fifteen per cent of the available crop land." This increase gains added importance when we consider the loss of acreage in some areas because of large housing developments; the shrinkage in potato acreage in Nassau and Suffolk Counties in New York, as new suburban housing took its place, may be cited as an example. The general trend toward rural living and only part-time farming may result in a further decrease which must be offset.

In the ten-year period from 1941 to 1951, there took place an increase of twenty-nine per cent in farm production, despite a decline in farm population of over 5,000,000. These figures truly summarize the achievements of mechanization of the farms and modern agricultural science.

In an earlier chapter the initial educational phases of the land-grant college and activities which mostly stemmed from that movement were mentioned. It was not until after World War I that organized extension teaching of scientific agriculture got under way. The Smith-Hughes Act of 1917 led to the development of courses in the rural high schools, and the county agent system began to make great strides about that time. By 1928, at least seventy-five per cent of the counties of the United States had agents whose duty it was to visit the farmers with technical advice and assistance. The American Farm Bureau Federation and local and private associations have offered assistance and have fostered the many farmers institutes.

* * *

Much of our present knowledge of human and animal nutritional needs and the ways of satisfying them may be traced directly to the work of Stephen Moulton Babcock and his earnest younger associates at Wisconsin, E. B. Hart and G. C. Humphrey. Their experiments in the feeding of cattle proved that some seemingly adequate foods lacked a certain something necessary for performance of a healthy life cycle. De Kruif, in his *Hunger Fighters,* has said: "Babcock was first of all modern men to discover this hunger that doesn't gnaw at men and beasts but only strikes them down with strange ills, maybe kills them."

From the work of Babcock and his group came impetus for other scientists to take up the search for the vital "somethings." Discovery of vitamins which are necessary for proper health and growth marked a turning point in our whole concept of nutrition. Adequate consumption of the fruits, vegetables and milk products which contain important amounts of these substances has been made possible by progress in canning and freezing of seasonal produce; together with improvements in refrigeration methods it became possible to provide a year-round supply. Because of Steenbock's discovery that the sun's ultraviolet rays generate vitamin D, we now buy irradiated foods and so benefit by his research. As other necessary elements are discovered to be lacking in our diet, scientists are finding ways of reinforcing commonly-used foods to provide these elements.

Meat From Hoof to Market

Railways and refrigeration changed the whole picture of the American meat industry. Until swift transporation became possible, fresh meat was limited to sale in local markets entirely, at all seasons of the year. The packers could process pork and ship it, but other meats, not so well adapted to salting or smoking, were sold only to local consumers. As the West opened up, cheaper lands and the growth of population in the industrial East pushed the meat industry westward. The dealers followed the herds, but cattle had to be shipped alive because of the perishability of fresh meat.

Chicago became the center of the meat industry when it achieved importance as a principal railway hub, for cattle and hogs could be shipped in from all points to be sold there. Its role as a meat supplier in the Civil War gave it, as we have seen, a great boost. In 1851 the city had only eight packers but by 1864 there were fifty-eight. With the opening of the railroad-financed Union Stock Yards in December, 1865, Chicago acquired a dominant position from which it has not been budged. As the railways advanced still farther westward, cattle drives to railheads established lesser centers, and the meat industry erected plants in other cities to care for the western growers, but Chicago has remained the chief area of operations.

Cincinnati, with its river port facilities and location in the midst of the corn belt, had been the leader of the hog packing industry, justly deserving its title of "Porkopolis." Its first packing house was established in 1818. Here this large industry began to use modern practices—mechanical aids and methods of waste recovery. Lard rendering was the first subsidiary industry. The manufacture of other by-products soon followed. The pig was king in the packing houses until the railroads and modern refrigeration made possible the shipping of dressed meats. This animal performed a dual service. It has been said that "the task of the American hog was to market the American corn crop."

Natural ice was first used in 1857 to permit some summer slaughter. In 1868 G. H. Hammond shipped beef east in early refrigerator cars using natural ice, not with complete success. Around 1870, the packing houses began to adopt mechanical refrigeration in their plants. Meat was being shipped into Boston in 1872 on schedule, and the packers were moving toward successful transporting. Early attempts, without proper knowledge of air currents, were failures. Hammond of Detroit and Gustavus F. Swift of Chicago were the most persistent experimenters. Finally, a Boston engineer named Chase, working with Swift on the problem, designed a satisfactory refrigerator car, and so revolutionized the industry. Working on the theory that cold air, being heavier than warm, would fall, he iced his car at the top and provided space for circulation of the cooled air, with ventilators for the escape of the warm air. Earlier cars were iced only at the ends. By 1880, dressed meat was being shipped east regularly.

Opposition came from both local butchers and the railroads. The former feared loss of business and the latter foresaw loss of the heavier revenue from live cattle which of course weighed more than the dressed meat. Since the railroads claimed that they could not afford to buy the expensive refrigerator cars, a deal was made whereby higher rates were charged to compensate the railroads for transporting dressed meat, and refrigerator companies were formed by the packers and the cars leased to the railroads. Despite these arrangements, the prejudice against western beef was so strong that towns and even some states legislated against it.

Swift led in overcoming the opposition, by taking into partnership some of the leading eastern butchers, whom he had known either in his youth when in business in Massachusetts, or later as a supplier of live cattle. Family business connections were called on to help. Armour and Company established branch houses in the East to gain markets; Morris and Swift both followed this example.

Because the cost of cleaning, icing and otherwise maintaining the new cars was high, the packers started to transport other perishables in them to keep their fleets in maximum use. Soon local and seasonal fruits, dairy and poultry products were being shipped to the cities, where they brought good prices. It has been claimed that refrigeration, extended from the meat plant to transportation, to retail markets, and to homes, has been the most significant factor in the evolution of the meat industry in the last six decades. Seasons have been practically abolished and markets expanded to national dimensions by the use of refrigeration. One may go farther and say truthfully that refrigeration changed the meat industry into an international business.

The 1880's saw the start of another very important development in the industry: complete utilization of all parts of the animal. Science joined hands with the packers when chemists were employed to find uses for all meat by-products. Many were already being processed for leathers, wool, soap, glues and so forth. But much remained as waste material or fertilizer.

The Armour Laboratories, first opened in 1886, deserve untold credit for their work in this field. They have led in developing products to utilize the former slaughterhouse waste, and so keep down the price of meat while producing new chemical offshoots and other materials to enrich our lives. At the same time, and by the creation of some of these very products, they have given to the world pharmaceuticals of life-saving potentialities, such as acthar and tryptar. Twenty-five years ago, these laboratories were using only part of the supply of glands from Armour's Chicago plant. Now they use all that can be supplied from their own plants, and still more from others.

Other research groups, such as Hormel Institute, are working alongside Armour Laboratories to produce new discoveries and improve products already known. The National Live Stock and Meat Board was established in 1903 "to initiate and encourage education and research on meat and its products." It turned the sights of scientists on meat, partly by means of seventy odd fellowships established for college and university research.

In 1944, the American Meat Institute was founded, with Thomas E. Wilson at its head. Its building, opened in 1949, was financed by the Institute, built on land donated by the University of Chicago, and partly equipped by allied industries. It already is known for its valuable work on spoilage of pork products, and on nutritional values of meat proteins. It has established an advisory service and a meat educational program for the public.

Canning of beef began in Chicago about the time of the Civil War. It was a business of specialists mostly, with few packers engaged in it. For years, its domestic market was small, with the armies and navies of the world providing the chief outlet. At first, the meat canners made their own tins. The pyramid shaped tin of Wilson's beef was widely known. In our own time, George A. Hormel and the American Can Company cooperated in developing the pear shaped ham can in which canned ham was introduced to the American public in 1926. Its success made other canners realize that the market was ready to accept good quality canned meats. Very quickly, the Illinois Meat Co. introduced its "Broadcast Brand" corned beef hash. Other companies followed this lead.

Unsanitary methods and other conditions in the stock yards caused much public indignation. Upton Sinclair's *The Jungle* is credited with arousing the public to demand an investigation and appropriate action to force the companies to observe practices in the best interest of the people.

The first federal meat inspection occurred in May 1891 at the Eastman & Co. abattoir in New York. Years of struggle on the problem followed, before the Federal Meat Inspection Act was extended in June of 1906 to cover sanitary conditions in all companies engaged in interstate trade. At that time, 162 houses in 58 cities had inspection. The American Meat Packers Association urged full cooperation of the packers who had sat too long in silence while the public demanded reforms and government committees investigated, without accomplishing any immediate good. The first two companies to apply for federal inspection were Armour & Co. and Morris & Co., later a part of Armour. Over eighty per cent of all meat animals slaughtered commercially in 1950 were slaughtered and prepared under supervision of the Federal Meat Inspection Service. In 1951, over a thousand establishments in 367 cities and towns had federal inspection. State and local inspection services, patterned on the federal, have been established in many places to check on the local meat supplies. The constant federal supervision and inspection to eliminate any diseased animals or parts, and the familiar purple stamp indicating federal approval, give the American public by far the safest meat supply in the world.

The animals are inspected before and after killing. Glands and flesh are examined to detect any dangerous conditions, and all rejected carcasses or parts are promptly segregated to be used for fertilizer or other commercial purposes. Seven laboratories are maintained by the Fedeal Meat Inspection Service to check samples of meat, meat products and their ingredients. The Service also passes on labelling for products and on plans for remodelling and building of structures for slaughtering and processing, to be sure that all regulations are met.

The modern plants have gone a step beyond refrigeration, and since 1930 have been installing air conditioning. New insulating materials, such as Pittsburgh Corning Corporation's new foamglass, are used on walls, floors, and ceilings to increase the efficiency of refrigeration. Operations are still largely manual, and because of the nature of the work, probably will continue to be so, but machines have been introduced wherever possible—mechanical hoists, transit wheels, belt conveyor lines, hog scrapers and so forth, to speed work. Electrical saws and sausage fillers eliminate much handling of products. A banding machine, developed by Dennison Manufacturing Co., permits one operator to band 25,000 frankfurters in a single day.

The Perfect Food

It has long been said that milk is the perfect food because it contains all the elements necessary to support life. The production and distribution of this one food and its by-products render it one of the largest industries in the United States. In 1951, there were close to 24,000,000 milk cattle in the country. To support this huge national dairy herd, the farmers have to keep their pasture lands in good condition, with proper replanting and drainage. This in turn is a means of conserving and improving our soil. In *Social History of American Agriculture,* Joseph Schafer has said that dairying "is helping to guarantee the permanence of American agriculture."

Starting with the essential basis of the industry, the cow, let us consider what science has contributed to give us the present productive animal which furnishes us with milk in fair supply and of high quality. Research, both private and governmental, has striven to breed sturdier and more productive stock. Again, the work at Beltsville is of untold importance—the developing of cross bred animals with superior records in production of both milk and butter fat, and in production of high grade heifers. Recently, experiments in cross breeding with Red Sindhi stock have been undertaken to produce animals which will have both good heat resistance and high production records.

The problem of feeding the dairy herds has been under study for many years. Its importance goes beyond the initial concern of satisfactory nutrition for production of good milk in large quantities, to the problem of the utilization of waste products from other industries. A rather surprising example is the discovery that a large percentage of the Florida citrus cannery pulp waste, when dried, can be used for feed. In 1940, ninety per cent of Florida's grapefruit cannery waste was used in this way. Before 1900, Fred C. Pillsbury discovered on his experimental farm the value of milling waste for cattle feed.

The various dairy herd improvement associations have done remarkably effective work in carrying information on breeding, feeding and management to the dairy farmers. In 1952, 2,109 of these associations had 1,166,297 cows in 41,105 herds in all states and Hawaii and Alaska. Working in cooperation with state extension agencies, they raised production levels in their herds in 1953 to an average of 9,195 pounds of milk and 370 pounds of butter fat per year. These results are bound to be noted by other farmers who will then seek to follow the methods used to obtain the high averages and, consequently, the greater profits.

Studies have been made, and others are still under way on

methods of cutting and storing of feed and relative resultant nutritional levels. Types of silos, barn construction and sanitation, methods of drying feeds, and of storing it have been improved by experimentation at Beltsville and on other station farms.

A hundred years ago, the average production per cow was 2,000 pounds of milk per year. In 1951, this figure had risen to 5,326 pounds, and 211 pounds of butter fat. Considering current costs of feed and maintenance, a cow must produce over 5,000 pounds of milk and 200 pounds of butter fat per year to show a profit for the farmer. The striking advance in these production records may be attributed directly to improvements in breeding and feeding, especially as sponsored by the dairy herd improvement associations. The importance of this advance may be still further highlighted by the fact that in 1951 our total milk production of 115.6 billion pounds was just about the same as that for 1941, when we had approximately twenty million less people to feed. Obviously, any means of increasing our supply of dairy products is of great value. Better production rates seem the most likely way because farmers who secure higher returns from other produce are not likely to increase herd size or to go into the dairy business.

Another method of increasing our supply of dairy products is the use in foods of the available non-fat nutrients in milk. At present, we use about seventy per cent of these, an improvement of twenty per cent over our record of twenty-five years ago. The Bureau of Dairy Industry researchers and other groups are searching now for ways of utilizing the remaining thirty per cent as food for humans.

The dairy belt which crosses the northern part of the United States, starting with Vermont, provides the great bulk of our dairy produce. The butter states are Minnesota, Iowa and Wisconsin. The latter is also our greatest producer of evaporated and condensed milk, while Pennsylvania and New York are the greatest ice cream producers. Besides natural conditions which render these state suitable for dairying, there is an added incentive in the heavy concentration of population in the large cities and in the generally heavily populated Northeast and Middle West.

The first notable innovation in the industry was Gail Borden's successful production of evaporated milk. He adapted the vacuum pan method used in early fruit canning by the Shakers, and in 1856 patented his method. Its success started Borden and his family on their way to the outstanding position they occupied in the whole industry. Other similar processes were discovered and soon several companies were making this form of milk, which has fed soldiers in many wars, gone the world over with those who are removed from supplies of fresh milk, and is used extensively as a baby food as well as for the customary household purposes. At present, over 500 brands are produced by forty-three companies in 132 plants, using six and a half billion pounds of milk annually.

The period from 1880 to the present has seen notable progress in methods and machines, with the adoption of silos for food storage, development of new feeds and new forms of old feeds, the invention of cream separators, milking machines, pasteurizers and milk coolers, and the perfection of condensing and drying processes. Babcock's discovery of a test to measure the butter fat content of milk was one of the great milestones in the industry, as was his discovery of nutritional deficiencies of foods. Recently the Bureau of Dairy Industry has developed a new test for fat content which is claimed to be as rapid and accurate as the other; it uses the same equipment, and is easier and safer for the operator.

In 1935, International Harvester introduced mechanical refrigeration of milk in response to demands for quick cooling to hinder bacterial development. This method was soon adopted and required by most state health departments. It led to the evolution of farm "walk-in" coolers for storing all perishable farm products. The new McCormick parlor milker has made possible greatly improved sanitation, as well as much saving of time and labor. This mechanical milking system strains the milk as it comes from the cow and pipes it directly from the cow to the cooler, eliminating much danger of contamination and spoilage, and saving considerable labor.

The Bureau of Dairy Industry, separated in 1924 from that of Animal Industry, conducts research on by-products as well as on all phases of milk production. Recent studies have brought to light the effect of rusty equipment in increasing milk spoilage, a discovery of much economic importance, especially in sections where milk must be handled a lot, as in cheese factories.

Our largest cheese manufacturer, the Kraft Foods Company, was started in 1904 by James L. Kraft in Chicago. From his tiny enterprise, with his new ideas and methods, he built up a huge business based upon speedy delivery of good packaged cheeses, well over a hundred varieties, and a blending method which assured purity and standardized good quality and taste. The Kraft packaging methods quickly attracted the housewife who recognized their economical value.

The cheese industry has benefited recently by research conducted in cooperation with the Bureau of Dairy Industry at Wisconsin and Ohio State universities, on the making of brick and Italian type cheeses. Experiments on utilization of the whey of Cheddar and Swiss type cheeses have produced a seemingly profitable product useful in spreads and other foods.

Research in the use of milk solids in baked goods has shown that their use improves both nutrition and appearance as well as the keeping quality in bread and other baked goods. Addition of these solids in the preparation of other foods has proven a means of increasing their nutritional value and flavor.

The larger dairy companies have also invested heavily in research activities which they in most cases have made available to the smaller companies. National Dairy Products Corporation and its subsidiaries, with their well-equipped laboratories afford a good example of this service. They carried on extensive tests to determine the best metals for use in equipment at various stages of the processing of milk. The large distributors such as Sheffield Farms, Borden and Hood maintain spotless laboratories where milk is constantly tested for its purity and butter fat content.

While advances have been under way in breeding and feeding, in production of more and better milk, and in use of its by-products, the marketing angle has not been ignored in any sense. At the turn of the century, in the towns and cities, milk was brought to the consumer's door daily in a horse drawn wagon. It was in large metal containers, poorly iced, from which the milkman ladled it into the buyer's receptacle by the pint or quart. Pasteurization was new and still under suspicion by many people. Sheffield Farms Co. was the first dairy in the country to adopt the new process. The rural family which did not have its own cow went to a dairy farmer to buy milk which was carried home in a pail. Inspection on the farm and in the towns was often only sporadic. In some areas it was nonexistent. The glass milk bottle, in round quart size, was just coming into use. The space-saving square bottle came only recently, and the larger two-quart bottle came into use because war shortage of labor and fuel made daily delivery impractical.

When the National Dairy Products Corporation was formed in 1923, it recognized the need of a central laboratory service to assist its member companies where their laboratory equipment was not sufficiently extensive. It established the now famous Sealtest laboratories, and with them a system of plant inspection and control. Plants operating under their control are permitted to use the Sealtest emblem, nationally recognized as an assurance of quality. The ice cream industry was the first to become so standardized.

The disposable paper quart container of today, and the careful pasteurization and icing are a far cry from the clumping horse which knew where to stop for each customer's house while the driver measured out the milk. Now we may have a silent truck stop while its driver delivers to us safe, clean, bottled milk, or we may go to a store and buy the equally safe container milk in its single use carton. Some apartment dwellers have only to go to the lobby and put coins in a slot machine to secure their cold, fresh and pure quart of milk from a carefully stocked, refrigerated dispenser. The cow is still necessary, but how many intermediate steps have disappeared!

The forms in which fluid milk is marketed have increased from fresh, sweet milk and buttermilk to include homogenized milk and, most recently, concentrated fluid milk which is easily reconstituted, and which cuts the cost of transportation and storage by as much as two-thirds. As yet, this form is largely experimental in markets. The public has not been generally introduced to it. Dried whole milk and non-fat milk solids have been widely accepted and are of great nutritional value as a means of enriching other foods in domestic as well as commercial preparations.

In the cheese business, one of the largest divisions of the dairy industry, great changes in method have occurred. Processed cheeses and spreads have been developed successfully and have been well received by the public. The old-fashioned method of ripening cheeses in large rounds or other shapes has been losing ground to the newer, more efficient and economical consumer-sized portions packaged in transparent plastics. The ripening process is carried on in these smaller containers. Thus loss from spoilage and drying is avoided. Gone too is the smaller loss from tasting, once the chief advertising method of the local grocer for his cheese.

Milking machines and improved refrigerated transportation are probably the big stories in the mechanization of the dairy industry. With the new parlor milkers, large refrigerated tank trucks, which in 1950 hauled close to eighty per cent of the milk, and superior producing breeds of cows, we have come a long way in our search for a perfect supply of the perfect food.

The Staff of Life

In a previous section, credit was given to the mechanical devices which helped to encourage our great grain production—the harvester, and later the combine, and to the scientific research which developed improved, rust and blight resistant strains of seed.

In the story of the milling of our wheat, two companies stand out from all the others. These are Pillsbury and Washburn Crosby, later General Mills, Inc. Both were started at the Falls of St. Anthony, now Minneapolis, in the 1860's. Cadwallader C. Washburn put his first mill in operation in 1866. Three years later, the Pillsburys, John S. and his nephew Charles A., bought an interest in a small mill. From that, with other members of their family, they built up a large chain of mills, constantly working to improve methods and products. Washburn too was eagerly seeking better ways of using the hard-kernel spring wheat grown in the Northwest.

The returns on this spring wheat were at best fair, because much of it, rich in nutrition, was bolted out as middlings, a waste. The problem was to find a way to recover this. In 1871, a method was discovered and called the "new process." It was a combination of screening and brushing of the middlings, which resulted in recovery of a large percent of the flour. The mills adopted it. Then, about 1880, another great innovation came in with the use of steel rollers to replace the age-old mill stones. These rollers speeded up the whole milling process. Production in 1871, before introduction of the "new process" was approximately 200,000 barrels, and by 1890 it had grown to nearly seven million barrels. Spring wheat production increased in leaps, as the new methods produced a fine grade of flour.

The roller process was imported from Hungary where its secrets had been carefully guarded. Both Charles A. Pillsbury, who went to Budapest to watch operations, and who secured some of the rollers to install in his mills, and Washburn Crosby Co., claimed the introduction of the system to the United States. The latter company entered its new white flours in the Millers' National Exhibition in Cincinnati in 1880 and, receiving gold medals for each grade, adopted the name "Gold Medal," now a world-wide trademark.

In 1903, Washburn Crosby, to combat high freight rates, opened a mill in Buffalo, the first completely electrified mill. Since then, changes in method have been minor. The greater innovations have been made in packaging machinery and method, in converting to cereals, and in manufacturing of prepared mixes. Washburn Crosby, realizing the need of nationwide markets for its products, merged with several other milling companies in 1928 to form General Mills, Inc., the world's largest flour milling company.

Our wheat crop reached a peak in 1947, with nearly 1,365 million bushels of which 694.7 million were converted into flour, mostly for our own consumption. Waste from this huge operation is made into animal feed, for which the demand exceeds the supply. The nutritional value of this mill waste was discovered, as stated before, by Fred C. Pillsbury. In 1892, it was first packed for export, and six carloads were shipped abroad. Local demand used all that could be had.

While the millers were improving methods and their products, the United States Department of Agriculture was busy with basic research on types of wheat. In 1889, it introduced into this country Durum wheat, and distributed it to farmers in the Dakotas the following year. By 1906, a production of fifty million bushels was reached, but the domestic market was small for this wheat which was not suited for bread flour. The bulk was shipped abroad to be converted into semolina, used for the several paste products, such as spaghetti. American millers did not have the equipment to produce this flour, so wheat was shipped abroad and returned as semolina. John S. Pillsbury decided to end this double transportation cost, so went abroad to study the special milling process. When he returned, necessary changes were made in production methods to produce the high quality semolina quickly accepted by the American paste manufacturers.

In the refined white bread flour produced by the roller process, much of the nutritional value, which had remained in the coarser stone milled product, was removed. The American public wanted the white flour for its bakery goods and worried little about nutritional losses, some of which were not then known. But with the discovery of vitamins and their great importance, and of ways of reinforcing foods to add the missing elements, some of the values lost in milling can now be restored to flour and bakery products.

Machinery kept pace with the growing milling industry. New grain elevators were developed and special grain-carrying ships were built. Weighing and dumping devices were invented for use at transfer points where trucks delivered grain to the elevators. We now see floating elevators in seaports. Whereas once the flour barrel was a familiar sight in the kitchen corner, pantry or country store, we now see the same flour packed in bags up to twenty-five pounds for household use, and larger for commercial purposes. This packaging in bags specially constructed to keep the flour dry and free from contamination is all mechanically done.

Some large baking companies, like the National Biscuit Company, have their own mills to insure uniform quality for their needs. At Toledo, Ohio, the N.B.C. mill processes over twelve million bushels of wheat per year. The large baking companies afford a fine example of complete mechanization of an industry. The company-tested ingredients of the products are mechanically measured, mixed, cut, and fed on rollers to the baking ovens, where electrical controls bake them to perfection before discharging the finished goods to be packaged, all without a single touch of human hands. They give us pure, attractive and good tasting breads and other baked goods, reinforced with important vitamins for superior nutrition. Packaging research has removed the unwrapped loaf of bread and the cracker barrel, in favor of the double-wrapped, moisture-proof loaf of bread and the attractively boxed or cellophane wrapped crackers. We may mourn the picturesque and politically important cracker barrel, but we have gained more hygienic foodstuffs.

Shortenings and Spreads

Butter, our oldest known spread and shortening, is far beyond the pocketbook of millions of people, even for a bread spread, and for many more as a shortening for cooking. Lard, an old standby, until recently did not keep fresh very long without refrigeration and the quality varied greatly. The taste of the early margarine was unpleasing to many, and its quality was also of uncertain grade. The need for new shortenings was definite.

While searching for new methods in the soap industry, American chemists heard of European experiments leading to the hydrogenation process. By this method vegetable oils, heated and mixed with hydrogen, can be converted into hard fats. They saw implications here for our cotton seed oil and our demand for more fats as food products. Proctor & Gamble brought to the United States an Englishman named Kayser who had been working on industrial applications of the hydrogenation process. His discoveries were patented here and in February, 1909, a commercial hydrogenation plant was put into operation. As a result, an entirely new shortening product, Crisco, was introduced. Its success made other manufacturers quickly adopt the hydrogenation method to produce other shortenings. Research developed uses of other oils. The soap maker was in a new business.

The new process, successfully applied to vegetable oils, forced the meat industry to turn its research efforts toward improvement of meat fats. Their good shortening quality and high nutritional value needed long shelf life and uniformity of appearance and taste to compete with the all-vegetable oil products which had these desirable features. Meat chemists searched for anti-dioxidants to retard rancidity, while leaving no objectionable flavor, color or harmful substance. Careful tests by the Meat Inspection Division of the Bureau of Agricultural Industry had to be met. Gum guiac was the first used with commercial success, in 1933.

The labor and the time-saving prepared mixes, which the American housewife has gratefully accepted, were made possible by improvements in the new types of shortenings. Lard still holds its own, especially in pie and bread making, but the use of vegetable oil, in solid or liquid form, with its new refinements, has boomed. In 1951, civilian consumption of non-butter spreads and shortenings included 1,853 millions pounds of lard, 993 million pounds of margarine and 1,364 million pounds of oils for cooking and other food uses. In that same year we also consumed 1,466 million pounds of butter. With legislation against colored margarine removed by most states, and the federal tax repealed, its use is growing rapidly. The American public can look forward to a plentiful supply of low cost, highly nutritious and pleasantly flavored spreads for its bread, and again give thanks to our food chemists.

Preserving and Packaging

Our modern packaging methods, particularly the tin can, and our new refrigeration facilities, are largely responsible for our high standard of living because they assure us of purity, high nutritional quality and honesty of weight. Important foods, once definitely seasonal or local, are made available in good state of preservation the year round, either protected in tin cans or glass jars, frozen in paper containers, or dried and packaged.

The preserving process, introduced soon after 1800 in France by Nicolas Appert to feed the French armies, was quickly adopted by the English, who invented the iron can in 1818 in place of Appert's glass jars, in provisioning their navy. William Underwood started America's first canning operations in Boston in 1819. In both World Wars, to say nothing of the Civil War, the tin can played an enormous part, helping us to feed our fighting forces, those of our allies, and the hordes of refugees. In World War II, it was adopted to transport general equipment.

By the close of the nineteenth century, the growth of the industrial East and the city areas in general created constant demands for food at all seasons. Canning was still a young industry, with crude methods and much manual labor. The fastest machine made but sixty cans per minute. The factories were mostly local enterprises, serving small areas, with little capital to use for experiments looking toward improved methods in manufacture or use.

The American Can Company, founded in 1901, absorbed over a hundred small companies to form a national enterprise. It set out both to expand and improve the industry, goals shared with independent companies. The research laboratories of the large companies have spent millions on experiments to secure speedier machinery and to perfect new types of containers, as well as to free the industry from dependence on foreign sources of tin. We now have machines which make 450 cans per minute, and in the same operation pressure test them. Within two hours after the tin plate starts through the process, cans are being loaded for delivery to the canneries. Our researchers reduced our tin requirements by a third between 1940 and 1950, when curtailment of supply loomed as an extremely serious problems. Our tin needs for canneries had increased fifteen fold in the first half of the century. The shortage in World War II and the Korean War led to American Can Company's "Operation Survival," which is a search for materials and methods which would remove our dependence upon imported tin. So far, limited success has been achieved, and the goal is in view, but still rather distant. For some products, a container can be made using only materials found on this continent. The use of steel with lacquer coatings, and of fibre and cellophane containers is most significant. A possible revolution in the container industry is in sight.

The vacuum pack can, the slide top can, and specially shaped cans are important innovations in our canning picture. Mechanical production line procedures, roller belt trays, new washing and sorting machinery, all have helped toward better quality of goods and speed of preservation. The filling methods in 1920 gave fifty cans per minute. Now close to 400 cans are filled and closed in a minute. Modern cooking machinery is a far cry from Shriner's closed retort, invented in 1874, which required hand loading and hand removal of cans. Now the filled cans move in at one end to be processed as others emerge on a conveyor belt from the other end, ready for labelling and packing. Can closing methods developed by the American Can Co. have been of decided value.

The canning companies, to bring their products to the consumer at the very peak of quality and nutrition, now often take their plant to the fields. The foods are preserved as soon as picked, fresh and full of their natural flavor. To insure steady flow of good crops, the canneries as well as governmental agencies provide technical assistance to growers, and even operate their own farms in some areas. The fruits and vegetables which we buy in cans often are actually "fresher" by days than those we buy unpreserved in our markets, despite fast rail, truck and plane delivery.

The frozen food industry has brought still further changes in our food picture. Items not successfully canned have been found to be suitable for freezing and distributing in consumer sized packages and tins. Many of them have their natural flavors and their original textures well preserved. In canning, these are often lost. Home size deep freezers and refrigerated display cases in markets have done much to popularize frozen produce. Some articles, such as pies and fried fish, are suited to freezing but not to canning, so the newer method of preservation has enlarged the range of our prepared, or partly prepared foods.

Together, canning and freezing have revolutionized the food preparation habits of the American home. The working wife is enabled by them to serve wholesome, attractive meals, purchased on her way home from office or factory.

Dehydration of foods was practiced to a limited extent during the Civil War. In *The Army Ration,* E. N. Horsford in 1864 wrote that "a block one foot square and two inches thick weighs seven pounds and contains vegetables for a single ration for 112 men." The process gained importance during the two world wars, but the industry still is young. In the United States, its products are not generally popular. They have been extremely valuable in feeding refugees and the people of war ravaged areas. The most accepted products are soups and beverages, particularly coffee. Powdered milk and eggs have long been used by bakers.

Improved transportation conditions, speed and refrigeration, permit transfer of fresh crops to wider areas in all seasons. Canning, freezing and dehydration utilize the remaineder of the crops in season, and so preserve our food supply, giving us a richer and more varied diet throughout the entire year.

The great change in packaging methods and materials came in 1920 with the introduction of cellophane, first of the transparent films. The Visking Corporation experimented for fifteen years before it introduced the first practical sausage casing in

1925. The Sylvania Division of American Viscose Corporation also put much energy into searching for transparent packaging materials. Machines were devised, to slice, weigh and package foods in these new films, saving time and lessening handling and chance of contamination. Other transparent plastics, as they were discovered, were tested for packaging use. Pliofilm was adopted soon after 1930.

Coated cardboard containers, moisture-proof, were adopted for use with automatic weighing machinery for lard, replacing the old tin pails. The same type of container proved desirable for margarine. A special wood pulp had to be created for use in milk containers to avoid any possibility of affecting taste. American Can Company, after much testing and experimentation, developed the flat-top fibre milk container and adapted the can assembly line to its production.

Modern packaging problems demand constant research, new methods, new materials and new machinery. Their results assure clean and attractive merchandise and generally permit self-service, which saves time and money. The buyer can learn the exact contents by reading a label or by looking through a film, and so make a more satisfactory selection.

Soaps and Cleaners

Once a luxury of the rich and a product of hard home labor, soap has long been accepted as a necessity. No longer need the housewife save every scrap of fat and make lye from the ashes to convert the two into precious bars of soap. Until the Civil War, soap making remained a crude, hand industry. Then the pressure of the war's heavy demands forced a change to mass production.

The Proctor & Gamble Company, started in Cincinnati in 1837, affords a good example of the evolution of the American soap industry and of its research activities. It operated in a crude fashion, with huge vats of fat and soda; the tasks of mixing, stirring and pouring were performed manually, although the business had grown rapidly as the West developed and river steamers opened up markets and means of easy transportation. Before the Civil War, the company was doing over a million dollars annually, still with hand labor.

A chance occurrence led to the discovery of floating soap, and the first cakes, called Ivory, were sold in 1879. Their immediate success gave incentive to the establishment of one of the earliest industrial research laboratories in the United States. The first great contributions were in connection with development of the hydrogenation process for industrial use, with results already described. Once again, in 1931, American chemists, scouting European progress, found that a new cleansing agent, usable in even the hardest water, was being developed. Proctor & Gamble chemists worked on the idea and in 1932, having obtained rights to the formula, began to produce synthetic detergents. Researchers have continued to work on methods, on applications and on new products. Hundreds of patents have been granted, and some of these have been licensed to competing concerns to give us a large supply of good soaps, synthetic detergents and other products of the industry for personal, home and commercial use. The three largest companies, Colgate-Palmolive-Peet, Lever Brothers, and Proctor & Gamble, their subsidiaries and the smaller companies, supply us with a wide variety of fine quality soaps, soap powders and grains, detergents, and dentifrices as well as household and commercial abrasives.

The introduction of the new detergents has been a great boon because it makes possible easier and more complete home cleaning of utensils, furnishings, clothing, and even the person. It promotes quick and easy sanitation in the food industries and in the hospitals, and it has revolutionized some commercial processes. No longer does the housewife take up her soap stone to scour pots and cutlery. From the soap concerns and the packing houses come improved cleansers with their pleasant scents and their deodorizing properties.

The scientist in the laboratory has been called upon by industry to create special purpose soaps such as those needed as emulsifiers in the production of buna synthetic rubber, or those used in processing textiles, in metal working and in leather preparation. He has sought ways of speeding processes. Hydrolization, developed by Proctor & Gamble, eliminates use of the huge kettles, and also produces, as a by-product, very high grade glycerine, a vital commodity in both war and peace. He has found uses for other by-products such as cotton linters and seed cake. These are now used in manufacture of textiles, film, paints, plastics, explosives and in many other ways. Seed cake feeds live-stock.

The petroleum industry also has had a hand in lightening the task of cleaning our homes and clothing. Dry cleaning, once chiefly the job of the cleaning shop, and a dangerous process even there, may now be carried on in the home with relative safety. Improved, less combustible fluids, odorless and less hard on the user's skin, permit the housewife quickly and with little effort to keep small articles clean and her household fabrics spotless. The beating and shaking, the scrubbing and rubbing have mostly disappeared from the task of housecleaning.

Clothing the Masses

It is possible, if not probable, that the sheep and the cotton ball, like the spinning wheel and the carder, may become antiques, and of historical interest only before long. Our clothing, based now on the products of agriculture, may come from the test tube and the air above us or the elements of the earth beneath us. The chemist will clothe us.

In the memory of many of us, the silk worm, the sheep, and

the cotton ball, with the hides of cattle, clothed us with little aid from the chemist. Then methods of tanning the leather we use came under his eye. And from Europe came word of the first successful artificial textile fibres, based on a patent by Hilaire de Chardonnet in 1884. Rayon quickly assumed importance. The first American company to make the new textile fibre was the American Viscose Corporation, which established a rayon plant at Marcus Hook, Pennsylvania, in 1910. Since then, progress has come with startling rapidity. Celanese, a discovery of two Swiss brothers, Camille and Henri Dreyfus, was brought to this country in 1918 by Dr. Camille Dreyfus. Synthetic fibres, one after the other, with amazing qualities of wearability and strength, have proven practical both in manufacturing and in personal and household use. Today, American Viscose makes a considerable amount of the man-made fibres produced in the United States. Its 1950 production was 416 million pounds.

Machinery for converting the old style natural fibres and for making the new ones has been vastly improved. Now industrial science is showing us ways of combining the old and new fibres into materials which retain the best qualities of each. Here may be the answer to our search for the ideal. Old agricultural crops will not be discarded. Science will have combined the natural and the synthetic to create a newer and better product.

The application of plastics to clothing is a venture on which we are as yet scarcely embarked. Such products as neolite may revolutionize the leather industry. Protective plastic coatings on our outer garments are still in a developmental stage. They promise to be a great aid to our national health and comfort.

Research activities of the larger companies, du Pont for example, are of amazing scope. It maintains several laboratories devoted to research aspects of its fibre department. It has a Pioneering Research Laboratory where only "fundamental investigations in the field of high polymers and the development of new types of synthetic fibres" are carried on. Its discoveries are turned over to industrial research groups in other laboratories for further work. Here both nylon, the invention of Wallace Crothers which was introduced in 1938, and orlon, were discovered. Another du Pont laboratory at Deepwater, New Jersey, works on dyes, water repellent finishes and shrinkproof agents. Other companies are equally interested in searching for new improvements. Monsanto's merlon renders cotton fabrics more crisp and wear-resistant; its resloom is used to give textiles both shrinkage and wrinkle resistance.

One of the newest materials, acrilan, introduced in 1952 by the Chemstrand Corporation, has many desirable features and may be combined with natural fibres or worked alone to resemble them; it is moth-proof, wrinkle and spot resistant, quick drying, and bulky and warm without weight. The newest story is that of du Pont's dacron, introduced early in 1953. In its laboratory tests, it has proven remarkably resistant to both water and wrinkling, and seems to be very well suited for tailored apparel.

Quality as well as quantity is of great importance in clothing as in industry. The engineers of Industrial Rayon Corporation set out in 1932 to produce a finished yarn of high standard and in 1938 introduced the continuous process, by which yarn is spun, washed, treated, dried and twisted as thread in one operational sequence. It is the greatest development in the industry since the basic methods were introduced at the end of the nineteenth century.

The great versatility of rayon has made possible a wide variety of fabrics, strong and comparatively inexpensive, suitable for many personal and industrial uses. The success of these fabrics is indicated by production figures. In 1928, 97 million pounds were produced in the United States, and in 1948, 1,124 million pounds were made. The public has accepted rayon in its several forms.

CHAPTER TWELVE

PRECISION IN AMERICAN INDUSTRIAL SCIENCE

Great emphasis has been given to the vast material resources, the mass production methods and the speed with which American industrial products are made available to the public. Such stress is accurate, because available materials and rapid processing methods are basic advantages of our industrial society. We must not forget, however, that the elements of precision and accuracy are vital to the successful standardization without which we could never have become a nation of such large productivity.

Without the delicate instruments which measure the flow and the characteristics of electrical current, the optical devices which have been basic in evolving modern microscopes, telescopes and cameras, or the measuring gauges which provide accurate standards for thousands of industrial parts, we could hardly function industrially. Most important of all, no extensive and permanent group of metal industries could have been established in this country without those "master tools of industry,"—machine tools, for they are not only the basic industrial machines, but are also the only "self-perpetuating" ones. Planers, lathes and milling machines help to build items such as automobiles and airplanes, but they also bring into being other planers, lathes and milling machines.

Before the advent of mechanized industry, the colonial craftsman had only his keen eyes, his skillful hands and his sense of proportion, with such tools as the time afforded. Yet he made excellent watches and clocks, amazingly accurate firearms, substantial farm and carpenter's tools, beautiful furniture and sturdy dwelling houses. Our success in a crude industrial age was furthered materially by the rapid immigration to this country of skilled workmen from across the Atlantic; watchmakers and clockmakers from Switzerland and other countries, glass blowers and lens grinders from Germany, weavers and metal workers from England and Ireland, machinery designers, especially in the textile business, from the British Isles. The political troubles of central Europe in the mid-nineteenth century sent overseas hordes of skilled immigrants.

A basic reason for the use of machines is to attain greater accuracy, as well as increased speed of operation. No human eye is good enough to match machine or instrumental accuracy, even in so simple an operation as the measuring of an exact length of a bit of material, or in the amount of time needed for a mechanical operation. Without precision machines and precision instruments, no accurate controls could be exercised over physical, chemical or electrical reactions; without such controls it would be impossible to assure industrial scientists and machine operators the correct conditions needed to attain satisfactory results.

Precision is thus a necessity in the development of a modern, scientific industrial society. Machine tools of amazing accuracy and scientific instruments of the utmost delicacy provide this precision, whether the device may be a pressure gauge, a microscope, a camera lens, an automobile speedometer, a turret lathe, or any one of a thousand others. By instrumental precision the size and contents of all natural objects can be ascertained, the appearance and color of distant objects can be recorded, the sounds which please or which may be useful may be preserved, and the records of business calculated and economically filed for future reference.

In the twentieth century we entered a new era of instrumental precision foreseen but not attained in the century before. Our huge industrial machinery—units as large as the engines of ocean liners, the great power plants of irrigation dams and factories, and the electrical communication apparatus which reaches to most parts of the known world—is moving rapidly to a state in which instruments in panels or on a switchboard exert maximum control. To a large extent, the mid-twentieth century might be termed "the instrument age;" the work bench of the ancient craftsman has given way to the control panel of today.

Business Machines

One of the most impressive developments of the times is the progress which has been made in the application of precision machines to business, particularly to its clerical and administrative features. For many years, even centuries, the details and records of business concerns were carried on by slow and laborious manual operations. The business office had little or no equipment to achieve dispatch and efficiency. Clerks copied letters, bills, and invoices and added up long columns of figures; the records of the company were packed away in drawers or piled in dust-covered corners. A bank clerk was identified by his ink-stained fingers and a bookkeeper by his consumptive cough, acquired no doubt from breathing the dusty air from the littered records among which he lived.

As the machine began to work changes in the manufacturing processes, it became vitally necessary for methods to be found for speeding up business correspondence and for the keeping of accounts. Experimentally the device which can calculate or record data is not new. The French scientist Pascal produced an adding machine by 1642, and the German philosopher and mathematician Leibnitz evolved a gadget some thirty years later which could not only add and subtract but could also multiply, divide, and extract a root. It was not until 1820, however, that a commercial product was produced which could apply some of Leibnitz's principles to the performing of basic mathematical operations. The inventor was Charles Xavier Thomas, a Frenchman. About 1870 an American, Frank S. Baldwin, saw a version of a Thomas machine in a St. Louis office and was inspired to attempt improvements upon it. At almost the same time, a weary bank clerk of Auburn, New York, determined to find some way of avoiding the long hours of adding columns of figures to arrive at a balance in his accounts. He abandoned his bank position and oddly enough, went out to St. Louis and took a job in a machine shop where, by a trick of fate, he came in contact with Baldwin, who was still laboring to make an adding machine. By 1886, William S. Burroughs, the former bank clerk, formed the American Arithmometer Company to manufacture his own adding machines.

About the same time another device appeared which was of great aid to merchants who wished to find an accurate way of balancing the cash received with the merchandise disposed of. In Dayton, Ohio, James J. Ritty worked at an idea which had been suggested to him, while on an ocean trip, by observing that a dial in the ship's engine room indicated the number of revolutions of the ship's propeller. His device, completed in 1879, was the first cash register. Five years later John H. Patterson, a Dayton business man who had used several of Ritty's machines, formed the National Cash Register Company, which after ingenious and aggressive promotion became one of our large industrial concerns.

Even though Queen Anne granted a patent to one Henry

Mill for what may have been the grandfather of all typewriters, it was not until after the American Civil War that a device appeared which seemed to promise a method of "impressing or transcribing letters singly or progressively one after another, as in writing, whereby all writing whatsoever may be engrossed in paper or parchment so neat and exact as not to be distinguished from print," as the Queen Anne patent phrased it.

Christopher Sholes, a Milwaukee printer, after six years of hard work, produced in 1873 a machine which wrote well and rapidly. A friend and backer of Sholes, James Densmore, took Sholes' device to the officers of E. Remington and Sons in Ilion, New York, where after some discussion, the experienced Civil War gun makers agreed to manufacture it. In 1874 the "Model I Remington" typewriter appeared. It was a very difficult task to secure $125 from the public for a complicated machine to do the work which many thought could be satisfactorily done with a penny pen. It is likely that the final commercial success of this indispensable gadget was accomplished only with the invention of shorthand systems and the advance of business schools which offered, among other types of instruction, training in shorthand and rapid operation of the typewriter. Only thus was the time-saving element proven beyond all doubt. Mark Twain, an early purchaser, wrote a classic testimonial to the makers of this machine:

> Please do not use my name in any way. Please do not even divulge the fact that I own a machine. I have entirely stopped using the Type-Writer, for the reason that I never could write a letter with it to anybody without receiving a request by return mail that I would not only describe the machine but state what progress I had made in the use of it, etc., etc. I don't like to write letters, and so I don't want people to know that I own this curiosity breeding little joker.

But business machines had come to stay, and in the years following, the adding machine, the cash register, and the typewriter took hold of the business imagination, as the old work habits of generations of office personnel were gradually broken down. Leon Bollee in 1887 produced a machine which performed multiplication by a direct method instead of repeated additions, as in the early adding machines. Printing devices were combined with adding machines, which became the basis of an entire series of billing, accounting and bookkeeping devices.

In 1880 in Washington, statistician Herman Hollerith was finding it extremely difficult to assemble and combine the information for the Federal Census. As the decade advanced, it seemed distinctly probable that in the next census the data gathered so laboriously would be outdated before they could be processed into useful form. After much work, Hollerith came up in 1889 with a method of compiling and recording data by punching holes in cards, the holes being arranged in various positions to indicate different classifications and varieties of detailed facts. These cards were then run through machines in which, by electrical means, adding and counting operations were performed, governed by the particular arrangement of the holes in the cards. Since identically punched cards would record identical mechanical results, totals could be computed and data recorded with great speed.

Many years of refinement and improvement of this basic method have resulted in the production of some amazing calculating machines. Hollerith himself formed a company to develop his device commercially which in 1911 became a part of the Computing-Tabulating-Recording Company, known later as the International Business Machines Corporation. In the early days, four machines were used in the mechanical accounting system: a mechanical key punch to put holes in cards; a sorter, for grouping the cards; a tabulating machine to add the amounts punched in the cards; and finally, a lever set gang punch, with movable levers set to represent several digits.

In 1928 machines were devised to subtract results; in 1931 mechanical multiplication was attained and in 1946 appeared the first card-controlled machine capable of dividing. Refinements of great variety were constantly added. Alphabetic machines were made which would print word information in addition to numerical facts and a machine so delicate that it would detect marks made on IBM cards with a soft graphite pencil and translate them into punched holes. Speeds of all the IBM machines were increased. In the early 1940's electronic tubes were introduced into the machines to speed the calculating and recording processes.

In designing its new Electronic Data Processing Machine, familiarly known as the "701," the International Business Machines Corporation has achieved the greatest marvel in the realm of precision recording and tabulating. Composed of eleven compact and connected units, the first of this 701 series will be installed in the New York World Headquarters of IBM. It attains a speed twenty-five times that of its Selective Sequence Electronic Calculator, but fills less than one-quarter of the space of the latter. Eighteen of these calculators, which are planned for production within a year, are consigned to government agencies or defense industries. They will be used to calculate radiation effects in atomic energy, to make aerodynamic computations for planes and guided missiles, to aid studies of the effectiveness of various weapons and many other important and complicated projects. The figures for the speed and capacity of this group of machines, which will be rented for $11,900 per month or more, almost stagger the imagination. The 701 can multiply and divide 2,000 times per second; print 180 numbers or letters per second; add and subtract 16,000 times per second and can store on magnetic tape over 8,000,000 digits without changing tape. It was designed, IBM states, "to shatter the time barrier" which confronts engineers in working on complicated industrial or defense projects.

Meters, Gauges and Telegraphic Instruments

In our highly complicated civilization, it is necessary to ascertain accurately how much electric current, water, gas or other commercial utility is being produced or is currently passing through a given channel. Considerations of safety, economy and comfort have led to the development of thousands of instruments to measure or gauge the flow of power. Some of these are very small, others quite large. A basic number of them is used by almost every householder, car owner, or operator of any sort of machinery. Practically, the producer of the service which the individual or industrial company uses, must know exactly just how much of the service is being used at a given time, in order that the proper charge may be made for it.

A watthour meter is a very common object, situated in the basement or on the outside of nearly every dwelling; more than 45 million of them are in current use in American homes and few people regard them as of any scientific importance. Yet they must be built with watch-like precision. The Westinghouse Company calls them "gold-plated scales balanced on sapphire wheels" which is a literal fact. Working on the principle of magnetic induction, a current passes through an electromagnet in the meter, causing a thin disk to rotate. As the flow of current is increased, the disk revolves faster. When 500 revolutions have been made, a kilowatt of electricity has been measured.

An important feature of the wattmeter is precision, as exactly the amount of current used has to be determined. To insure such accuracy, the disk rotates in a jewelled bearing, made up of two cup-shaped sapphires surrounding a steel ball about the size of a pin head. In the common house meter such as those made in the Newark plant of Westinghouse there are 265 parts; it takes 600 labor operations to make one meter. They must be made to operate in any weather; hence the gears of the meter's register are plated with gold to prevent corrosion. In testing, they are salt-sprayed, roughly handled and overloaded to prove their performance capacity. To assemble the tiny parts of the delicate little instrument, girls must use jeweller's glasses to magnify the parts sufficiently to put them together. Yet they can be made at a cost of little more than that of a good fountain pen.

The power station which serves even a small community has to be equipped with several hundred measuring instruments—voltmeters, ammeters, watt meters, frequency meters. In a modern plane the instrument panel is indispensable, for it carries instantaneous information of every condition, both within and without, needed for safe flying. In modern industrial plants, the control panel, with its array of instruments, is the vital nerve center of the entire organization.

Precision also serves to protect property and to save lives. In the electrical communications industry, protective relays detect the outbreak of power line trouble; a short circuit can be located and isolated in a fraction of a second, before expensive equipment can be damaged. In one instance, a workman struck a high tension cable with his heavy pick; almost instantaneously, the power was automatically removed from the line even before the pick was burned. In the earthquake in Seattle in 1949, relays isolated the sections where the shocks were felt, permitting the rest of the system to function normally, and to allow safe and quick repairs to the places affected.

The communications industry, many features of which have been described in a previous chapter, is a distinguished example of the extent to which precision instruments have come to play a dominating role in present-day industry. In 1914, eighty per cent of all telegraphic message traffic was transmitted over manually-operated circuits. Things are vastly different today. Western Union has largely mechanized its entire telegraphic system. In the early years of the twentieth century mechanical or printer telegraphy appeared in answer to the need for greater message handling capacity without the expensive construction of thousands of pole lines and additional use of miles and miles of wire. In the multiplex system, adopted and extended on all Western Union trunk lines between 1910 and 1930, four messages could be received and transmitted on one physical wire; a simpler and less costly method than multiplex, designed for interchange between the trunk relay office and surrounding branch offices, became available in the "simplex" or start-stop printer, now usually called the teleprinter or teletypewriter. In 1937 the reperforator system was developed, whereby messages were received on combination printer-perforators, which printed the characters and punched corresponding five-unit code combinations in the same tape. This tape ran continuously from each perforator into a nearby switching transmitter, electrically connected to a nine-conductor cord and plug.

The meaning of these and many more automatic methods, which soon get beyond the detailed comprehension of the average layman, is that manual telegraphy has all but disappeared in the United States. In its place has come an automatic system in which the message operations are handled on an inter-continental network through fifteen relay centers; each center can connect with any other by automatic and reperforator switching methods; plug-in switchboards have been replaced by push-buttons which speed the message on its way. If a message originates at Boston, destined for Nebraska, the originating operator types the code character for the Kansas City relay station; in Kansas City the message in received on a printer-perforator assigned to Boston traffic, and "switched," by pushing a button, to the proper branch office in Nebraska. Thus instrumental precision has transformed Western Union telegraphy into a rapid, automatic affair.

Other wonders are being added to speed the pick-up and delivery of telegrams from patrons. A small "desk-fax" instrument, not much larger than a telephone, may be mounted on a user's desk. The patron writes or types a message on special Western Union blanks, places the blank on the drum of the instrument, presses a button, and the message is received in the telegraph office through corresponding apparatus. The desk-fax also may receive messages direct from the WU office. Several other types of automatic transmitters are designed to give quick pick-up of telegrams from office buildings and apartment houses. An ingenious device is the "telecar" delivery service in which a message recorder is built into a specially-equipped motor car; this equipment is in radio-telephone contact with a transmitter at the nearby office. A message is transmitted, received in the telecar recorder, which scans the message, prints it on "teledeltos" paper, cuts it from a roll and ejects it in front of the apparatus. While the messenger is delivering one telegram thus received, another may be awaiting him as soon as he returns to the telecar. In this way it may soon be possible for a person to write a letter and have it delivered anywhere in the country, in his own handwriting, in a matter of a few minutes. He may drop it into a telefax machine slot and leave the rest to Western Union, which by the wonders of electrically operated automatic communication precision, sees that it reaches his friend hundreds of miles away, delivered by telecar! The details of this process are for the engineer to puzzle over. The results are meaningful for every American.

Precision in Glass

The element of precise accuracy is not an elusive factor in American industry, for its results are everywhere. Wherever and whenever care must be exercised in the use of just so much and no more of an industrial material and in all cases where parts of hard material, such as metal, have to be fitted together exactly, precision must be attained. Precision underlies the standardization of parts and assures fast production methods. Even mechanical precision is not enough, in many cases; we must have electrical and electronic precision. We must also have microscopic precision, to a very high degree, particularly in the manufacture of delicate instruments, such as those of the optical industry.

The making of glass is one of the oldest of human industries; its precise beginnings are shrouded in mystery and date back to ancient Egypt and Phoenicia and perhaps even farther. For centuries glass-making remained in the hands of a comparatively small group of craftsmen who mastered the difficult art. The glass houses operated by Wistar and Stiegel, the great American glass makers of the eighteenth century, did not differ markedly from those of medieval Europe. Colonial American glass-making, which began with the Jamestown settlement in 1607, was little more than the introduction of the European handicraft to a new environment. The glass industry was almost the last of all great American industries to be modernized and as late as 1900 was in about the same state as it had always been. Yet in the past fifty years amazing progress has been made; the glass industry has been revolutionized. Intricate machinery and abruptly-new manufacturing methods have done this in large part; the very nature of glass has been changed as glass chemistry, a new branch of science, has begun its transforming work. Only in the past half-century have glass makers realized that the same methods of precision which have built all the great industries may also be applied to this versatile material, with the result that glass of the 1950's is a material of many diversified uses, most of which were not dreamed of a generation or two ago.

The movement of glass making to modern industrial importance owes much to pioneers like Michael J. Owens, a glass blower from West Virginia. After years of experimentation, Owens invented in 1903 a bottle making machine. This device had fifteen arms which revolved about a verticle axle. Each arm, carrying a mold, picked up enough glass to make a bottle as it passed a pot of molten glass; the glass was blown into the mold by compressed air and while the machine still revolved, the mold opened to release the finished bottle. The capacity of this fantastic device was a million bottles a week, all of the same dimensions and wall thickness; thus the way was opened for the devising of other precision machinery to fill and cap them.

Shortly before Owens' epochal invention, the failure of the New England Glass Company induced its head, Edward D. Libbey, to strike out to new fields. Locating in Toledo, Ohio, Libbey's company began to manufacture tableware of cut glass, which became immensely popular after being featured at the World's Fair of 1893 at Chicago. Another pioneer, John B. Ford, struggled for years after the Civil War to develop a successful plate glass factory. He set up factories at New Albany, Indiana and subsequently at Jeffersonville in the same state and at Louisville. These enterprises, though they produced good quality plate, were all failures. Competition from better-established foreign makers, the Panic of 1873, and the great distance of his plants from the eastern market, contributed to his lack of success. Yet, nearly seventy years of age, he was not defeated. Without any funds, Ford borrowed $100 from a former employee, and went to New York to see Peter Cooper, himself an inventor and, moreover, a man with funds to invest and with wide influence. Cooper helped Ford sell a patent on glass sewer pipes for $30,000 and with this cash, plus $20,000 in commissions earned in selling California land for General John C. Fremont, Ford returned to western Pennsylvania where in 1880 he established a factory which grew into the Pittsburgh Plate Glass Company.

The firms which these three men established became merged in 1930 with the establishment in Toledo of the Libbey-Owens-Ford Glass Company. Edward Ford, son of John, upon retiring as president of the Pittsburgh firm in 1896, determined to establish a glass works of his own. He purchased 173 acres on the Maumee River near Toledo, constructed a model town, and built the largest plate glass factory in the country, with a capacity of 6,000,000 square feet per year, a quantity which has since grown to many times that figure. The project was increased to giant size by the combination of Ford with Libbey-Owens later. In twenty years, eighty per cent of all plate glass used in America was domestically produced.

Before the merger took place, Owens, as Libbey's right hand man, was busily looking about for better methods of forming window glass. In 1912 a public bankruptcy sale of the assets of Irving W. Colburn occurred. He had exhausted his fortune and his credit in trying to draw window glass in a flat, continuous ribbon from a melting tank, or lehr. Owens persuaded Libbey to buy the Colburn patents; then Colburn and Owens went to work to perfect the process. After four years of effort and the expenditure of more than a million dollars they were able to draw out window glass 1/16 of an inch thick from the annealing oven in a continuous sheet four feet wide at the rate of better than five feet per minute. Window glass had formerly been made in cylinders and later flattened; this had usually led to distorted vision; the Colburn method, with later improvements, changed the entire production system of this type of glass.

So far Americans had merely improved window glass and established a domestic plate glass industry, important in themselves but not outstandingly new. The growth of the automobile industry, with its demand for huge quantities of plate glass, presented an opportunity and later a challenge to glass makers, for the vehicle, bearing several sheets of shatterable material, presented grave risks to the motorist. This was accentuated of course by the wide popularity of the closed car, glass-enclosed. The answer, in large part, to this critical situation, was laminated or safety glass. This material, now a familiar part of all motor vehicles, is actually a sandwich of two layers of glass, with a tough plastic film between them. The search for the middle layer was the hard part, for though many tough materials were available, none existed which would not discolor or weaken with age. Not until chemical engineers found a method of producing plastic cellulose acetate from wood pulp or cotton at low cost could the problem be solved. After grinding and polishing the plate glass blanks, coated with invisible cement, and assembling them with the layer of acetate between them, the three layers are welded together under great heat and pressure into a transparent sandwich. Other plastics are also used for the inner layer.

In 1879 the Corning Glass Works perfected a method of making glass bulbs for Edison's incandescent lamp. The improvement of this machine method through the years has resulted in the cheapening of bulb costs to a few cents each. In 1915 the same company produced Pyrex heat-resistant glass cooking ware, and twenty-one years later a much tougher "top-of-stove" ware. In 1934 Corning cast the 200 inch glass mirror for the immense Mt. Palomar telescope in California. This mirror had to be made from a glass disk seventeen feet in diameter, twenty-six inches thick and weighing twenty tons, the largest single piece of glass ever cast. The problems which were successfully met included not only its size, twice as large as the glass disk in the Mt. Wilson observatory, but also the fact that it was made of telescope glass which would not flow unless kept at a very high temperature. This made it necessary to place the mold inside an igloo-shaped oven to make action by gas burners render all parts of the glass fluid enough to fill every part of the mold. The tremendous heat made it difficult to prevent melting of the metal rods which held the ceramic cores at the bottom of the mold. The work took nearly two years and after its trip of three thousand miles in a special railway car to Pasadena, it had to be machined into a shallow paraboidal surface at the shop of the California Institute of Technology. This involved removing two and three-quarters tons of glass by grinding, and the accuracy demanded was within two-millionths of an inch (or better!) at all parts of its 36,000 squares inches of surface.

* * *

From such basic developments in the field of American glass making, it is possible to turn to a wide variety of uses to which glass has been put, resulting from the researches of more than a generation of glass chemists. Almost incredible toughness has been found possible in certain types evolved in industry. Libbey-Owens-Ford has developed a type called tuf-flex from sudden cooling of hot plate glass. A pane of this glass, resting on a cake of ice, remains unharmed when molten lead is poured over it. It will resist the bounding of a two-pound steel ball dropped from a height of six feet; a three-ton elephant, standing upon a sheet of tuf-flex which was supported at two ends a foot from the ground, succeeded only in bending, not breaking the glass sheet. This "tempered glass" may find uses in port holes of ships, in oven doors and kitchen range ports.

Vitrolite, a structural flat glass, another LOF product, is opaque and may be colored or lacquered or enameled. Thermopane, an insulating glass, has been made since 1937 by LOF by sealing up the edges of parallel plates with a metal-to-glass seal, to close in an air space between the plates. Admiral Byrd took lights of thermopane to Antarctica for windows in his expedition's laboratory in 1939; these served to maintain an inside temperature of 75 degrees above, with an outside temperature reaching 75 below, with perfect visibility. It seems to be solving many problems of condensation and blurred vision, as well as heat or cold-sealing.

Then there are heat-absorbing plate glasses, invaluable for many-windowed dwellings or office buildings, such as the new headquarters of the UN in New York; a mirror which is transparent, called mirropane (as one looks from a dark room into a light room, it is a window; looking from the light room into the dark one, it becomes a mirror); electrapane, with a thin metallic film over it, through which an electric current can pass, developed during the war, to keep the windows of airplanes and ships de-iced and de-fogged; fiber glass, made from molten glass filaments, of use in textiles, insulation, in metal substitutes and other industrial materials.

Photosensitive glass, developed in 1947 in the laboratories of the Corning Glass Works, makes it possible to produce colored photographic images within objects made of glass. This is done in two steps: first there is exposure with ultraviolet light through conventional negatives, and then heat treatment. The photographic image may be reproduced in a variety of colors, is three-dimensional, and often gives a stereoscopic illusion. This new photographic glass is expected to find use in portrait work, murals, decorative windows, church windows, advertising displays and so on.

As the mid-century arrived, new and intricate combinations and forms of glass continued to appear. Optical glass has been manufactured automatically by Corning Glass Works since 1944 and television tubes since 1947. The same concern announced in September, 1951 the successful production of a new type of photosensitive glass in which by a method of "chemical machining" lace-like patterns are formed, at one time regarded as impossible to attain in glass. A design is first printed in the glass by use of an ordinary photographic negative and ultraviolet light; it is then heated to 1200 degrees Fahrenheit for about two hours, causing a milk-white image of the design to appear. Upon immersion in hydrofluoric acid the white areas are eaten through and removed, leaving the remaining unexposed glass in the form of the original design. Since this chemical machining involves no mechanical stress, patterns too complex to reproduce on any other material except by long and arduous skilled work, can be quickly reproduced in photographic accuracy. This machining is suited to perforating holes of any diameter, from the tiniest to the largest needed. It has also been used in making "printed" electrical circuits for electronic instruments. After the circuit is photographically reproduced on the glass with holes for fastening to a chassis and the glass treated as above, conducting metal may be placed in the glass container to form an electric circuit, of high precision and durability.

* * *

The science of optics is one of the most important branches of physics. Steady progress involves constant research in glass chemistry to obtain the best types and forms of glass for correcting defective vision, to improve the instruments used by the ophthalmic profession, and scientific equipment for various branches of engineering and experimental science. Bausch and Lomb Optical Company finds it necessary to maintain a staff of fifty chemists, in addition to many optical physicists and other technical personnel. While the optical specialty is but a small part of the entire industry, research in this field represents a greater chemical variety than any other type of glass chemistry and the product represents, moreover, a very high dollar value. The same company maintains a close relationship with the Mellon Institute where a full-time fellowship in annealing is maintained for a Bausch and Lomb scientist; likewise there are close ties with scientists in university departments of physics and chemistry.

Glass destined for optical use takes very careful handling. In melting, platinum containers and platinum stirrers are used to avoid contaminating the mixture. The glass is tested for its optical properties, its expansion characteristics (with a dilatometer), its chemical composition (with a spectograph), its annealing rate, its qualities of light absorption, its behavior under high energy radiation and many other conditions. It is of course difficult to allocate directly to chemists and optical physicists the credit for new departures in optical equipment. In some cases, however, there is no doubt that commercial success of a product proceeds directly from a long physico-chemical search for certain needed properties. Such a case is the attainment of exact color identity between the lenses of each pair of Ray-Ban sun glasses—in hue, color saturation and in color brightness. Another achievement of the glass laboratory men is the production of a supersensitive type of glass which can detect the presence of atomic energy. This dosimeter glass was the joint project of United States Navy scientists and Bausch and Lomb chemists.

Among important new equipment items developed by Bausch and Lomb are the deep vitreous attachment for ophthalmic examination equipment, which enables the specialist to examine the posterior eye for tumors or cysts or other deep eye affections; the dynoptic microscope which has a number of mechanical improvements such as roller-bearing focusing adjustment and "below-stage" controls to avoid fatigue in operation; and the "30 mile eye" an adaptation of the forty inch focal length B&L lens to a standard movie camera, used by the Navy to take long range pictures of moving aircraft.

* * *

From only a few of many possible industrial examples, it is clear that precision is basic in all large-scale industry. Machining must be to the minute part exact, or parts will not fit or operate in moving assemblies. The most delicately made instruments are needed to enable us to keep accurate track of such common things as our gas and electric bills. Business itself, on the scale to which it has advanced, could hardly be conducted without the use of the most intricate and accurate of business machines. Science has built the vast industrial system and throughout the entire process of production, instruments of precision must perform measurements and exert constant control over a broad range of mechanical, electrical, chemical, and optical operations.

CHAPTER THIRTEEN

INDUSTRIAL SCIENCE AND NATIONAL DEFENSE

In the troubled twentieth century the United States has engaged in two major wars and is at the present time engaged in helping to restrain the expansion of communism by force. In case another full-scale war breaks out, implication of the United States is almost certain, since the only predictable form such a struggle could take would be some major challenge of the defensive preparations underway by the free nations, of which the strongest is this country.

In past centuries wars could occur with only brief interruption of the normal life of the contending nations. At the conclusion of those contests, few far-reaching adjustments were necessary, sometimes none at all. Immediate effects upon the population were almost entirely evil, but the general results were limited and not irreparably disastrous. Only the Napoleonic Wars in Europe and the Civil War in America resembled in any important way the titanic struggles of the twentieth century. Even the great nineteenth century conflicts by-passed large segments of the population and left undamaged great stretches of territory.

Those days of limited, temporary conflicts have passed

forever. In the present century, with applied science delving into all kinds of destructive possibilities, with communications so complete as to leave no sector of the globe unaware of a new crisis, and with transportation so rapid as to allow no isolation, war can hardly be started without the involvement, sooner or later, of most of the nations of the world. Those countries not industrialized, and hence incapable of building any strong military defense, may become at any stage of the potential conflict the victims of the stronger, more highly developed industrial nations. "Total" war, the type inherited by this century, involves the entire economy of a nation; domestic political, social and economic matters become for the time subordinate. To wage war on any lesser scale is to court disaster. It is by way of illustration to point out that had the Axis Powers attained their final objectives in World War II, very little of what we regard as the way of life of the western world would remain.

Wars of the intensity of the two great contests of 1914 and 1939 have been won by trained manpower with superior equipment. Time in which to prepare, and space in which to absorb temporarily the attack by better-trained enemies could, in other times, be vital. While both space and time still have tactical value for the side possessing them, they are no longer decisive for total war. There is no longer space enough or time enough to stand off the weapons of modern war for a decisive period.

One consideration, therefore, outweighs all others in twentieth century defense: the strongest military nations are, in the long run, those with the best industrial "know-how." This implies that only a nation which possesses the best possible industrial system and the most alert industrial scientists will be able to erect a strong defense when attack comes. If in addition to an advanced position in scientific research can be added access to great supplies of raw materials, availability of large capital funds, and a strong group of industries willing and able to convert their plants to military purposes, we have the best hope of national survival in total war. The United States is fortunate indeed to possess all these prime requisites in considerable measure. In an emergency their mobilization, placed at the disposal of trained man-power, would provide the essentials of a successful defense and would make an attack by any power, however large or well-prepared, costly and probably disastrous. However, merely to have strong military potentialities is quite different from making the best use of them in an emergency.

The Wartime Capabilities of the United States

The early military history of the United States sheds very little light on the problem of how to conduct the kind of war which has come to be characteristic of this century. For one thing, most of the wartime emergencies which have confronted the nation have been of the same limited type already mentioned. In addition, most of our early wars were not professionally planned or conducted. Speaking broadly, we first became involved in a situation, and then began to consider ways and means of getting out of it; we were never adequately prepared. The fact that we have fortunately avoided complete disaster in our wars has blinded many to important circumstances.

The American Revolution was won with the aid of France, Spain and the Netherlands; until foreign aid came there was no decision. The War of 1812, a minor side-show of a greater European struggle, was not won at all; the parties simply called it off, on a *status quo* basis. The Mexican War was fought against a weak, badly-divided nation. The Civil War took four years to come to a decision; only gradually did the Union swing its superior naval power into action and begin to utilize fully its overwhelming manpower and industrial potential. The War with Spain was a masterpiece of ineptitude and inefficiency, in which the Navy acted a deciding role, but against very little opposition. To quell the Philippine insurrection required nearly three years time and more than 1,000 armed actions by a large military force. Of all these conflicts, not a single one was fought against a major power by the United States alone; only one of them, the Civil War, was a major struggle and that, of course, was among ourselves; only in the last two years of the War of the 1860's was there a detailed, comprehensive strategy. Not until 1903, after the completion of the Philippine War, did we create a General Staff. Yet nations around the world already had large standing armies and superior navies, professionally led and staffed; many of them held periodic large-scale military maneuvers and had worked out detailed mobilization plans. Without taking any of these precautions, in a world seething with hostility, we had, without undue anxiety, established ourselves in the Caribbean and had extended our colonial control into the middle and far Pacific, following our easy victory over Spain.

Our entry into World War I, after a serious effort for two and a-half years to stay out, was scarcely less casual than our previous record of involvements. The principal European armies had several millions each in their active military establishments, while the largest expeditionary force ever to leave the United States previous to World War I would scarcely have equalled in strength a single British or German division of 1914. The greatest battle ever fought in North America had been won by the Union Army at Gettysburg with less than 100,000 men. We are not dealing here with the moral justification of our declaration of war against Germany. The plain fact was, that without proper military or industrial preparation we declared war upon two great military nations who had, since 1914, balked the efforts of millions of troops of Great Britain, France, Italy and Russia to drive them out of territory which they had won almost immediately. Indeed, at the very moment of our entry, Germany and her allies were planning a blow to win complete victory, with a strong prospect of success. We had no military or naval force strong enough, at the time of our entry, to exert any marked influence upon the course of the struggle. We had wealth, great industries, and a large pool of potential military manpower, yet hardly a move had been made to prepare us for what had been, ever since 1914, a distinct possibility.

Not all Americans, it is true, had been blind to the threat clearly implied in Germany's decision to use her submarine arm to its greatest possible advantage. The *Lusitania* crisis, and emergencies precipitated by other sinkings had set off a mild preparedness wave in the country, which was vigorously opposed by pacifist and various pro-German and anti-British groups of citizens. With great reluctance and marked hesitation President Wilson, a man of peace, endorsed the preparedness movement. After strong bi-partisan support of the President became evident, it was possible, in 1916, to secure from Congress a program of mild preparedness legislation. It was agreed, on paper, to increase the regular army and the national guard; an appropriation was made to increase the Navy; under the supervision of a new Shipping Board, an Emergency Fleet Corporation was set up to build and operate a merchant fleet; a Council of National Defense was authorized, to coordinate industry and defense.

The German High Command regarded the re-election of Mr. Wilson in 1916 as a guarantee that no immediate or effective intervention by the United States would hinder their speedy winning of the War. With Russia on the verge of revolution, with German, Austrian and Turkish armies retaining their large gains on all fronts, it seemed possible for the Central Powers, by concentrating their forces in the West, to knock out the hard-pressed French and British before American aid could become decisive. With the military situation clearly in their favor at the end of 1916, a larger and more aggressive U-boat program might stop the flow of American supplies to the Allies. This would probably bring the United States into the War against them, but the military decision could come too soon for that to make any difference.

How nearly correct the German calculations were is now a matter of sober historic fact. In the spring of 1918, with Russian collapse an accomplished fact and that of Italy a matter, apparently, only of time, Germany was able to mount in the West, from March to July, the greatest offensive of the War. In a series of terrific lunges, Ludendorff drove back French and British forces along a wide front to a depth of some thirty-five miles and reached Chateau-Thierry on the Marne River, hardly forty miles from Paris. Here, as we know, the trend changed as the great Allied counter-offensive began, coordinated under Marshall Foch and bolstered by the presence and support at the battle lines of more than a million fresh American troops. The Allies did not stop their offensive until the decisive battle of the Argonne, in which a force of 1,200,000 American troops, well-equipped with planes, tanks and heavy artillery, blasted their way to the key railway terminal of Sedan and forced the Germans to request an armistice.

This rough outline of the last months of World War I is a familiar one. Not so well-understood is the reason why American intervention with, at best, only a part of her potential military and industrial power, could reverse the course of military operations and quickly change an apparently hopeless situation into a brilliant victory. Did the United States learn so quickly the lessons of total war, which so many nations had worked long and hard to acquire? What did the victory, in which the United States had played so important a part, teach us which would be valuable in the future?

The Lessons of World War I

It is fair to say that the United States, though it started to prepare late in the game, almost too late to be of aid to its associates, supported the Wilson administration with enthusiasm and loyalty, once the issues were clear and the magnitude of the nation's task became evident. The undivided loyalty of millions of Americans of foreign birth or of foreign descent was astonishing. Enormous loans, five of them, were patriotically subscribed by the people. Successfully dramatized by President Wilson, the democratic, moral and humanitarian goals of the United States were ably propagandized by George Creel's Committee on Public Information. More than 4,000,000 men were processed into the armed services, largely through the agency of the Selective Service Act of May 18, 1917. Relief agencies, headed by the Red Cross, did a magnificent job of welfare, sanitation and medical aid to soldiers, sailors and their families. The American Expeditionary Force was ably led by General John J. Pershing; the corps of officers under his command was of high calibre, partly because of a better system of officer training than in former wars, in which the political general and the elected field officer had proved more often liabilities than assets.

But it was not merely these efforts, important as they were, which enabled the American contribution to hasten German military defeat. The war problem, when the nation entered the conflict, was primarily administrative, industrial and logistical. We had first of all to create a war administration with comprehensive and far-reaching powers. To cope with the supply problem, a large part of American industry had to be geared immediately to war production. The front on which the army would have to operate was three thousand miles from the American coast, in a nation wracked and torn by years of war; one which, moreover, had few transportation and industrial facilities which could be placed at the disposal of our convoys and troops; shipping to take the men to France was lacking and all the paraphernalia of supply had to be taken to an unused area behind the lines of combat, on a scale which far exceeded anything in American experience. These were the vital and difficult problems to which the President and his war administration had to devote themselves, if victory were to be won. The logistical problem was probably the most difficult of all, for it meant that a large part of the American war economy would have to operate in a foreign land, at great distance.

The serious military situation of the Allies was driven home by missions of Allied leaders which immediately began visits to the United States upon our declaration of war, and by the constant deterioration of the Allied positions on all fronts. Without full economic and military participation by the United States, it became evident that little hope of reversing the tide could be expected. Complete mobilization, however, could hardly be expected inside a period of two years. In the planning which ensued, therefore, the year 1919 was expected to be the year of victory. Actually, due largely to the desperate urgency of the situation in early 1918, this schedule was shortened, the War finally falling into three phases, so far as the United States was concerned: six months of planning, eight months of putting the plans into motion, and five months of full scale operations.

When the new Congress met in December, 1917, the initial efforts to probe into the conduct of the War to date were offset by optimistic promises from Secretary of War Baker that the spring would see a host of a million and-a-half men dispatched across the Atlantic, as fast as supplies and shipping could be made available. Convinced by this and similarly hopeful predictions by the administration, Congress proceeded to pass a series of acts granting the President enormous war powers, capped by the Overman Act, which made him substantially a dictator for the duration, plus six months thereafter.

The creation of a Council of National Defense in 1916 had provided a mechanism through which a real war economy could be evolved, particularly by means of the Council's Advisory Commission of seven civilians, representing manufacturing, transportation and labor and headed by the able Daniel Willard, president of the Baltimore and Ohio Railroad. In view of the fact that the Council of National Defense was made up of the Secretaries of War, Navy, Interior, Agriculture, Commerce and Labor, all of them heavily burdened with the stepped-up activities of their departments, the Advisory Commission assumed an important role, becoming in actual fact a "civilian general staff." Through the plans which the Commission formulated the government was reorganized for the purpose of winning the War. Specifically, six great agencies were set up before the end of 1917 and were granted large control over industry, shipping, fuel, food, trade and labor.

At the center of these war agencies was the War Industries Board, under Bernard M. Baruch, fully organized as an executive agency by March, 1918. It assumed control of war purchases for our government and for the Allies; it could fix prices, standardize products, determine priorities of production and delivery, control conservation of facilities and materials. Mr. Baruch became in effect the economic dictator of the nation for the duration of the War. Prices were fixed on the following raw materials, among many, by the Board: iron and steel (with their products), wool, foreign and domestic hides, aluminum, domestic manganese ores, lumber, sulphuric and nitric acids, copper, hemp and cement.

In December, 1917 a Railroad Administration was set up under Secretary of the Treasury McAdoo and for eleven months the railroads of the nation were operated as a single system; the addition of inland waterways and the railway express companies to Mr. McAdoo's agency gave him complete control over the internal transportation of the nation. The Shipping Board of 1916 was converted into a war agency by 1917; the Board's Emergency Fleet Corporation, largely because of the leadership of Charles M. Schwab of Bethlehem Steel, did a magnificent job of building, purchasing and commandeering a merchant fleet, as related in a former chapter. By September, 1918 the Shipping Board had a fleet of more than ten million tons; it carried nearly half of our Army and all of its military supplies to France. The War Trade Board, under Vance C. McCormick, exerted close control over exports and imports by a licensing system; it forbade the export of necessary commodities and put upon a black-list all companies suspected of trading with the enemy.

The Food Administration, under Herbert Hoover, obtained price-fixing powers under the Lever Act of August, 1917. As the War went on, the Food Administration gathered large powers over the purchase, importation and exportation of foods; its licensing system affected 250,000 firms by the end of the War. In addition Mr. Hoover made an effective contribution by his encouragement of economy and conservation, the use of food substitutes, and by the instituting of "beefless" and "wheatless" days. Most interesting of all were the agency's operations in the grain market and its stimulus to greater grain production through organizing a special Grain Corporation. By means of pegging the price and by guaranteeing purchase of the entire crop, wheat production was increased in three years by nearly 200,000,000 bushels.

Fuel also came under the strict purview of the administration; scarcities due to tremendous industrial and commercial consumption, accentuated by the extremely severe winter of 1917-1918, confronted the administration. Under Harry A. Garfield a Fuel Administration functioned in several strategic ways to alleviate the worst of the deficiency. A series of "fuelless" Mondays were ordered in the winter until March 25, 1918. A power shutdown of all manufacturing plants east of the Mississippi was ordered for five days; the opening of submarginal mines, encouraged by the Fuel Administration, increased the production of bituminous coal to a point more than fifty per cent greater than in 1914. Daylight Saving Time, instituted as a war measure, aided in the conserving of coal.

In spite of all the controls and price fixing of the government and its special agencies, the cost of living mounted steadily; yet real wages and actual earnings in business mounted as well, though not as rapidly. The year 1917 was one of strikes and mounting industrial unrest; in 1918 there was considerably less labor difficulty, due to two reasons: wages were increased in shipyards, munition plants and for railroad workers; in addition, new machinery was set up for the conciliation of labor disputes. In January, 1918 the President created a special labor agency under the Secretary of Labor; at the advice of the latter a National War Labor Board was set up by the President in April, 1918. In addition to the lessening of strikes by this policy, the sympathetic governmental attitude contributed to the growth of organized labor; the A.F. of L. added half a million new members by the end of 1918.

These, in rough outline, were the essential adjustments made to our domestic economy to gear it to produce in greater volume and at greater speed. The special controls exercised over the economy functioned well and served to supply and equip our greatly enlarged military and naval forces in time for them to make their decisive contribution. This was the purpose, and the only one, for which the extensive emergency alterations in our governmental system were made. At the conclusion of the emergency, the controls were relinquished and American industry was free to resume its course as before. It is interesting to note that in only one particular did the government take over bodily a strategic phase of our economy; that was in the case of the railroads. The wisdom of the course has been vigorously debated; in World War II the experiment was not repeated.

The National Defense Act of 1916 gave clear-cut powers to the government to take over possession of any plants refusing to furnish war material to the nation in time of war or of imminence of war. Mr. Bernard M. Baruch, in commenting upon this section of the Act many years later, stated that he knew of no instance in which an important industrial firm was commandeered, though such action was on one occasion urged. It was a case in which an industrial plant was not giving full cooperation to the war effort. The War Industries Board considered the matter, and vetoed taking such action. The argument which apparently clinched matters ran about as follows, according to Mr. Baruch:

> Who will run it (the commandeered plant)? Do you know another manufacturer fit to take over its administration? Would you replace a proved expert manager by a problematical mediocrity? After you had taken it over and installed your Government employee, what greater control would you have then than now? Now you can choke it to death, deprive it of transportation, fuel and power, divert its business, strengthen its rivals. Could any disciplinary means be more effective? If you take it over, you can only give orders to an employee backed by threat of dismissal, and with far less effect than you can give them now. Let the management run the plant and you run the management.

Symptomatic of the sharp reaction against the internationalism of the war period, were the discussions which went on later regarding certain of the practices carried on by industry during 1917 and 1918. The high profits secured by industry were called excessive; bonuses and other incentives to employees were deemed unpatriotic; such behavior was compared unfavorably with that of the soldier who sacrificed more and received only his ordinary pay. In the 1930's the journalist John T. Flynn suggested a limitation of wartime profits to three per cent of the investment, and a maximum personal war income of $10,000 per year. Mr. Baruch, in a section of his complete report, *American Industry in the War,* combated this view and called attention to the fact that the industrial system, a vast and complex machine, was "built and geared to run on investment and profit. There is no proof that it will run on psychology and there is much that it will not. Certainly we should not select an hour when the enemy is at the gates to find out whether it will or not."

In testimony before the Nye Committee of the Senate, which in 1935 held many hearings regarding the armaments industry and various other aspects of our policy regarding World War I, an interchange took place between Senator Clark of the Committee and Mr. Eugene Grace of the Bethlehem Shipbuilding Corporation. The Senator called into question the incentive system, while Mr. Grace stated that it was necessary to secure increased production; that it worked well in time of peace, and he believed that it applied just as well in time of war. Before the Nye Committee was set up, General Douglas MacArthur, Chief of Staff of the Army, made a statement regarding Public Resolution No. 98 (June 27, 1930), which set up a War Policies Commission to study ways of "equalizing the burdens and removing the profits of war." Said the General:

> In our attempts to equalize the burdens and remove the profits from war, we must guard against a tendency to over-emphasize administrative efficiency and underestimate national effectiveness. . . . It is conceivable that a war might be conducted with such great regard for individual justice and administrative efficiency as to make impossible those evils. . . . It is also conceivable that the outcome of such a war would be defeat. With defeat would come burdens beside which those we are considering would be relatively insignificant.

It is worth considering whether General MacArthur did not here sum up much of the discussion which took place about profits, incentives and commandeering industry in wartime. The crisis and all that it actually portended was the important problem; the winning of a quick, effective victory was the prime objective. Matters of how much or how little profit was necessary, how many or how few incentives were needed, and whether industry should be seized or discreetly managed "with a club in clear view," are details which every war administration would have to work out for itself.

This is not to say that the life of the War Industries Board was in any way a rosy one. A growing shortage of steel occupied the attention of the WIB during much of spring and summer of 1918; it was necessary in the summer to prepare the automobile industry for an immediate cut in production, and to urge its members to enter war work as soon as possible, and no later than January 1, 1919. Upon the submission to the WIB of inventory and production information, the Board calculated the amount of steel needed to maintain each factory at fifty per cent level; during the last half of 1918 the auto manufacturers were free to buy these alloted quantities of steel, if they could get them. The sudden end of the War made complete conversion unnecessary.

Problems poured in upon the industrial administrator as he met the representatives of one or another of the nation's great industries; he and his colleagues heard their troubles, listened to their objections, and tried to find proper answers; often there were none. On one occasion an industrialist stated that he saw no reason for the "Government's building stuff and storing it, as I was told yesterday, that couldn't be shipped for five years" to which Mr. Baruch replied patiently:

> Of course we hear a good many statements, but we would like to have the facts on it. It may be true. The Government has a military and naval program which it must carry out. The program is a very, very large one . . . if there is any possible way that you can suggest to this Board that we can speed up the production of steel mills,

or anything that particularly interests you . . . we would be very glad to hear it.

Extended discussion of the trials and tribulations of the busy war agencies would go far beyond the limits of a single chapter. One thing seems fairly clear in regard to the WIB: its organization was well-conceived, with sufficient authority direct from the President. Five of its members were salaried civilian officials of the Government; the other two represented the Army and the Navy, respectively. It was able to deal with many of its problems effectively and might have solved other problems, such as the steel shortage, had the War continued longer. Able and clear-headed, Mr. Baruch steered a solid course between industry, with which as a financier he was sympathetic, and the United States Government, to which he was unqualifiedly loyal.

It is misleading, in many ways, to speak of direction of war production by the government. While the center of the vast war economy was Washington, the actual direction, when the needs were understood and the amounts and specifications disclosed, came into the hands of the nation's business men. These were in many cases the heads of great corporations like Ford and General Motors, United States Steel, Westinghouse, General Electric and many others. Their experience, and that of thousands of others, in large and small industrial corporations, had been gained in years of production for the civilian economy. Under such a system as ours, the initial disadvantage in war of extensive individualism was soon counterbalanced by the zeal and speed with which these massive combinations forged the weapons of victory.

Much nonsense has been written about the greed of the industrialist, of his scheming and grasping for power, and of his desire to become a "merchant of death." The opposite is certainly true of many. The industrial dislocations which come in case of war serve to terrify, rather than to thrill, the average employer of labor. The spectre of loss or damage of a profitable business appears; the raw materials once in plentiful supply become almost unobtainable; if he can convert his plant to war purposes he usually does so. But conversion is a slow, painful process. He may not receive a large allotment of war orders. The war will end, and he must convert his plant back again. If he happens to be a socially-minded person, and thousands of manufacturers are, he feels responsible for the continued employment of hundreds, perhaps thousands of people; an entire community may depend upon his plant for its existence. Why should he, for purely selfish reasons, wish to embark upon the dangerous and disagreeable task of making instruments of destruction? If he is able to continue his industrial operations by securing war contracts, he is under constant governmental scrutiny. If profits rise during the process, along with costs and wages, he risks investigation after the war; his motives are constantly questioned and his patriotism often under fire. Yet the industrialist is the key to our entire defense system; if he fails, the war effort bogs down.

The logistical problem of the nation in World War I was complicated by a very real submarine menace in the Atlantic. This portion of the problem was solved by the United States Navy in collaboration with naval forces of the Allies. The relative success of the merchant fleet program enabled us to transport a substantial part of over two million men overseas, three-quarters of them in the last six months of the War. To accomplish this feat, it was necessary to assemble more than three million tons of shipping for that purpose alone; the loss of life in this large operation was almost negligible. The northern ports of France were crowded with the shipping of the British and the French; it was impossible for the United States to use them. Furthermore, inland through western Belgium and northern France was the vital British operational area; to the east, radiating from Paris to the front, was the long French operational zone. Since the armies in these vital regions monopolized all the port facilities, roads and railways of northern France and western Belgium, no additional foreign army could be properly supplied if stationed in those sectors. It was necessary, therefore, to station the principal American headquarters farther east,—at Chaumont, as it turned out. From this point, the American lines of supply and communication had to traverse the middle of France, to the less congested seaports of Brest on the west coast, St. Nazaire and Bordeaux in the southwest, and Marseilles on the Mediterranean. At Tours, a little more than halfway from Chaumont to St. Nazaire, was set up the Services of Supply, through which funneled the immense tonnages brought into the seaports alloted to the A.E.F.

The Services of Supply, so well described by General Johnson Hagood, was a gigantic and successful experiment in the coordination of civilian industrial production, wartime emergency shipping and naval convoy, to distribute millions of tons of supplies to a large army operating more than three thousand miles from its home bases. In order to achieve the objective, port facilities had to be reconstructed, a large section of the French railway system reorganized, vast storehouses and hospitals constructed. To a large extent it was a triumph of American technology, business organization and engineering skill, adapted to a novel situation. The experience was of untold value in the meeting of future emergencies.

Thus it was that the United States gained valuable background in conducting total war, from its rather brief and incomplete mobilization of 1917-1918. The critical military situation of the spring of 1918 called for haste in the training of manpower, and the use of large quantities of field artillery and airplanes already manufactured by the Allies, in order to make the American military effort available in 1918, the critical year. Solution of shortage problems and the complete development of newly-designed American weapons had to wait, as it happened, because of the desperate urgency created by Ludendorff's "Big Push." There is every reason to believe that the year 1919, could the situation have been saved to that time, would have seen the full productive capacity of the United States thrown into the conflict.

What, after all, had the War established, regarding the wartime adaptability of the United States? It proved, in the first place, the willingness of the American people to accept a partial dictatorship over our economy in the face of an emergency. It demonstrated the ability of our industrial system to begin a far-reaching conversion to a total war economy. It established the importance of industrial science, business leadership and advanced scientific technology as indispensable to military success. It proved, in the totals of shipping built in rapid time and in the beneficial controls exerted over the vital areas of food, fuel, transportation, manufacturing, foreign trade and labor, that in a more extended emergency our economy would probably survive; that, moreover, our productive power could be made a chief instrument of eventual success.

The United States did not *win* World War I. Germany, after its desperate gamble in 1918, with its last reserves of manpower and most of its remaining material committed to the effort, had *lost* the War as soon as its offensive failed to reach its final objectives; for Germany lacked the means to replace either the men or the munitions so committed. To this failure and to a quicker victory than would have been otherwise possible, United States manpower, food, shipping and raw materials made a vital contribution. Had it been necessary, an overwhelming amount of American men and materiel would have accomplished the same result in 1919.

Industrial Science Faces a Supreme Test

Following World War I, demobilization of men, industry and shipping was almost immediate and complete, with disastrous effects upon our national economy, to say nothing of the effects upon our armed forces. Despite the persistence of disorder and revolution beyond the official end of hostilities, the American military and naval establishments were quickly reduced to peacetime levels.

What followed was a regime of so-called "normalcy" in which our government and others were forced to deal with many difficult postwar problems. To a large extent, the political events of what is more accurately termed "the long armistice," from 1919 to 1939, followed two opposing courses. On the one hand, many leaders sought earnestly to discover the true meaning of World War I and to undertake measures to provide against a new conflict; opposed to these internationalists were those who followed a blind, ill-timed policy of resuming those activities interrupted by war; they sought to turn back the clock and return to the "good old days." The collective security which might have been established after such a ruinous conflict was not to emerge; between impractical pacifism on the one hand and selfish economic nationalism on the other, the postwar world was allowed to drift into a second great crisis. What happened is known to all. Into the power vacuum created by the defeat of the Central Powers and the disintegration of orderly economic life in Europe, rushed new forces to challenge peaceful society. Hitler, Mussolini and Stalin emerged with their panaceas for their respective states; worse, their policies involved the disruption of the settlements after the War. No decisive measures were taken to hinder the rise of a powerful new combination, headed by Germany and including the disgruntled Italians and the insatiable Japanese, with the Soviet Union, now grown to great potential military size, lurking insidiously in the background, ready to make terms, no doubt, with the winner.

The challenge to undo the settlements of World War I was made and accepted, and the world was once more plunged into global war. The United States steered a traditional, though somewhat less naive, neutral course for two years; then Pearl

Harbor catapulted her into the cauldron along with the others. The new war, at first glance, seemed rather familiar. The old allies of 1914, Great Britain, France and Belgium, were together again; France was seemingly better prepared than before, with her great Maginot Line of underground pill boxes, bristling with men and artillery, from Belgium to the Alps; Italy hesitated, as she had in 1914.

But certain changes in the line-up soon appeared. Italy, instead of joining the Allies, cast in with the Axis; the former overwhelming sea power of the Allies was noticeably absent; Russia, a one-time ally of Britain and France, preserved a watchful neutrality. The pattern was soon to change even more abruptly. After contributing to the general complacency of the winter of 1939-1940, Germany suddenly launched a "blitzkrieg" or "lightning war" against the West in the spring of 1940. Following a pattern begun in its attack upon Poland in the previous September, Germany, disregarding formal declarations of war, sent great masses of bombing planes against nations with which it was at peace, to disorganize the cities and to spread terror among the civilian population; upon the heels of the air attacks, fast-moving columns of motorized infantry swooped upon key objectives, bypassing the fortified points which they were expected to attack along the Belgian and French frontiers. By June, Norway, Denmark, the Netherlands and France had been defeated and Germany was poised to repeat her performance against a resolute Britain, the only nation which still opposed her. Failing to accomplish the conquest of Britain by "air blitz" in the last half of 1940, the Axis Powers turned their attention to the East; on a series of battle fronts from the Baltic to the Mediterranean, the Axis after 1941 overran hundreds of miles of Soviet and Balkan territory; in collaboration with the Italians, Balkan resistance was crushed and the Mediterranean nearly closed to British shipping.

Geared closely to the swift successes of the western Axis in

Europe in 1940 and 1941, Japan, which had already advanced into southeast Asia, launched a sudden attack in the Pacific against British, French and American positions. Completely successful in her many-pronged naval and air campaign in late 1941, and in her advance into the Dutch East Indies, the Island Empire was ready to proceed to a conquest of India; from such a vantage point she could readily form a junction with her triumphant western allies. Between them they would thus intersect and destroy the British line of empire; in firm possession of all the sea lanes leading to western Europe and eastern Asia, they would then dictate terms to the rest of the world.

The situation facing the United States in 1917 had been bad enough; in December, 1941 it was infinitely worse. No convenient landing zone for men and supplies was available to an American force in France, for that nation was securely in the grip of the Nazis; without the French and Italian fleets to aid them, the Allies could no longer blockade Germany as before; submarines roved the seas at will, seriously endangering American shipping along the Atlantic coast and menacing the Pacific coastal area as well. By the terms of the Berlin-Tokyo Axis agreement, we had gotten into not one war, but two, by the attack on Pearl Harbor. That part of our Navy which could have helped defend the Philippines and our other Pacific Islands, lay at the bottom of the sea in Hawaii.

It is no part of our problem to analyze the manner in which this desperate war was fought and won, after the United States became a party. It is necessary, however, to call attention to the foregoing and some other military facts in order to make clear the massive task which American leaders, and American industry particularly, had to face. Logistically we confronted an almost impossible situation, even had we possessed a two-ocean navy, an immense army, and a full arsenal of military weapons, none of which was ours at the impact of the 1941 crisis. Instead of safe bases in the rear of strong military lines in France, we had, in the West, only the British Isles and the

Arctic ports of Russia where ships could land safe from the Axis. To supply the hard-pressed British in the Middle East, meant a supply route of 9,000 miles around South Africa; very soon there might be no British in Egypt to send aid to. On the Pacific side, only a few islands besides Australia remained outside Japanese control, and at the rate she was proceeding, it did not seem that they could remain free much longer. To hold off Japan in the Philippines was not possible, though a courageous rear-guard action was staged there by General MacArthur through the early months of 1942, with what he had. Even Hawaii could have been Japan's in 1941, had she grasped the military weakness of the Islands.

Back in the United States the news of Pearl Harbor was received with a mixture of shock, anger and determination. An attack by Japan had not been unexpected, though the exact time and placed had not been effectively foreseen. However, it must be admitted, we were not as completely unprepared as in 1917. Selective Service had been in force for more than a year; helpful precautions had already been taken, in relieving the British of protection of their Atlantic bases; we were occupying Iceland and Greenland; Lend-Lease, aid "without the dollar sign," had been in force since March, 1941 and shipments to Britain and the Soviet Union had already begun.

* * *

Long before World War II, some preparatory moves had been made in the direction of industrial mobilization planning. An Act of Congress in 1920 had placed the planning responsibility upon the Assistant Secretary of War; four years later the Army Industrial College was founded to train Army and Navy officers in the function of military procurement, and to carry on studies under the Assistant Secretary's direction which would assist him in mobilization planning. By 1931 the Joint Army and Navy Munitions Board formulated the results of these endeavors, as the Industrial Mobilization Plan of that year, a plan which underwent subsequent revisions. As a detail of the last revision of the I.M.P., in 1939, a top-level agency was urged, one which would have complete authority over all phases of industrial mobilization; it was urged further that it should be directed by outstanding business leaders.

Much controversy has ensued over the failure to put this plan into effect, after so many years of effort had been spent upon it. The most likely reason is that the plan assumed a sudden change from peace to war. Since our involvement in World War II was really a gradual one, despite the dramatic nature of Pearl Harbor, several part-way measures were first employed to aid our future allies, outside the limits of official wartime conditions. When the War came in 1939, Assistant Secretary Johnson appointed a War Resources Board headed by E. R. Stettinius, Jr., Chairman of the Board of the United States Steel Corporation, containing members drawn from military, business and scientific circles; its report, drawn up in October, 1939, was not acted upon. The urgency of the situation had not become clear, either to the American people or to their representatives in Congress.

The sudden change in tempo in the spring of 1940 brought Presidential action, although no one could say precisely that we were threatened with attack by the Axis. By use of an Act of 1939 which permitted him to reorganize the executive branch of the Government, President Roosevelt created on May 25 the Office of Emergency Management; making use of the defense legislation of 1916, still in force, he appointed a few days later a Council of National Defense, with a National Defense Advisory Commission. This move was actually a re-creation of the defense mechanism of Woodrow Wilson's time. The Council even repeated the Wilsonian formula to the extent of naming the same cabinet members as before: the Secretaries of War, Navy, Interior, Agriculture, Commerce and Labor. Secretary Morgenthau, as key man in the treasury aid to the democracies, which was later to blossom forth into Lend-Lease, and Secretary Hull, the senior member of the cabinet, might have been added. But this would have required new legislation. In search of a cabinet of "national unity," the President called to the War and Navy secretaryships two distinguished Republicans, Henry L. Stimson and Frank Knox, with Robert P. Patterson and James Forrestal as their respective under-secretaries.

The Advisory Commission drew heavily upon financial and industrial leaders for its personnel. It included Ralph Budd, President of the Chicago, Burlington and Quincy Railroad for transportation counsel; William S. Knudsen, President of General Motors Corporation, as advisor on production; Chester C. Davis of the Federal Reserve Board, as consultant on agricultural and food problems; Sidney Hillman, President of the Amalgamated Clothing Workers, to advise on labor and manpower problems; E. R. Stettinius, Jr., of the United States Steel Corporation, to work on raw materials; Leon Henderson, of the Securities and Exchange Commission, to aid in the control of prices of munitions and foodstuffs; and Harriet Elliott, Dean of Women of the University of North Carolina, as advisor on the participation of women.

Our Government was not unaware of the task which faced American industrial mobilizers for at least three years before we went to war. In 1938 Ambasador Joseph P. Kennedy had reported from Great Britain that British arms production was lagging far behind that of Germany; he added also that the Nazis had made such strides in rubber and petroleum production as to make her blockade-proof in those items. These reports were checked and found to be correct by Bernard M. Baruch, who was sent to Europe by the President to investigate them. The Military Intelligence Division of the Army backed up both civilian diplomats as to Germany's air power at the same time. It reported that Germany had 3,353 heavy and medium bombers; that Russia had between 1300 and 1900; that

France came third with 916; Britain fourth with 715; and Japan fifth with 660. At the same time, while the President was being attacked in some centers as a war-monger, the United States had but 301 bombing planes of all kinds.

A measure to gear industry to possible war needs was passed by Congress in 1939. The Educational Orders Act, for which the War Department had lobbied in vain for many years, appropriated five million dollars a year for five years to aid in the familiarizing of key manufacturers with the kind of war materials they might be expected to put into production on Mobilization Day. Previous to this and again in 1939, the President and the Assistant Secretary of War had tried without much success to secure the passage of bills which would allow the Government to stockpile strategic materials, those which would be in heavy demand in wartime but not usually obtainable in the United States. The arrival of the serious emergency in the spring and summer of 1940 provided some stimulus to the defense effort which nothing else could do, as will appear.

The NDAC, while far from the effective defense general staff which it might have been, having no coordinating head and hence in a confused relationship with other agencies which also had defense responsibilities, did perform some useful acts in its short history, from May to December, 1940. It brought into the picture Donald M. Nelson of Sears, Roebuck as coordinator of defense purchases; he worked out a method by which the Army and the Navy could contract with manufacturers without adherence to the usual requirement of competitive bidding. By this system production could be spread over a wide variety of companies according to the availability of factories and labor supply. Leon Henderson and Nelson together formulated the famous five-year amortization plan, whereby companies which made defense contracts could write installation costs off their tax returns at twenty per cent a year; at the end of five years, therefore, the facilities would be free and clear and be the property of the companies. Since such plants would obviously be useful for a much longer period than five years, it was a great advantage to them to "deflate taxable earnings, while increasing their earning facilities." Even the large profits taxes which would be collected from defense contractors, could be in part recovered, by claims for refunds, to balance losses incurred by them in executing the contracts.

The disaster which the British suffered at Dunkirk in 1940 served one useful purpose. It swept the slate clean, for the British Army had lost all its equipment in the evacuation. It taught them what would and what would not do, in attempting to defend their Island against the Luftwaffe and perhaps the Wehrmacht, should actual invasion come. Out of the bitter experience came improved designs for planes and tanks, and enormously stepped-up requirements. They estimated that by 1941-1942, if Britain should last that long, they would need 3,000 planes per month to fight Germany in the air. Against this total, the United States in August, 1940 was scheduling 895 planes, with only 236 ear-marked for the British. To expedite the production of fighter planes British authorities sent Alex Taub to this country; Taub had worked miracles in helping to re-equip the British Army after Dunkirk; it was he who designed the Churchill tank in sixty days; his mission here was to urge our manufacturers to go into mass production of the British Sabre engine for the very practical reason that they were needed in vast quantity in the air defense of Great Britain. It is possible that this might have speeded up airplane engine production for British fighters, but the project, after some consideration, was not put into operation. It appears that the War Department, which was enthusiastically developing a four-engine bomber program, did not favor large allocation of facilities for British fighters, lest this endanger the long-range American program.

It is fair to say that the NDAC and our military men were heavily involved, in the summer of 1940, with our many problems of manpower and materials. We had begun to contract, in very large figures, with industry; we were confronted with the need to expedite the production of rubber and steel. Powers were given the Reconstruction Finance Corporation to create defense production facilities, either through private companies or by creating government corporations. The RFC set up the Rubber Reserve and the Metals Reserve Corporations to deal with shortages, and a Defense Supply Corporation, to expedite the production of high octane gasoline. As the Selective Service Act of September 16th went into effect and began to withdraw large numbers of men from industry and business, new shortages appeared. As the manpower destined for the armed forces grew to large totals, lack of building materials, clothing, food, fuel and ordnance became greater in the areas in which new cantonments were situated. Since these were scattered in many of the less-congested sections of the South and West, the regional markets did not exist to furnish materials for the training programs. Housing was soon a problem, as thousands of workers moved into the vicinity of airplane and ordnance plants where they were employed; the many agencies charged with responsibilities over housing were not coordinated into a single one until 1942.

The NDAC gave way in late 1940 to the Office of Production Management, headed by Knudsen, and in the following year was set up the Office of Price Administration and Civilian Supply, under Leon Henderson. Before its demise, the NDAC had contracted for more than ten billion dollars worth of defense materials; ships were to require a third of this sum, about $3,300,000,000; defense housing, aircraft and munitions (including tanks and trucks), a billion and-a-half each. The contracts, which were added to earlier British and other foreign orders, called for the production of 50,000 planes, 130,000 engines, 17,000 heavy guns, 25,000 light guns, 9,200 tanks, 50,000 trucks, immense numbers of automatic rifles, machine guns and regular rifles, with ammunition for all of these.

Nearly 600 naval and merchant vessels, forty government arsenals, more than 200 camps and cantonments, some 80,000 miles of road construction (both new and repaired), and clothing and other equipment for 1,200,000 men made up what might be called the first procurement list for total war.

The initial year of defense preparation resulted in rather more than setting the requirements and starting the orders going; it also got some actual production started. By August 80,000 workers had been drawn into shipyards, 50,000 into airplane plants, 18,000 into the machine tool industry. Some 80,000 defense workers were recruited by the defense training program initiated by Sidney Hillman of OPM, and Isadore Lubin of the Department of Labor. This program, set up as a division of OPM, was headed by Channing Dooley of Socony Vacuum Oil and Walter Dietz of General Electric. General Motors' Saginaw Steering Gear Division took up a contract for 25,000 Browning machine guns in June, 1940. It was necessary to study the process at the Colt Arms plant in Hartford, then build an arsenal. The contract called for the delivery of the first gun in eighteen months. The site of the arsenal was picked in November, 1940; production began in April of the next year. The first gun was delivered in eleven instead of eighteen months; by the twenty-first month (March, 1942), when 280 guns were due, the firm delivered over 28,000 and cut the cost per gun from a calculated $667 to $141 each.

In the process of getting ready for the production of 1,000 medium tanks, awarded to the Chrysler Corporation, it was soon evident that while industry was willing to make what was wanted, OPM was not equipped to give exact specifications. The result was not a superior tank, but some improvements, such as increased weight and thickened armor plate, over the tank which had performed badly at the Battle of Flanders in the early spring of 1940. There was trouble with the tank treads and the cooling mechanism; springs were specified which, it was later discovered, had been discarded years before in railway car equipment. Eventually the engineering design was accomplished by the Chrysler people them selves.

Trouble came with the order to cut automobile production twenty per cent beginning August 1, 1941. The discussion here revolved about the matter of converting automobile plants into munitions factories, which the automobile makers did not want, though they were quite willing to execute munitions contracts outside their auto plants. There was a question, argued on both sides, as to whether the machinery of such plants was suitable for plane and tank production.

Shortages had by early 1941 become so acute that priority orders began, issued by OPM's division set up for that purpose, on March 23, 1941; the first order required producers of aluminum to put defense orders at the top; this order was soon followed by others, as contractors, loaded with defense orders, began to besiege the OPM with requests for assistance in securing needed raw material. Copper, iron and steel, cork, chemicals, nickel, rayon, rubber, silk were soon placed on the list; this responsibility of assigning priorities was transferred on August 28, 1941 to a new agency, the Supply, Priorities and Allocations Board, which was to decide which of the scarce materials would go to the Army, the Navy, the Maritime Commission, for Lend-Lease and for civilian necessities.

* * *

When Pearl Harbor came in December, 1941, the United States had already been alerted to the possibility of war; moreover, it had begun a considerable overhauling of its civilian governmental organization; it had made a start in industrial mobilization and production. The picture was confused, for many organizations and many individuals were exercising seemingly helpful functions, yet not much seemed to happen. Shortages were nation-wide, industry was arguing with the government agencies, which were in turn arguing among themselves. Mobilization plans had been made, revised and discarded. Defense agencies had been set up and soon demolished or superseded. Central authority for gearing the economy to a giant war effort was lacking. Many of the administrative problems of the defense period would remain unsolved right through the war period.

But the time for arguing was now over, for we were in the War up to our necks, with three major powers. There was no longer any question as to whether it would be better to make tanks inside or outside of a plant equipped to make automobiles. The answer was provided by an Island Empire in the Pacific: we were to make them in both places and in lots and lots of other places. Thus Pearl Harbor did one good thing; it stopped hundreds of silly arguments, and reminded one and all that the time for action had come. Organizational changes continued: OPM evolved into the War Production Board, under Donald Nelson; WPB approached more nearly the status of a general staff of industrial warfare than anything that had appeared; under its supervision was accomplished a large proportion of the production of World War II. On May 27, 1943 the President centralized this function still further by creating the Office of War Mobilization under James F. Byrnes. But before we had reached the end of 1943 war production had attained its peak.

The productive phase of the War was won between 1940 and 1943, in spite of frequent bad management on the governmental level, which, it is true, was affected by the political necessity of moving only as fast as the people of the United States allowed their Government to move; the tempo was slow between September, 1939 and June, 1940; the pace quickened after the fall of France; after Pearl Harbor the war effort raced, not always efficiently but on the whole effectively.

The automobile industry was hard at work by the beginning of 1941. By January of that year Packard, General Motors,

Studebaker and Ford were working on plane engine projects. This represented a break-down of the giant plane program, in which the motor car corporations contributed to the flow of sub-assemblies to the great bomber plants. The Ford Company worked with Consolidated Aircraft; Chrysler, besides carrying on its tank project, worked with Goodyear and Glenn Martin. General Motors furnished engines for North American. The Willys Company worked on the "jeep" and the auto industry in general did considerable work on ordnance, armored cars and army trucks. The Timken Roller Bearing Company turned much of its capacity over to the making of armor plate. Houdaille Hershey, makers of shock absorbers and other automobile parts and supplies, did an important job on machine guns; they eliminated costly tooling which the army had believed would be necessary for the operation, and developed a new gun handle which saved aluminum, an extremely scarce item. Substitutes appeared: plastics for steel, stainless steel in place of chrome, copper instead of brass. Sometimes these substitutes were helpful; often they caused shortages in the substitutes themselves.

The steel shortage was an early and a long-continued one. It seems likely that OPM, in its period of control, did not have adequate figures as to the flow of steel or as to inventories on hand, for its officials insisted well into 1941 that supplies were adequate. Priorities were too readily granted; some steel manufacturers found that all steel orders carried some kind of priority. It was this situation which led to the creation of the Supplies, Priorities and Allocations Board in August, 1941. Steel requirements, already under severe strain from the tank and plane programs, also received great pressure from the Maritime Commission, which was planning a program of 1200 Liberty Ships to add some 13,000,000 tons to our merchant fleet. By the spring of 1941 nine new shipyards and 131 new shipways were being built. In January, 1942 the President set our ship-building requirements at eight times the 1941 figure, a total which was met. The time needed to build a ship was reduced from 355 to fifty-six days. One of Henry J. Kaiser's yards built a ship in fourteen days, an astonishing feat, which may have had something to do with Kaiser's ensuing difficulties in obtaining steel. This deficiency disappeared when the RFC aided Kaiser to set up his own steel mills in the West. Kaiser had the foresight to transform ship-building largely into a production line operation when he caused the cutting, shaping and fabricating operations to be done behind the shipways, as the steel moved up to the hull.

The dislocations occasioned by the cutting-back of civilian automobile production were largely remedied by the contracts awarded to Ford, General Motors, Chrysler, Hudson and others; yet there were hundreds of communities which were faced with "priority depressions" just before Pearl Harbor because their plants, often small ones, were not getting contracts or sub-contracts from the defense authorities; those who received them could often not get parts or materials. A Division of Contract Distribution was organized in OPM under Floyd B. Odlum of Atlas Corporation to deal with this difficulty. Manufacturers in distressed areas made pools of plants or of supplies to handle defense orders. Machinery in limited use was placed at the disposal of other firms; travelling exhibits were sent out to acquaint prospective contractors with what was needed. In one case thirty washing machine manufacturers took over a twelve million dollar machine gun project; three of the firms became prime contractors, the others sub-contractors. Stove companies, tool firms, jewelry manufacturers and many other groups organized industry-wide pools. At one time more than a hundred communities had distribution pools in operation. One artillery contractor asked for more than five million dollars to re-tool his plant for gun manufacture. The OPM found that in the area where this manufacturer operated, were other commercial facilities not yet involved in defense work which all together could make 118 of the 121 parts needed for the gun. So the prime contractor was tooled up for the three remaining parts and asked to sub-contract with the others for the rest.

Lend-Lease, in effect by March 1941, put heavy pressure upon the requirements of our own armed forces, which were growing fast. The Soviet Union, which was made eligible to receive Lend-Lease equipment immediately after the attack on her by Germany in June, 1941, presented very large requests for high octane gasoline, along with other items. About the same time the British discovered that they had underestimated their own requirements for the same item. Oil companies began installations for refining the high octane grade, and refiner equipment firms joined the rush to secure material priorities. Harold Ickes, as Oil Administrator, began to push shipments abroad as fast as they could be gotten together. Ickes engaged in a tremendous controversy with the railroads over his pipeline project, but Ickes won and the pipeline to the East Coast was completed, supplementing in a strategic manner the efforts of the railroads, which were choked with coal and food shipments.

Immediately upon the declaration of war, the President stipulated what the year 1942 would mean in war production: 60,000 planes, 45,000 tanks, 20,000 anti-aircraft guns, and 8,000,000 tons of shipping. The response of industry was notable. In 1941 only about fifteen per cent of our production went for munitions; by the end of 1942 more than half of a much greater total was for war production; while this went on, manufacture of civilian goods fell by more than twenty-five per cent. Whereas OPM the year before was contracting at the rate of about $1,500,000,000 a month, the military appropriations for 1942 reached the colossal total of one hundred billions, and war production for that year, at controlled prices, actually exceeded $30,000,000,000. Momentum had begun with a vengeance, but with momentum came confusion. The promised coordination of our production effort did not appear under the War Production Board, and the position of Nelson of WPB was seriously weakened by the instituting of the Office of Economic Stabilization in October, 1942. James F. Byrnes, as administrator of the new agency, was a "director of a director," since his new office placed him over Donald Nelson of WPB.

In late 1942, North Africa was about to be invaded, and scrutiny was being given the defects in our mobilization by the Truman Committee of the Senate. In mounting our first large-scale invasion, we had to have extensive and continual support from industry. The production effort, well-started, was involved in a mass of requirements, priorities and shortages which combined to hinder the delivery of war material at a designated spot in a given time, and gave little assurance that the invasion could be heavily supported for the period needed. Eliot Janeway, in his *Struggle for Survival,* gives marked credit for improving the domestic industrial situation to Ferdinand Eberstadt, head of the Army-Navy Munitions Board; Eberstadt came up with a Controlled Materials Plan, promptly adopted on November 2nd, which concentrated upon control of the flow of three essential materials: steel, aluminum and copper. As a result, according to Mr. Janeway, orders were placed "for finished products in full and precisely calculated balance of all the parts, components and materials involved in producing the finished product in question." At least, 1943 beheld a doubling of the munitions output, to nearly sixty billions; 1944 carried the total well over that figure. By that time, of course, the "battle of production" had been won. The problem, by then, was reconversion and demobilization, problems just as knotty and vexing as the great build-up which had preceded.

No historian in short space could possibly analyze in detail the massive economic effort which was put into our productive machinery as it moved to a peak in 1943. Figures do not tell the story; nor does the record of the rise and fall of agencies and administrators shed very much light upon the why and the how of it all. Yet the totals give a fair idea of the tempo and the intensity of our economic mobilization. The task was not merely to raise and equip a vast Army, Navy and Air Force; it involved as well furnishing the means for keeping millions of British, Russians, Chinese, French and others fighting; there was also the responsibility of allocating enough of the wartime production to civilians in order to prevent the loss of morale at home. The fact that all these objectives were accomplished in time to save our nation and many another from destruction, is probably the greatest industrial miracle in history.

By the middle of 1944, the production level attained by the United States and the British Commonwealth together was four times that of the Axis Powers. At the time of Pearl Harbor we had 1150 combat planes and about the same number of tanks. In five years, from 1940 to 1945, American industry produced 86,338 tanks, 297,000 planes, 17,400,000 rifles, 315,000 pieces of artillery, 4,200,000 tons of artillery shells, and 41,000,000,000 rounds of ammunition. In the shipping field we built 64,500 landing vessels, 6,500 Navy ships, and 5,500 cargo ships. Our synthetic rubber program was producing 836,000 tons per year, fifty per cent more than our pre-war crude rubber importations. Lend-Lease aid was furnished to the total of more than forty billion dollars; in reverse we received from our allies goods and services valued at over five and-a-half billions. Even these few figures prove the effort so huge as to make all other production statistics seem puny by comparison. Nowhere had people ever accomplished such a productive feat in anything like a similar period. Very few persons, in government or out, really believed that such results could come anywhere near realization.

There has been a large amount of speculation on this matter. Much of it has centered about the leadership or lack of it by President Roosevelt; other historians have directed attention to this or that individual in the productive picture, and have showered high praise on some and focused condemnation on others. One point is sometimes missed by historians, even by those who seek to determine the forces which enabled us to do this amazing productive job:—World War II, for all its confusion and turmoil was a people's war against oppression. One would need to talk to the millions of Americans who worked and fought and sacrificed throughout its grinding course, to find the real reason why we won. If the people of this nation had not thoroughly believed that this was everybody's war, yours and mine, the effort which opened the eyes of all, which shocked and ruined dictators, would never have been possible. It was the men and women of America, thoroughly aroused and determined, who filled the airplane plants, toiled long hours in the munitions factories, pushed the job along in the shipyards, and put their savings away for the future. Fortunately we had a mechanism, the most powerful and successful ever to be developed by man, to put to service for war. That mechanism was American industry. The workers of America, manning this great machine, provided the spirit and production which won the war.

But one more element was needed, and without it we might have failed in this supreme test. Experimental science had to work at top speed, for World War II was not only a battle of production, but also a test of the scientific skill and ingenuity of the contestants. As Arthur M. Schlesinger well phrased it, scientific achievements which "would otherwise have taken decades" had to be telescoped into a few years. It was well-known that Germany had made sensational advances in new weapons; dire threats of newer and more deadly secret weapons had been made, with what we know to have possessed strong possibility of success. While it is a serious error to simplify the story of the war into a race for scientific superiority, we are quite aware that German rocket development was so well-advanced in 1944 as to endanger the success of the Normandy invasion.

Probably the most important part of our scientific mobilization in the War revolved about the amazing activities of the OSRD, the Office of Scientific Research and Development, headed by Dr. Vannevar Bush and enjoying the assistance of dozens of the most brilliant scientists of the nation. Crucially necessary devices such as radar, submarine-detecting devices, flame throwers, radio proximity fuses, means for jamming enemy radio and radar were developed or adapted to American military use by scientists of the OSRD and the many cooperating agencies which worked with them. A vast and very necessary project was carried on by technical units in our combat forces, in collecting and sending back to this country thousands of items of captured equipment; these were studied and analyzed and from them came many improved designs. Less spectacular, but just as important as the work of the OSRD and its army of scientists, was the work done continually in the laboratories of industry, as engineers designed better airplane and marine engines, more rugged tanks, and vastly improved land and marine transportation equipment.

The OSRD also performed wonders in military medical aid, by putting penicillin into quantity production, in building up production of the insecticide DDT, so important in alleviating malaria in the Pacific and the Italian campaigns; it helped, in collaboration with the Red Cross, to make available 5,000,000 pints of blood plasma in 1944. Most sensational of all scientific achievements of the War, and perhaps of the century, was the production of the atomic bomb in time to have important military use in the Pacific. The atomic project was a joint enterprise, financed by the United States to the extent of two billion dollars, participated in by Americans, British, Canadians and exiled European scientists. A brilliant piece of American scientific intelligence was the discovery in late 1944, that Germany, despite frantic efforts, had made no practical progress on an atomic bomb; by that time, of course, the success of the American atomic project was almost assured.

The lessons to be gleaned from our economic mobilization effort are many. One which is seldom high-lighted is that it is possible to initiate defensive measures of far-reaching importance in time of peace. We did learn to plan, and in many ways to plan well, even though it took us until 1942 to put plans in workable form. Another lesson which should have been learned is that the capacity of American industry to produce should not be underestimated. Our production limits had not been reached in World War II when cutbacks and reconversion began; this is a fact for engineers, politicians and potential enemies to ponder on. What seems to be the most valuable lesson of all is the final proof, if proof were needed, of the effectiveness of a team composed of the scientists, the industrialists and the military, supported by the determined workers of America. This is the team which won the War, regardless of what happened in Washington; it was a team of men and women who understood the partnership which must exist in wartime between the laboratory, the production line and the fighting front.

CHAPTER FOURTEEN

A YEAR OF INDUSTRIAL SCIENCE

One of the most arresting qualities about our industrial system is that it is never completed. It cannot be, because the material universe from which its strength is drawn is in a constant state of change. Moreover, techniques which seem excellent today will in many cases be replaced by others tomorrow. The scientific miracles of 1953 will become standard procedure a few years or months from now. Even as we pen these words, industrial science is pushing into the background many of those matters which we have been describing as current and new. In the drafting rooms, the laboratories and the pilot plants of industry, new processes and products are constantly appearing; from these sources will come new plastics, better chemicals, stronger metallic alloys, improved transportation devices, new sources of power. Industrial science is dynamic, not static; nothing is so good that research may not uncover something a little better. That is the way the science of industry functions.

When World War II came to an end in 1945, many business men feared that a sharp recession would follow the demobilization of men and materials. Since the American phase of the War was infinitely larger than in 1917-1918, the removal of the vast markets provided by global war held the possibility of a serious jolt to our economy. But this did not happen, for a number of reasons. In the first place, demobilization did not take place with the rapidity of the earlier war, and it was in no way as complete as before. Furthermore, hardly had the partial conversion to a civilian economy gotten under way, than international political conditions halted it, as the United States and the other free nations were compelled to deal with the facts of the "cold war." Even before this novel situation arose, military occupation in Europe and the Pacific, with the supply system which it entailed, rendered necessary the continuance of a part of our military industrial economy. In addition to these factors, the broadening of the cold war into a shooting war in Korea brought actual military remobilization and stimulated the production of ships, airplanes, automotive units, and all the supplies needed to maintain a large combat army; raw materials to back up the manufacture of finished products were likewise produced in greater amounts. In other words, we now live in about as much of a war period, so far as the gearing of a large proportion of the economy to military items is concerned, as we did in 1917-1918, though this has been carried on with less fanfare than in those days.

This is not the entire story, however. In the planned demobilization of 1945 and after, there were signs that we had begun to appreciate the danger of stripping ourselves of protection immediately after a major war. Research in atomic weapons, in supersonic aircraft, in guided missiles, all continued at a commendable pace after 1945. The changeover to civilian production between 1945 and 1950, was accomplished, in great part, without grave disturbance to our industrial system. Civilian goods and services have continued to improve in quality and to increase in volume, despite the persistence of international danger, accentuated since 1950.

Take, for example, the year 1952, with the unprecedented

situation of having two economies, a peacetime civilian production system and a partial wartime production system, operating side by side, often in the same industrial plant. Since this situation had existed for nearly three years, it was being accepted and planned for by government and industry alike, because no industrialist, or political leader either, could tell positively how long it might continue. In many ways this kind of industrial effort presents greater problems than a total war economy, for the production of civilian goods must be constantly increased at the same time that the levels of military production have to be kept high, and possibly increased as well. To accomplish such a task requires exceptional skill by management and poses heavy problems at the high government level.

The General Industrial Picture

The year 1952 was one of marked business activity. Retail sales were high, the total of 165 billions being an increase of three and a half per cent over the previous year. Foreign trade reached a level close to the record high figure of 1951; in the export field, of course, military shipments bulked large and served to offset declines in purely commercial exports. The steel industry, hampered during the year by strikes and difficulties in obtaining materials, was expected to reach its defense expansion goal of 120,000,000 tons annual production capacity about the middle of 1953. All of which would have meant that by that time the industry would have added nearly 20,000,000 tons of steel-making capacity since the Korean crisis began in 1950. Even during 1952, though production fell below the two previous years, steel manufacture exceeded any World War II year, or any year since the War, except 1950 and 1951. This was accomplished despite the longest and most expensive strike in the history of the industry.

In the production of electric power, it has been estimated

that the United States produced nearly half of the current generated in the entire world during 1952. The total of 400 billion kilowatt hours represents two and-a-half times as much as the nation generated the first year of World War II, and 150 times the amount generated by the electrical industry a half century ago. It is expected that by 1955 the private electric utility companies will have doubled their capacity in the decade ending that year. An interesting feature of this enormous increase, speeded up since the beginning of the Korean War, is the increase in the size of power plants, some of which are of 12,000,000 kilowatt capacity. Three power projects, each of 3,000,000 kilowatt potential, are being constructed to serve the nation's atomic energy installations. Meanwhile the average cost per kilowatt to residential users has been decreased from five to three cents over a period of ten years.

The gross earnings of the railroads of the nation in 1952 passed the ten billion mark for the second consecutive year. Net earnings also increased to the highest point in twenty years, except during the exceptional war years of 1942 and 1943. An important factor in the increase in net earnings, achieved despite wage increases, has been the high efficiency of the Diesel-electric locomotive. According to the Association of American Railroads, Diesels were used for 65 per cent of the freight service, 70 per cent of the passenger service and 75 per cent of the yard and switching operations, during 1952. In addition to modernizing railroad equipment, improvements were made in freight service. The Atlantic Coast Line and the Seaboard have each cut a day off their fruit and vegetable forwarding schedules from Florida to the Eastern markets. Passenger service showed a serious deficit for the year; remedies for this may lie in discontinuance of trains which do not pay, and improved service on long distance runs, which are usually profitable; modification of ICC rate and operational controls would be helpful, railroad executives believe.

The expansion of communications continued at a fast rate. The American Telephone and Telegraph Company and its associated companies added one new telephone during every four seconds of each working day, on the average, during 1952. It is planned that ultimately all telephone users will be able to dial their own calls; eight of each ten of the Bell telephones were dial-operated during 1952 and forty per cent of the long distance calls made were being dialed directly by the original operator to the telephone called. New major switching centers were added during the year, making eighteen such centers from which operators can dial some 1700 towns and cities. Fourteen more cities were brought into the national television network by the extension of coaxial cable and radio relay microwave systems. Telephone service was extended from the United States to ten points in the world which had formerly been without it. It is said that 96 per cent of the world's telephones can be connected through the Bell overseas service. Of great importance in the field of communications were the extensive contracts being filled for the armed forces, which included facsimile, radio carrier and automatic transmission projects.

The automobile industry showed the effect of our partial change to military production. The Automobile Manufacturers' Association estimates that the total production for 1952 was just over five and-a-half million passenger cars, trucks and buses, a cut of 18 per cent below 1951, which represents two and-a-half million fewer vehicles than were turned out in the peak year of 1950. Government restrictions on the purchase of steel, copper and aluminum held down production of civilian automobiles. This stress was alleviated somewhat by cutbacks in the delivery dates of military items, which permitted release of strategic metals; the use of imported steel, and a liberal government policy of allowing manufacturers to carry over unused allotments from quarter to quarter, and to borrow against forthcoming allotments, also eased the situation. By these means, serious unemployment which might have struck the great industrial centers of Detroit, Flint and Pontiac was partly prevented; such layoffs as did occur were due chiefly to the steel strike, which stopped nearly all automobile manufacture for six of the eight strike weeks. Total earnings, however, held up well because of the income from defense contracts, retooling for which was practically complete by 1952. General Motors reported defense sales of more than a billion dollars for the first nine months of 1952 and the Chrysler Corporation increased its defense sales in the same period to more than four times the total for the corresponding period of 1951.

Aluminum production was greater during 1952 than in any previous twelve months in history. The total production of 1.87 billion pounds might have reached a higher total but for a costly drought which cut heavily into available power production. In the expansion program of the industry, which has necessitated the building of new smelting plants and the enlargement of others, it is expected that output will climb to 2.5 billion pounds in 1953 and to three billion in 1954. Even with this expansion, some of the effects of which were already shown in 1952, the metal was in short supply for the year. It was necessary to "borrow" British aluminum from Canada and to halt for four months the domestic stockpiling program. The Defense Production Administration has asked the country for a third installment of aluminum for defense purposes in the amount of 400,000,000 pounds; it has contracted with Olin Industries and the Harvey Machine Company to enter the aluminum business in order to secure a part of this added output. Reynolds Metal Company and Kaiser Aluminum and Chemical Corporation are both developing large new bauxite ore resources in Jamaica, British West Indies.

Agriculture had a favorable year, with grain crops and the materials for meat, poultry and dairy products all registering high production levels. In quantity the crop total was the second largest in United States history; the acreage under production was smaller, but average yields were higher. The cattle herd increased to 93 million head, a gain of five million for the year. Farmers earned a net income of about fourteen billion dollars, about the same as in 1951; the rise in volume of production offset slight rises in operating costs and small declines in prices. The main problems which face agriculture are a continued decline in commercial exports, and, closely related to that unfavorable trend, an expectation of mounting surpluses. A factor still unknown is the question of what sort of price support may be expected from the subsidy policies in Washington. Leaders of the industry are said to be viewing the future with moderate optimism, with ample supplies of foods available resulting from the bumper crops of the year.

Construction projects reached a total of 42 billion dollars and there appears to be an expectation of an increase in this type of activity. An official of the Johns-Manville Corporation has stated that this huge total can be divided into four almost equal categories. Private residential construction accounted for about one-fourth, in the building of over a million new homes; private non-residential projects provided another large section; public buildings a third part; and finally, modernization of existing structures, or maintenance and repair construction, was about equal to the others. An important factor in the continued high level of construction work was the decision of the government to stretch out defense expenditures; this policy removed partially the pressure on materials and labor, and allowed housing capital to finish much construction which could not otherwise have been completed in 1952.

The rubber industry had a successful year. All controls except those on pale crepe rubber were lifted in April, 1952, rendering the industry almost completely free of government restriction for the first time in eleven years. Consumption for the year was estimated at 1,250,000 long tons of new rubber, a figure only slightly under the year 1950, when the Korean crisis brought on a wave of "scare buying." The above figure included about 800,000 tons of the synthetic type and 450,000 tons of natural rubber. In addition the industry used about 280,000 tons of reclaimed stocks.

One important phase of reconversion in the rubber industry remains to be accomplished: the transfer to private hands of the government-owned synthetic rubber plants. Action on this matter has been delayed by Congress, which has insisted that the nation must first have a stockpile adequate for any new emergency, and also a synthetic industry which can maintain itself. The Rubber Manufacturers' Association maintains that these conditions have now been met, that the synthetic capacity is now near 1,000,000 tons annually, and that the national stockpile is sufficient for a five-year war without resorting to importation. Action on the matter began in the spring of 1953, after a report to Congress by the Reconstruction Finance Corporation.

As would be expected, tires led the rubber group in 1952, with 73 million units for passenger cars, a rise of seven million over the previous year. High demands for truck and passenger cars still existed during 1953. Belting and hose are also in high demand, as are other mechanical rubber goods. The total sales for the industry were established at five and-a-half billion dollars, with indications of increase in 1953. The high production of synthetic rubber has continued to provide insurance against the instability formerly characteristic of the rubber market. Gone forever are the days when rubber could vary in price from three cents to a dollar and-a-quarter per pound.

The direction of the chemical industry can hardly be other than upward, in the opinion of Charles S. Munson, Chairman of the Board of the Air Reduction Company and head of the Manufacturing Chemists' Association. He stressed the probable increase of production levels, announced the end of the sulphur shortage, and predicted the appearance of many new products. Chemical industry, he said, "must process wastes to stretch our wealth, lengthen the life of materials we now use with coatings and preservatives, help solve the mysteries of disease, and comb the sea and the earth and the soil for new resources which are as yet unknown." With the past record of the chemical industry in mind, one can be relatively sure that Mr. Munson is not overstating the mission or the expectations of chemical industrial science. The index of chemical production stood at 574 in October, 1952, with 100 as the average of 1936-1939; this level compares with 412 as the peak of World War II production.

Wages were pushed to new highs during the year 1952, a period in which more than twice as much industrial work time was lost through strike activity than in 1951. Net profits averaged about four per cent of each dollar of gross income for the year, the lowest rate since 1932. Most business prognosticators believe that no very startling developments are visible to change the general trend of well-being which pervades business. Government spending will doubtless continue at about the same rate, consumer purchases may increase somewhat, and the customary investments in new plant and equipment will continue. Improvements in profits might occur if the remaining price controls are lifted and if excess profits taxes are eliminated. Income and excess profits taxes took about sixty per cent of corporate income for the year. Certain factors, the exact direction of which is still unknown, will affect the future of American industry and business. Of these the most important are the international situation and the future policies of the administration governing our destinies. If increased efficiency in public administration and cutting of waste in expenditures can be made, substantial cuts in taxes can soon be accomplished, with beneficial effects to business and the general public. If war comes, or if the present war tension situation sharpens, it is impossible to gauge with any certainty just what the effect will be.

There is a wider and healthier distribution of industrial

activity throughout the United States than ever before. The South and the West are becoming large centers of industry, a trend which has helped to distribute the national income better, and which has strengthened the economic position of many of our states, some of which were for many years in a backward and undeveloped state. Gone are the days when New England, the Eastern Seaboard and the Great Lakes area dominated the financial and industrial activities of the nation. Among metropolitan industrial areas, the Los Angeles region, once ninth in rank, has climbed to a position to rival Philadelphia; in other words, close to fourth among industrial areas; in number of workers employed in manufacturing, it is slightly ahead of Philadelphia. Florida has doubled its manufacturing totals in the past seven years. South Carolina has set up nearly 1000 manufacturing plants in the same period.

Milestones of 1952

Technology races on with dizzy speed. In no field was this more evident than in air transportation. For several years the United States Government, in collaboration with aircraft companies, has been working on the modernization of the World War II military airplane. Such plans have coincided with a revolution in aircraft science which makes it difficult to carry on a long-range air mobilization, for engines designed for the air are changing so rapidly that models effective a few short years ago now approach the obsolete stage. The very heavy bombers developed at the end of World War II, especially the B-36's, are exceedingly expensive machines; to discard them would be dangerous and would necessitate a large and very quick program of replacement with more modern machines. Furthermore, strategic bombing has not yet advanced to the supersonic stage, and even bombers of less than 500 mile per hour speed are still extremely important as offensive military weapons. However, it is necessary, even as modernization of the heavier bombers goes on, to develop, as fast as possible, modifications of these bombers, with more modern power units. The Consolidated Aircraft Corporation has built and tested for the Air Force the YB-60, a jet-powered adaptation of the B-36 bomber. This supersonic bomber was received by the Air Force and official tests upon it begun before the end of 1952.

The same company, in cooperation with the Navy, has also developed the F2Y Sea-Dart, with delta wing and jet propulsion, having also retractable skis; this is a striking attempt to combine the versatility of a water-based plane with the performance of a land flyer. The F2Y was first tested in December, 1952. New engine development has marked the recent activity of Continental Motors Corporation; it has acquired American rights to the Turbomeca group of gas turbines; its Marbore 352 engine will furnish power for the new Cessna jet trainer which is intended to help fill up the speed gap between the present trainer, powered by the conventional reciprocating engine, and the jet fighters which the Air Force fighter pilots will eventually fly. Continental has also designed types of jet engines to furnish supplementary power for commercial transport planes. In this way one jet engine can power a twin-engine plane while its standard engines are cut off. The extra engine also is said to improve take-off performance, especially when fully loaded, and in hot weather, when the thinner air gives less wing lift.

Underlying all these developments, as we have seen, is the rapid increase in speed due to military requirements. As speeds from 1000 to 1500 miles per hour have become distinctly possible, and as other advantages of jet propulsion begin to attract the commercial field, the pilot has almost ceased to guide the machine, and has really become a monitor of the electronic instruments which do the guiding. In the field of guided missiles, in which many of the aircraft companies perform research in cooperation with the Armed Forces, the place of the pilot has been entirely taken over by electronic equipment. It is an interesting comment on the progress of air transportation to realize that if a box were constructed just large enough to hold a B-36 bomber, the first power flight by man could take place entirely within that space.

It is a mistake, however, to assume that air-minded passengers will soon be whirled through space at supersonic speeds. For the ranges of flight required by commercial aviation today, the sub-sonic reciprocating engine is still the most practicable power source. Turbo-prop and turbo-jet engines do not as yet have the effective durability of the reciprocating engine, and they cannot begin to match it in reasonable cost. The place of the jet engine is still mainly that of a power unit for the military fighter plane, and is now being adapted to the military bomber. Faster engines still are under development; the "ram jet" engine, a project of the Wright Aeronautical Corporation, is expected to be capable of speeds of 2,000 to 2,500 miles per hour; its niche will be as power for guided missiles.

In late August, 1952 the press carried an account of a gigantic aluminum project under consideration by the Aluminum Company of America. The locale is a mountainous area in the Chilkoot Pass region, made famous by the Alaskan gold rush of 1897-1898. At a cost of $400,000,000 the new facility would provide year-round employment for 4,000 workers and to produce 200,000 tons of aluminum each year. At an altitude of 2,200 feet above sea level, the water of five Canadian lakes would be dammed; from the huge artificial lake thus formed the water would pass through a thirteen mile tunnel. At the tunnel's end, the stream would drop down eight penstocks to an underground powerhouse where, it is planned, would be eight generating units of 100,000 horsepower capacity each; from this point the water would pass through a second tunnel just under eight miles in length. after which it would drop down 1,100 feet more to a second generating area at sea level, where 800,000 additional horsepower would be developed. The "Taiya" project, as it is called, will necessitate legislation by the United States, there being at present no legal method of acquiring the necessary 20,000 acres of Alaskan land near Skagway ;it will also require the approval of the Canadian authorities, since the water to be impounded lies in the Dominion, as does the entrance to Tunnel No. 1, although the aluminum factory and the underground powerhouses would lie in Alaskan Territory.

The answer to the international difficulties posed by the Taiya plans may be found in United States aid to a similar Canadian project at Kitimat, on an inlet of the Pacific almost due north of the upper point of Vancouver Island. Here the water of an immense inner drainage area will be led toward the Pacific by damming up a number of British Columbia lakes. At the western end of the reservoir thus formed, a tunnel ten miles long will be drilled through the Coast Range; the water will proceed to a point where by a drop of half-a-mile, an underground powerhouse will develop 1,650,000 horsepower for the gigantic aluminum plant which is rising at Kitimat; this tiny village will grow to a great city as the construction rushes forward. The relationship of these two projects to each other, though in some ways competitive, may be mutually helpful if, by a statesmanlike business diplomacy, Canadian water rights for Taiya may be exchanged for United States tariff concessions and enlarged aluminum orders for Kitimat; here is an opportunity for the United States and Canada to pool their respective advantages, not only for increased prosperity of Alcoa and Alcan (Aluminum Company of Canada), but for the security of both nations. The successful defense of the western world depends in no small degree upon a very large supply of this vital metal.

Important developments are afoot in the field of atomic energy, both as to its military applications and in its possible commercial industrial possibilities. Of primary importance, according to William L. Laurence, a leading authority, was the testing of the first hydrogen bomb at Eniwetok on November 1, 1952; this test appears to have established the truth of what scientists had suspected, that such super-weapons could be made of almost any desired power, even to that of bombs 1000 times more deadly than the atomic fission bombs used at Hiroshima and Nagasaki. To produce such stupendous weapons, three ingredients are needed: first of all, superior fission bombs, such as those of World War II, but better, as "triggers" or detonators of the hydrogen bomb; secondly, large quantities of double-weight hydrogen, or deuterium, which is obtainable from water; third, triple-weight hydrogen, or tritium, which is not found in nature and will have to be (or has been) prepared by modern chemical industrial means. In order to further this project, a plant has been constructed near Aiken, South Carolina.

As we have noted earlier, the keel of the first atomic submarine, the *Nautilus,* was laid in 1952; the craft will probably be delivered to the Navy sometime in 1954. Plans to apply atomic power to military aircraft and aircraft carriers are in the developmental stage. It has been generally agreed that since nuclear reactors produce enormous heat, this by-product can be readily used to develop large amounts of electrical energy; but it would not pay any power company at the present time to convert its power generating machinery to the nuclear type, because of its prohibitive cost. Yet it is interesting to note that studies are underway, administered by a section of the Atomic Energy Commission, to inquire into the possibilities of vast power development by utilization of the additional energy available from nuclear reaction.

For more than a year several industrial teams have been at work upon this problem, described in 1951 by the AEC as "the practicability of business and industry building and operating reactors for the production of fissionable materials and power." The initial groups of industries to begin these studies were: Monsanto Chemical Company with the Union Electric Company of Missouri, Dow Chemical with Detroit

Edison, Commonwealth Edison with the Public Service Company of Northern Illinois, and the Pacific Gas and Electric Company with the Bechtel Corporation. In May, 1952, the AEC set up an office to administer the "industrial participation" program; later in the same year the project was greatly expanded by the addition of eleven other corporations to the Dow-Edison group and by the setting up of a fifth study group. The continued broadening of the organized study of nuclear physics is a well-known trend in education and may be followed from day to day in the press. The Oak Ridge Institute of Nuclear Studies has announced that it will offer as many as seventy-five fellowships in radiological physics for the academic year 1953-1954, sponsored by the AEC and under the supervision of three leading universities.

A Glance at the Future

Sound economic thinking recognizes that the industrial progress in the United States will in the long run be strengthened by the accelerated advance of neighbor countries in their own economic well-being. The shrinking of the physical world into a single great community has come with the progress of technology, with the result that no people can expect ultimate profit to result from the misfortunes of their neighbors. It is of the highest significance that the Dominion of Canada attained a record national income of 23 billion dollars in 1952, and that the smaller, yet fast-expanding economy of the Mexican Republic indicates a strong trend toward industrialization; two-thirds of our southern neighbor's five billions of income came about through the proceeds from industry other than agriculture. Similar trends are in process in smaller or greater degree in the other states of the western hemisphere.

Competition with the other American nations should not harm the industrial economy of the United States. Basically, the move into a stronger and more prosperous economic situation should create greater stability within the various states of the hemisphere, and in time should remove, on the part of any of their neighbors, the temptation to interfere with one another's affairs, a temptation always present when raw materials of great value can easily be bargained away by a weak and short-sighted government. Hemispheric peace, both political and economic, with the fullest commercial interchange on a basis of equality, can benefit all the nations of the Americas as no other factor can. As the less-developed nations become more and more complex industrially, the living standards of their people should continue to rise. New needs will appear which the older economy of the United States will help to supply. The economies of the American states will continue to supplement one another. The United States will need the vast pulpwood, mineral and power resources of Canada, and that nation will continue to require the mechanical equipment, cotton, textiles, chemicals and fertilizers obtainable from the United States. Our processing and manufacturing enterprises absorb huge quantities of the copper of Chile, the petroleum of Venezuela, the sugar of the West Indies, the coffee and fruits of Central America. In the nine months from January to September, 1952, foreign trade between the United States and the Latin American countries reached a total volume of more than five billion dollars, about evenly divided between imports and exports. The volume of trade between Canada and this country equals, roughly, the total between the United States and all the other American states. To a very large extent, new industrial developments in the Dominion of Canada and in Latin America are financed by United States capital. According to a census taken at the end of 1950 by the United States Department of Commerce, the direct foreign investments of the United States in hemispheric projects reached a total of 8.2 billion dollars, or about seventy per cent of all foreign investments made by citizens of this nation.

It is quite important, in all this hemispheric interchange, to keep any tinge of political interference out of wholly natural, helpful and friendly economic cooperation. Wisely-drawn trade agreements can do much to adjust irritating competitive situations. It is likely, that as all the American nations become more and more industrialized, and decrease their export surpluses of food, this formerly difficult situation, which has been known to embarrass our relations with Canada and the other American states, will be alleviated or eliminated. The wise policy, for all governments and business men of Pan-America, is to proceed openly and fairly in building the strongest neighborhood economy in the entire world.

Such a principle ought to extend to the rest of the world; we could make great strides toward a cooperative world economy were it not for selfish and destructive political ideologies which for the time being enthrall the attention of great masses of European and Asiatic peoples. It is the tragedy of our age that a considerable part of the human race must follow through with the sorry experiment of communism in all its economic misery and social disaster, instead of uniting with western society in a worldwide program of constructive and peaceful development.

In June, 1952, the President's Materials Policy Commission, under the chairmanship of William S. Paley of the Columbia Broadcasting System, made a report of its eighteen-month study of what the future may bring. The Commission made the timely statement that there is no entirely domestic policy toward materials which every nation needs, but only "world policies which have domestic aspects." In its attempt to guide American thinking on material requirements for the twenty-five years from 1950 to 1975, the Commission warned that the drain on our resources, plus the increasing cost of raw materials, might imperil national security unless corrective steps are taken. It was pointed out that in 1975 the United States would have to import twenty per cent of its materials to support a total product, in terms of purchasing power, of 566 billions, as compared to 283 billions in 1950. While our raw materials will probably not be exhausted, in the twenty-one years ahead, the cost of extracting them and of processing them into useful goods may rise to dangerous levels.

In 1975, said the Commission, our increased requirements will call for a 90 per cent rise in our over-all mineral supply, including a 60 per cent increase in coal production (as petroleum and gas resources diminish); our needs for agricultural products will increase by 40 per cent, and our total energy requirements will rise by 100 per cent. Much must be done if the United States is to meet these goals of production; legislation must be passed to eliminate obstacles to our trade with other nations; there should be a scientific study of our taxation system; greater incentives must be given to the discovery and development of new sources of materials; there must be improvements in our forest conservation; technological advance should be encouraged to help reverse the rising costs of materials.

While no such set of industrial predictions can be more than approximately accurate, the dangers to which the Commission points are real, and the best industrial thinking must be applied to meeting them. It is likely, judging by the past history of American industrial science, that the necessary objectives, whether they are those arrived at by the Commission, or others equally impressive, will be achieved. The probability of a half-trillion dollar economy is not fantastic; such a level has been suggested before, by government officials and by business men. Such a peak of production and consumption, however, should be measured by productive, not deflated dollars. A one hundred per cent increase in industrial production in twenty-five years should bring with it not only more dollars, but also an improved national life on all levels and among all economic groups. There is no reason why it cannot be so, barring the criminal idiocy of a new world conflict.

BIBLIOGRAPHICAL NOTE

A great variety of material is available to the student of American industry. Vital to such research are the many excellent economic and industrial histories of the United States. More specific in their bearing upon this subject are studies of manufacturing, transportation, invention, agriculture and the development of scientific research. Still more detailed and frequently very useful are the histories of individual industries, as steel, railroads, aviation, electrical power and communications, and many others. At the individual company level, the author has found materials, published by the corporations themselves, of much value; more than sixty corporations have assisted the author with a wide variety of publications, including histories of their companies and annual reports of their operations. This has, it is hoped, given a contemporary character to the volume, since production policies and newer processes may be studied in this way, and coordinated with material in book form which was correct for its time, but has become out-of-date by the passage of even a few years. Government publications which deal with overall industrial matters have been used frequently, and the contemporary press has been used freely with respect to the latest developments in industry.

Basic to any study which concerns American business is Henrietta M. Larson, *Guide to Business History,* Cambridge, 1948; especially useful in this study is the section which introduces the student to the vast literature of the history of industry (Part V, pp. 242-731). Additional titles may be found in the general industrial and economic histories of the United States, and in the histories of specific industries.

General trends in industry may be found in the many excellent economic histories of the United States, such as E. C. Kirkland, *A*

History of American Economic Life, 3rd. ed., New York, 1951; Broadus and Louise P. Mitchell, *American Economic History,* Boston, 1947; and Harold U. Faulkner, *American Economic History,* 5th ed., New York, 1943. More detailed than these single volume studies are the volumes of the *The Economic History of the United States* under the editorship of Henry David, Harold U. Faulkner, Louis M. Hacker, Curtis P. Nettels and Frank A. Shannon, New York, 1945- , especially those by G. R. Taylor, F. A. Shannon, H. U. Faulkner and Broadus Mitchell. Several of the volumes of *A History of American Life* (13 vol., A. M. Schlesinger and D. R. Fox, eds., New York, 1927-1948) contain valuable accounts of the growth of American industry. The bibliographies of both multi-volume sets are well worth examining.

More relevant to industry alone are J. G. Glover and W. B. Cornell, eds., *The Development of American Industries,* New York, 1936; E. B. Altderfer and H. E. Michl, *Economics of American Industry,* New York, 1942; Walter Adams, ed., *The Structure of American Industry: Some Case Studies,* New York, 1950; N. S. B. Gras, *Industrial Evolution,* Cambridge, Mass., 1930; Thomas C. Cochran and William Miller, *The Age of Enterprise: A Social History of Industrial America,* New York, 1951; H. T. Warshow, ed., *Representative Industries in the United States,* New York, 1928; Arthur Pound, *Industrial America: its Way of Life and Thought,* Boston, 1936; E. L. Bogart and C. E. Landon, *Modern Industry,* 2nd ed., New York, 1937; Siegfried Giedion, *Mechanization Takes Command,* New York, 1948; and Burton J. Hendrick, *The Age of Big Business,* New Haven, 1919.

Manufactures are dealt with comprehensively in V. S. Clark, *History of Manufactures in the United States,* 3 vol., Washington, 1916, 1929; also useful are: Solomon Fabricant, *The Output of Manufacturing Industries,* 1899-1937, New York, 1940 (with more than 300 pages of technical notes and production statistics); Malcolm Keir, *Manufacturing,* New York, 1928 (lacks a bibliography); R. M. Tryon, *Household Manufactures in the United States,* 1640-1860, Chicago, 1917; Robert A. East, *Business Enterprises in the American Revolutionary Era,* New York, 1938; J. B. Walker, *The Epic of American Industry,* New York, 1949.

Invention is well-treated in Waldemar Kaempffert, *A Popular History of American Invention,* 2 vol., New York, 1924; for an earlier period, Edward W. Byrn, *The Progress of Invention in the Nineteenth Century,* New York, 1900, is excellent, with its many diagrams and pictures of mechanical, electrical, photographic and other devices which appeared by 1900. Roger Burlingame, *March of the Iron Men,* New York, 1938, is a popular history of invention to the Civil War; the same author, in his *Engines of Democracy,* New York, 1940, carried the story on after 1865. Holland Thompson, *The Age of Invention,* New Haven, 1919, is a useful summary, but is naturally dated. Also interesting is James Stokley, *Science Remakes our World,* rev. ed., New York, 1946.

Histories and technical studies of individual industries and companies are almost endless. The following have been found useful: Arthur Pound, *The Turning Wheel,* New York, 1934; Labert St. Clair, *Transportation: Land, Air and Water,* New York, 1935; Jeremiah Milbank, Jr., *The First Century of Flight in America,* Princeton, 1943; Fred C. Kelly, *The Wright Brothers,* Indianapolis, 1943; Matthew Josephson, *Empire of the Air,* New York, 1944; Frank J. Taylor, *High Horizons,* New York, 1951; John Moody, *The Railroad Builders,* New Haven, 1919; S. K. Farrington, *Railroads of Today,* New York, 1949; same author, *Railroading the Modern Way,* New York, 1951; G. L. Wilson and L. A. Bryan, *Air Transportation,* New York, 1949; William Haynes, *This Chemical Age,* New York, 1942; William S. Dutton, *Du Pont: One Hundred and Forty Years,* New York, 1949; M. Campbell and H. Hatton, *Herbert H. Dow, Pioneer in Creative Chemistry,* New York, 1951; William Haynes, *American Chemical Industry,* vol. 2 and 3, New York, 1945; H. Barger and S. H. Schurr, *The Mining Industries,* 1899-1939, New York, 1944; C. C. Carr, *Alcoa, An American Enterprise,* New York, 1952; Douglas A. Fisher, *Steel Serves the Nation,* 1901-1951, U. S. Steel Corp., 1951; A. T. Schurick, *The Coal Industry,* Boston, 1942; Max Ball, *This Fascinating Oil Business,* Indianapolis, 1940; Paul H. Giddens, *Early Days of Oil* (a pictorial history), Princeton, 1948; Dorsey Hager, *Fundamentals of the Petroleum Industry,* New York, 1939; Malcolm McClaren, *The Rise of the Electrical Industry during the Nineteenth Century,* Princeton, c. 1943; Alvin F. Harlow, *Old Wires and New Waves,* New York, 1936; C. N. and C. D. Mooers, *Electronics,* Indianapolis, 1947; Hugh Allen, *The House of Goodyear,* Cleveland, 1949; Alfred Lief, *The Firestone Story,* New York, 1951; Howard and Ralph Wolf, *Rubber: A Story of Greed and Gold,* New York, 1936; C. C. Furnas, *The Storehouse of Civilization,* New York, 1939; Alice Payne Hackett, *Fifty Years of Best Sellers,* 1895-1945, New York, 1945; Charles Morgan, *The House of Macmillan,* 1843-1943, New York, 1944; Thomas B. Lawler, *Seventy Years of Textbook Selling* (Ginn and Company, 1867-1937), Boston, 1938; William Miller, *The Book Industry* (report of the public library inquiry), New York, 1949; Stanley Unwin, *The Truth about Books,* London, 1947; R. L. Duffus, *Books: Their Place in a Democracy,* Cambridge, Mass., 1930; John A. Guthrie, *The Newsprint Paper Industry,* Cambridge, Mass., 1941; J. B. Bettendorf, *Paperboard and Paperboard Containers,* Chicago, 1946; B. C. Forbes, ed., *America's Fifty Foremost Business Leaders* (contains interesting material on present-day heads of corporations), New York, 1948; Frank Crane, *George Westinghouse: His Life and Achievements,* New York, 1925.

Also of importance are: Harlan Logan, ed., *How Much do you Know about Glass?,* New York, 1951; Jonathan N. Leonard, *Tools of Tomorrow,* New York, 1937; Ruth C. Christman, ed., *Industrial Science: Present and Future,* Washington, 1952; Joseph Schafer, *Social History of American Agriculture,* New York, 1936; T. S. Harding, *Two Blades of Grass: A History of Scientific Development in the United States Department of Agriculture,* Norman, Okla., 1947; S. E. Johnson, *Changes in American Farming,* Washington, 1949 (U. S. Dept. of Agriculture, misc. rpt. no. 707); Harper Leech and J. C. Carroll, *Armour and His Time,* New York, 1938; L. F. Swift, *The Yankee of the Yards: the Biography of Gustavus Franklin Swift,* Chicago, 1927; L. B. Jensen, *Meat and Meat Foods,* New York, 1949; "The Significant Sixty," (an historical report on the progress and development of the meat packing industry, 1891-1951), *National Provisioner,* v. 126, no. 4, Chicago, 1952; Paul de Kruif, *Hunger Fighters,* New York, c. 1928; C. H. Eccles, W. B. Combs, and Harold Macy, *Milk and Milk Products,* 3rd ed., New York, 1943; B. L. Herrington, *Milk and Milk Processing,* New York, 1948; E. O. Whittier and B. H. Webb, *Byproducts from Milk,* New York, c. 1950; A. W. Bitting, *Appertizing, or the Art of Canning, its History and Development,* San Francisco, 1937; H. W. Von Loesecke, *Drying and Dehydration of Foods,* New York, 1943.

General books on American science are not very useful for this subject, since they seldom emphasize economic and industrial implications. Bernard Jaffe, *Men of Science in America,* New York, 1944, has several suggestive chapters; by his comprehensive view of the subject and his inclusion of scientists from Thomas Harriot to Ernest Lawrence he touches fields important to American industry. J. G. Crowther, *Famous American Men of Science,* New York, 1937, features Franklin, Henry, Gibbs, and Edison as leaders of the early, middle and later periods of science and thus connects experimental science with economic utility. Other writers in this field, as Henry C.

Tracy, William J. Youmans and F. C. Hylander feature biological science in the main, and other topics too remote from the subject to be of appreciable value.

Government publications often contain helpful comparative figures for industrial production and narrate important phases of research related to industry. Especially useful were: U. S. Bureau of Mines, *Minerals Yearbook,* Washington, 1949; U. S. Department of the Interior, *Report of the Secretary,* Washington, 1951; U. S. Federal Trade Commission, *Report on Interlocking Directorates,* Washington, 1951; U. S. Department of Commerce, *National Income,* Washington, 1951; U. S. Department of Agriculture, *Agricultural Statistics,* Washington, 1952; same, Bureau of Animal Industry, *Report of the Chief,* 1950 and 1951; same, Bureau of Dairy Industry, *Report of the Chief,* 1950, 1951 and 1952; same, Bureau of Plant Industry, Soils and Agricultural Engineering, *Report of the Chief,* 1951 and 1952; same, Production and Marketing Administration, *Grain Production and Marketing,* Washington, 1949; same, *The Inspection Stamp as a Guide to Wholesome Meat,* Washington, 1952; same, *Technology in Food Marketing,* etc., 1930-50, Washington, 1952; same, *Research at Work from Farm to You,* Washington, 1949.

Light is shed upon the role of industrial science in wartime by: W. F. Willoughby, *Government Organization in Wartime and After,* New York, 1919; Johnson Hagood, *Services of Supply,* Boston and New York, 1927; Eliot Janeway, *Struggle for Survival: A Chronicle of Economic Mobilization in World War II,* New Haven, 1951; Donald Nelson, *Arsenal of Democracy,* New York, 1946; Bernard M. Baruch, *American Industry in the War,* New York, 1941; Seymour E. Harris, *The Economics of America at War,* New York, 1943; and G. B. Clarkson, *Industrial America in the World War: the Strategy Behind the Line,* 1917-18, Boston, 1923.

INDEX

A

Abrasives, manufacturers of, 209, 338
Acetate, 140, 141
Acetylene gas, 246
Acids, 138, 142
Acids, manufacture of, General Chemical Co., 134
Acrilan, 341
Acthar, 320
Adding machines, 347
Adhesives, 148
Ader, Clement, 99, 101
Advertising, use of airplane for, 111
Advisory Commission, World War I, 369-370
Aerial photography, 111
Aerodynamics, 98; Langley's treatise on, 100
Aeronautical Corp. of America, 121
Aeronautical exhibitions, 98
Agricultural education, 51
Agricultural experiment stations, 51, 307; see also State experiment stations
Agricultural machinery, 308, 310; electrical, 304; tractor powered, 312, 314; world market opened, 311, 312; see also names of individual machines (e.g. Reaper)
Agricultural societies, state, 29, 33
Agricultural Research, 309
Agricultural Research Administration, 309
Agricultural Research Center, see Beltsville, Md., Agricultural Research Center
Agriculture, 7, 15, 16, 21, 24, 33, 34, 49, 88; colleges of, 50 51; mechanization of, 22, 88, 303-304, 310-315; production in 1952, 405-406; scientific, 51, 303, 307-308, 309, 316
Air Cargo Inc., 118
Air conditioning, meat industry, 322
Air, control of, 90
Air defense, 387
Air express, 56, 118, 119, 123
Air freight, 56, 118
Air Mail Act of 1934, 116
Air mail compensation, 116
Air mail service, transpacific, 114
Air power, 91
Air Reduction Co., 133, 142, 143, 306, 407
Air transportation, 91
Air-atomic war, 91
Aircraft, 388; aerodynamic computations for, 348; for barrage, 115; for reconnaissance, 115; improved designs for, in World War II, 387; giant engine program, 392; industry, 145; manufacturing, 123; profits in making, 122; supersonic, 401; supply concerns, 121; World War II equipment, 409-410
Aircraft, see Airplanes
Airplanes, 387, 388; Boeing transport,

INDEX

122; engines, 86; engines, production of, 387; exhibition, 108; foe of isolation, 90; for private use, 108; industry, 120; motors, manufacture of, 111; parts, 86; passenger service, regular, 114; plants, 389; postwar makers of commercial, 120; speed of, 91; steam-driven, 98; uses for, 304

Airplanes, see Aircraft

Airports, 116

Air-races, 111

Airship disasters, 115

Airship industry, stimulus to, 110

Akron, 115

Alathon, 145

Alberta, Canada, 71

Alcoa, see Aluminum Co. of America

Alcock, John, 112

Alcohol as blending agent in gasoline, 223

Alcoholic beverages, distillation of, 34

Alexanderson alternator, 172

Alexanderson, E. F. W., 167, 169

Alkali manufacture and Solvay Process Co., 134

Alkaline products, 187

Allard Lake region, 214

Allen, Hugh, 266

Allied Chemical and Dye Corp., 133, 134

Allied counter-offensive in 1918, 367

Allies of 1939, contrasted with 1914, 381

The Allison Co., 121

Alloy steels, 140

Alloys, 41; recent, 199

Alternating current, 160, 164

Alternator, 169

Aluminum, 187, 206-210; control of, in World War II, 395; production in 1952, 405; shortage in World War II, 392; uses, 210

Aluminum Company of America (formerly Pittsburgh Reduction Co.), 207, 208, 209, 211, 213, 411-412

Aluminum Company of Canada, 412

Aluminum Limited, 208

Amalgamated Clothing Workers, 385

American Aeronautical Society, 99

American Airlines, 118

American Arithmometer Co., 345

American Bemberg Co,, 141

American Book Co., 290, 296

American Can Co., 321, 334, 335, 337

American Car and Foundry Co., 64

American Chemical Society, 245, 246

American Company of Booksellers, 289

American Cyanamid Co., 133, 138, 213, 230, 299, 300, 306

American Electrical Works (later Kennecott Wire and Cable Co.), 201

American Enka Corp., 132, 141

American Expeditionary Force in World War I, 368

American Farm Bureau Federation, 316

American Federation of Labor, 372

American intervention in World War I, 367

American Locomotive Corp., 66

American Locomotive Works, 64, 65

American Magnesium Corp., 212

American Meat Institute, 320

American Meat Packers Association, 321

American Paper and Pulp Association, 281-282

American Potash and Chemical Corp., 230

American Railway Express, 117

American Revolution, 5, 364

American Smelting and Refining Company, 203, 205

American Speaking Telephone Co., 153

American Strawboard Company, 283

American technology, 378

American Telephone and Telegraph Co., 154, 155, 170, 171, 172, 175, 177, 404

American Viscose Corp., 141, 340, 337

Ames, Oakes, 43

Ammonia, 225, 226

Anaconda Copper Mining Co., 201, 202, 205

Anderson, Capt. Orvil, 115

Anglo-Consolidated Investment Co., 202

INDEX

Anhydrides, 142
Aniline and derivatives, 256
Animal power, shortage of, 304
Anne, Queen of Great Britain, 345
Antibiotics, 139, 140
Anti-knock chemicals, 147
Appalachian Mountains, 48
Appert, Nicholas, 334
Appleby, John F., 311
Appleton-Century-Crofts, Inc., 290
Applied science, 363;—schools of, 52
Apprentice system, 2
Arc lights, 158-159
Architectural Forum, 295
Argon, 142
Arizona Chemical Co., 300
Arkwright, Sir Richard, 9
Armistice of 1918, 367
Armories, 30
Armour and Co., 306, 319-321
Armour Laboratories, 320
Arms manufacturers, private, 30
Arms, small, 30
Armstrong, Maj. Edwin H., 182
Army camps, 389
Army Industrial College, 383
Arsenals, 30;—government, 389
Asbestos, 226, 228, 229
Assembly line in motor car industry, 85
Associated Press, 168, 287
Association of American Railroads, 403
Atchison, Topeka and Santa Fe, 44
Atlantic, attempts to fly the, 96
Atlantic Coast Line, 72, 403
Atlantic Steamship Lines, 73
Atlas Corporation, 393
Atom smashers, 181
Atomic bomb project, 181
Atomic energy, 91, 181, 184; development of "A" bomb in World War II, 398; new installations, 403; postwar developments in, 421-424; radiation effects in, 348; research since World War II, 401
Atomic Energy Commission, encourages power research, 413-414
Atomic research, 184
Audion, 70
Audit Bureau of Circulations, 288
Auto-buggies, 314
Auto truck, 85
Automatic couplers, 62
Automobile, see Motor car.
Automobile Manufacturers Associations, 404
Aviation, 93; commercial progress in, 114; meet at Rheims, 109
Axis powers, 363; conquer Western European powers, 381; decrease in productivity in 1944, 396

B

Babcock, Stephen Moulton, 316, 317, 325
Babylonians, 271
Bag makers (paper), 283-84
Bags, production of paper for, 277
Baker, Newton D., 369
Baking industry, 331, 332
Balchen, Bernt, 112
Baldwin, Frank S., 345
Baldwin Locomotive Works, 64
Balsam fir, uses for, 275, 276
Baltimore and Ohio Railroad, 31, 45, 57, 58, 59, 67
Banks, 15, 18, 27, 37
Barber, Ohio C., 263, 264
Barclay, Henry, 274
Barn-storming, 111
Barrels, 4
Barrett Co., 134
Baruch, Bernard M., 243, 249, 370, 372, 375, 376, 385
Baruch Committee, 243, 250
Bausch and Lomb Optical Co., 359, 360
Bauxite ore, 207, 208, 212
Bayer aluminum process, 208, 257
Bayer Co., 256
Beater, uses for, 278
Bechtel Corp., 414
Beef, canning of, 320
Beef, western, prejudice against, 319
Beet sugar culture, 32
Behn, Hernand and Sosthenes, 177

INDEX

Bell, Alexander Graham, 151, 152, 157, 168
Bell overseas service, 404
Bell Patent Assoc., 153
Bell Telephone Co., 153, 154, 179
Belmont, August, 18
Belmont Park, New York, 109
Beltsville, Md. Agricultural Research Center, 307, 308, 309, 323, 324
Beltsville turkey, 308
Bendix Corp., 121
Beneficiation, 190
Bennett, Floyd, 112
Bennett, James G., 287
Bentley, Edward M., 162
Benz, Carl, 79
Benz motor car, 80
Benzol, 226
Bessemer, Henry, 39, 41
Bessemer process, 40, 41
Best, Jeremiah, 28
Best-seller books, 292, 293
Bethlehem Steel Co., 370
Bible, The Holy, 293; Revised standard version, 300
"Big Push" of 1918, 367, 378, 379
Bikini tests, 184
Binder, 311, 312; invention of, 304
Biologists, 139
Biplanes, 110
Bissell, George H., 31
Bituminous areas, 30
Black, Joseph, 93
Blackstone, R., 14
Blake, Francis, 153
Blanchard, Jean Pierre, 94, 95
Blandin Paper Co., 288
Bleriot, Louis, 109
Block caving, 191, 192
Blockade, 34
Boehm, 110
Boeing Air Transport, 117
Boeing Aircraft Co., 121
Boiler, multitubular, 14
Boll weevil, 51
Bollee, Leon, 107, 347
Bolton, Dr. E. K., 245, 246
Bomber program, 387
Bond paper, production of, 277
Book agents, itinerant, 289
Book clubs, 293
Bookman, 292
Book paper, production of, 277
Book publishing, 280, 285, 289, 294, 300, 301; cost, 294; paper books 300, 301; statistics 291, 293, 296, 300; See also, University presses and names of individual publishing houses
Book stock, raw materials for, 278
Bordeaux, France, 378
Borden, Gail, 325
Borden Co., 327
Boston and Albany Railroad, 57
Boston and Maine Railroad, 57
Bowker Book Guides, 292
Bowker, R. R. Co., 292
Boyle, Herbert, 132
Braden, William, 202
Braden Mining Company, 202, 203
Bradford, William, 273, 289
Brass button industry, 19
Brest, France, 378
Brew masters, 30
Bridge companies, 18
British Army, 95, 387
British Commonwealth of Nations, productivity of, by 1944, 396
British Marconi, 168, 169
Broadcast Brand, 321
Broadcasting, 174
Bromine, 227
Brown, Arthur, 112
Brown, Moses, 9
Brown-tailed moth, 51
Brush, Charles F., 158, 176
Brush Electric Light and Power Co., 158
Bryant, Gridley, 62
Buckman, W. D., 259
Budd Co.,66
Budd, Ralph, 385
Building materials, 28; lack of, 388
Building stones, uses of, 231
Bull Run, 27
Bulldozers, 88
Buna rubber, 244, 245, 247
Burlingame, Roger, 13

INDEX

Burr-Leaming corn, 307
Burroughs, William S., 345
Burroughs Adding Machine Co., 345
Burton Oil process, 221
Business machines, 344, 449
Business schools, 346
Butadiene, 248, 249
Butyl, 248
By-products, commercial use of, 134
Byrd, Richard E., 112, 357
Byrnes, James F., 391

C

Cable communication with Europe, 150
Cactus as rubber source, 244
Cadillac Motor Car Co., 81
Calatron, 184
Calcium, 229; uses of, 231
Calculating machines, 347, 348
Calendar machine, 237
California Institute of Technology, 357
California Zephyrs, 66
Camden and Perth Amboy Railroad, 31
Cameras, 342, 360
Camphor, synthetic, manufacture of, 300
Canada, Dominion of, 71; national income in 1952, 414; raw materials, 276; spruce forests, 275
Canadian spruce, use in paper manufacturing, 275
Canals, 16, 42
Candee, L. and Company, 251
Canning industry, 333-336; meat, 320-321; mechanization of, 336, milk, 325; tin needs in, 335
Cantonments, 389
Cape Cod Canal, 70
Capital, 12, 15, 22; amount of, 4; availability of, 9; foreign, 18; lack of, 7, 16; plentiful amount of, 37
Capitalism, commercial and industrial, 36
Car, see Motor car
Carbine, breech-loading, 14
Carbon dioxide, 306
Carey, Matthew, 289
Cargo subsidies, 74
Carnegie, Dale, 293
Carrier system, 156
Carrier-repeater system, 156
Carty, John J., 171
Caruso, Enrico, 171
Cash register, 345, 347
Cataract Construction Co., 163, 164
Caustic soda process in paper manufacturing, 273
Cavendish, Henry, 93, 132
Cayley, Sir George, 92, 97
Celanese, 340
Celanese Corporation of America, 306
Cellophane, 141, 336
Celluloid, 129
Cellulose, 141
Cellulose products, 129
Cement, uses of, 231
Census Report on Mines and Quarries, 191
Centennial Exposition at Philadelphia, 150, 151
Central Pacific Railroad, 43, 44
Central Railroad of Georgia, 59
Chaffee, Edward, M., 237
Chain-breast machine, 192
Champion Paper and Fiber Company, 299
Chanute, Octave, 99, 102, 103
Charcoal, 12, 39
Chardonnet, Hilaire de, 129, 340
Charles, Prof J. A. C., 93, 94
Chase Brass Co., 201, 204
Chattanooga-Birmingham region, 38
Cheese industry, 326, 328
Chemical companies, 305
Chemical dyes, 135
Chemical Foundation, 130
Chemical industry, 145, 187; diversity of, 132; pattern of, 132, production in 1952, 407, 408
Chemical machining, 358
Chemical manufacturing, 135
Chemical processes, businesses which use, 132

INDEX

Chemical processes in the paper industry, 274
Chemicals, heavy, 125
Chemistry in agriculture, 305, 307
Chemistry in food industries, 306, 317, 319, 320, 327-328, 332-333, 336-337
Chemistry in paper industry, 274, 299-300
Chemistry in textile industry, 274, 339-341
Chemists, 25, 139
Chemigum, 245, 247, 267
Chemstrand Corp., 341
Chicago and Eastern Illinois Railroad, 67
Chicago and Northwestern Railroad, 31
Chicago, Burlington and Quincy Railroad, 59, 385
Chicago *Times-Herald,* 80
Chicago *Tribune,* 288
Chicago, University of, 320
Chicago World's Fair of 1893, 161, 354
Chilean nitrates, 228, 229
China, aid to, in World War II, 396
Chinese cane, 32
Chlorine, 227
Chloroprene, 246
Christy, James, Jr,, 259
Christy, Will, 260
Chromium, 215
Chrysler Corporation, 86, 389-390, 392, 393, 405
Chrysler Motor Car Co., 84
Chrysler, Walter P., 81
Chuquicamata Copper Mine, 202
Churchill tank, 387
Cities Service Oil Co., 249
Civil Aeronautics Authority, 116
Civil Aeronautics Board, 116
Civil War, 26, 95, 126, 150, 364, 365; manpower shortage, 311, 362
Civilian goods, drop in, during World War II, 394; increase in production of, after 1945, 401
Clark, Bennett, 374
Clay, 234; uses of, 231
Clemens, Samuel L., see Twain, Mark
Cleveland Telegraph Supply Co., 158
Clincher Association, 261, 265, 266
Clipper ships, 22
Clocks, 18
Clothing, 8, 308, 339, 388; materials, 339, 340; ready-made, 29; scientific discoveries in, 303
Coal, 12, 21, 30, 47, 48, 187; anthracite, 48, 195; anthracite, role in industry, 218; bituminous, 48, 74; bituminous, role in industry, 218; breakers, 189; chemicals, 225-226; cutters, mechanical, 191; yard, 71; industry, 225-226; industry, bituminous, 195; mining, 185; mining, anthracite, 189; movement of, 68; place in American economy, 218; reserves, 225; role in industry, 218; soft, 48
Coal-tar dyestuffs, 130
Coal-tar products and the Barrett Co., 134
Coatings, 140
Coatings, heat-setting, 139
Coaxial cable, 156, 183, 404
Cobalt, 215
Coffee, 8
Coffin, Charles A., 163
Cogswell, W. B., 131
Coke, 48
Coke and by-products manufacture by Semet-Solvay Co., 134
Coke ovens, 41
Colburn, Irving W., 355
Colgate-Palmolive-Peet Co., 338
"Cold War," 401
Collier, Robert J., 110
Collodion, 128
Colorado, 52
Colt Arms Co., 389
Colt, Samuel, 14, 20
Columbia Broadcasting System, 416
Columbus Buggy Company, 258
Columbus, Christopher, 236
Combine, 315
Commercial capitalism, 36
Commissioner of Immigration, 33

INDEX

Committee on Recent Economic Changes, 257
Commonwealth Edison Co., 414
Communications, 363; telegraphic, 175; telephonic, 175
Communism in Europe and Asia, 416
Companies, integrated, 120
Competition, 61
Competition in motor car industry, 85
Comstock Lode, 188, 189, 193
Concentration method of refining, 190
Concentration of ore, 195, 197
Concrete, production of, 231
Conde Nast Publications, 295
Confederate States, 27, 35; credit of, 34
Congressional investigations, 44
Connecticut Experiment Station, 307
Connecticut R., 14, 42
Conrad, Frank, 174
Conservation, 69, 188, 198
Conservation and expansion program, 215
Consolidated-Vultee Aircraft Corp., 121, 392; new jet bombers and fighters, 410
Consolidation of railroad lines, 61
Construction industry in 1952, 406
Construction industry, needs of, 231
Container Corporation of America, 280, 299
Container industry, 280, 297, 327-328, 334-35
Container stock, production of paper for, 277
Container stock, raw materials for, 278
Continental Army, 5, 6
Continental Oil Company, 202
Continental Motors Corporation, develops new aircraft engines, 410
Continental telephone system, 157
Contract labor, 33
Controlled Materials Plan, 395
Conversion to defense industry, 390, 392
Convoys, naval, 378
Cooper, Peter, 354
Cooper Union, 52
Cooperative world economy, 416
Copper, 186, 187, 190, 192
Copper industry, 194
Copper ore, 197
Copper sheet, 19
Copper, boom in, due to rise of electrical industry, 201; cold rolled, 13; control of, in World War II, 395; mining of, 200-204
Copyright, 289, 293
Corn pickers, 314
Corn planter, 314
Cornell, Ezra, 19
Cornell University, 51
Corning Glass Works, 233, 356, 357, 358
Corporation, chartered, 17; joint-stock, 22
Corporation finance, 24
Corporation law, 24
Cotton, 34, 140; crop, 32; harvester, 314; industry, 139; manufacturing, 29; mills, 14; seed oil, 332; short staple, 7; spinning, 28
Council of National Defense of 1916, 366, 369; of 1940, 384
County agent system, 316
Couzens, James, 82
Cox, Charles R., 203
Cox, Tench, 16
Crease resistants, 139
Creel Committee on Public Information, 368
Creosote oil, 226
Crisco, 332
Crookes, Sir William, 166
Crop dusting, 304
Crothers, Wallace, 341
Crowell-Collier Publishing Co., 296
Crown Zellerbach Corp., 280, 282, 284
Cryolite, 206, 211, 215
Crypostegia, 242
Crypton, 142
Crystal radio detector, 169
Cultivator, invention of, 304
Cuneo Press, 296
Cuprammonium rayon process, 141
Curtis, Glenn, 101, 109
Curtis Publishing Co., 295

INDEX

Curtis-Wright Corp., 120, 121
Curtis-Wright engine, 111
Cutlery, 39
Cyanide method, 197
Cyclopropane, 143
Cylinder machine, use in paper manufacturing, 278
Cylinder paper machines, 278, 279, 281

D

DDT, 137
da Vinci, Leonardo, 91, 92
Dacron, 137, 145, 341
Dahl, Carl F., 273
Dahl's sulphate process for pulp manufacture, 273, 276, 277
Daimler, Gottlieb, 79
Dairy herd, national, 322
Dairy herd improvement associations, 323
Dairy industry, 322-328; distribution methods, 327; mechanization of, 325-6, 328
Dairy machinery, 325-6, 328
Dalton, John, 132
Davenport, Thomas, 14
Davidson, Richard, 92
Davis, A. V., 209
Davis, Chester C., 385
Davis, Harry P., 174
Davy, Sir Humphry, 158, 206
Daye, Stephen, 289
Daylight saving time, 371
de Forest audion tube, 156
de Forest, Lee, 167, 170, 172
de Kruif, Paul, 316
de la Vaulx, Count Henry, 96
Debt-funding program, Hamilton's, 18
Deepwater Chemical Company, 228
Deere, John, 19, 310
Deering, John, 311
Deering Harvester Co., 311, 313
Defense construction, 78
Defense mobilization, 364, 369
Defense production, for 1941 and 1942, 394; lack of coordination in, 394-95
Defense Production Administration, 204, 405
Defense Supply Corp., 388
Defense, twentieth century, 363
Degen, Jacob, 92
Dehydration, 336
Delagrange, Leon, 107
Demobilization, after World War I, 379-380; during World War II, 395; after World War II, 400-401
Dennison Manufacturing Co., 322
Densmore, James, 346
Depression of 1857, 26
Depression of 1929, 282
Desk-Fax telegraphic instrument, 352
Detector, 167, 168
Detergents, 138, 140; synthetic, 338
Detroit Automobile Company, 81
Detroit Edison Co., 413-414
Deutsche Lufthansa, 114
Deville, Henri St. Clare, 206
Diamond Rubber Company, 253, 254, 256,
Diary of George Templeton Strong, 300
Diesel engine, 49, 65, 88
Diesel locomotive, 66, 86
Dietz, Walter, 389
Dilatometer, 359, 360
Diode tube, 170
Direct current, 160, 164
Dirigible, 96, 111
Distillation in oil production, 223
Distribution of industrial activity, 408-09
Division of Chemistry, 307
Division of Contract Distribution, 393
Division of Soils, 308
Dodd Mead & Co., 290
Dodge brothers, 81, 82
Dolomite, 212
Dooley, Channing, 389
Doubleday & Co., 290
Douglas Aircraft Co., 121, 122
Douglas DC-4, 121
D'Oviedo y Valdes, G, F., 236
Dow Chemical Company, 133, 212, 227, 231, 228, 49, 413
Drake, E. L., 31, 48, 219

INDEX

Drew, Daniel, 45
Drexel, Morgan and Co., 159
Dreyfus, Camille and Henri, 340
Drills, compressed air, 189
Dry cleaning, 339
Dry ice, 306
du Pont de Nemours, E. I. & Co., 11, 133, 137, 138, 141, 144, 145, 213, 244, 245, 246, 249, 340, 341
du Pont de Nemours, Eleuthere Irenee, 126
du Pont, Irenee, 127
du Pont, Lammot, 126
du Pont, Pierre Samuel, 126
Duffus, R. L., 289
Duluth Range, 38
Dunlop, John B., 239
Dunwoody, H. C., 169
Durum wheat, 330
Duryea, Charles, 80, 82
Dutton, E. P. & Co., Inc., 290
Dyes, 129; azo, 137; chemical, 135; synthetic, 129; vat, 135
Dyestuff manufacture and the National Aniline and Chemical Co., 134
Dyestuffs and du Pont, 133
Dynamite, 126, 128, 190, 192
Dynamo, 151; building of, 158

E

Eastern Airlines, 118
Eastman & Co., 321
Eberstadt, Ferdinand, 395
Eckner, Dr. Hugo, 113
Economic dictatorship, partial, 379
Economic nationalism, 380
Edison Co., 161
Edison effect, 170
Edison Electric Co., 163
Edison Illuminating Co. of New York, 159
Edison, Thomas A., 153, 157, 159, 161, 162, 164, 169, 176, 244
Edison incandescent lamp, 356
Educational Orders Act, 1939, 386
Edward VII, King of England, 107
Eiffel Tower, 97, 109
Eiffel Tower experiments, 171
Ekman, Carl D., 273
Electrapane, 358
Electric-arc furnace, 180
Electric furnace, 41
Electric helmsmen, 180
Electric insulation, 145
Electric light, 176
Electric locomotives, 64
Electric motor, 14, 161
Electric motor cars, 80
Electric power in 1952, 403, 405
Electric street lighting, 158
Electric appliances, development of, 304
Electrical circuits, "printed," 359
Electrical equipment, 175
Electrical furnace, 180
Electrical generating plants, 53
Electrical industry, 19, 49, 146, 187; scientific discoveries in, 303
Electrical power, 47
Electrical scientists, 25
Electricity, progress in, 22
Electrification, extension of, 304; of industry, 48; rural, 304, 305
Electro-magnet, 14, 19
Electro Metallurgical Co., 147
Electro-Motive Division of the General Motors Corp., 65
Electro-Motive Engineering Corp., 65
Electron microscopes, 166
Electronic amplification, 155
Electronic compasses, 166
Electronic computers, 184
Electronic Data Processing Machine, 348
Electronic precision, 353
Electronic-welding, 166
Electronics, 151, 165, 166, 177; industry, 142
Elliott, Harriet, 385
El Teniente Copper Mine, 202, 203
Embargo, 10, 34
Emergency Fleet Corp., 74, 75, 366, 370, 371
Employment of women and children, 33
Emulsifiers, 339
Emulsions, 138

INDEX

Enamels, 139
Engine, atomic powered aircraft, 181; internal combustion, 49; internal combustion, development of, 304; multi-cylinder, 110; pusher type, 110; rotary, 111; tractor type, 110; twin, 110; water-cooled V, 111
Engineers, 139
Ensilage harvester, 312, 314
Entrepreneur, industrial, 8
Equipment, superior, 363
Ericsson, John, 14
Erie Canal, 42, 69, 70
Erie Railroad, 45, 69
Erosion, 49
Esch-Cummins Act, 60
Ethylene, 143
Evans, Oliver, 15
Expenditures, governmental, 27
Experiment Station Record, 309
Experimental science in World War II, 397, 398
Explosives, 129, 138
Explosives industry, 125
Exports, 73

F

FM, 182
Factory system, 9, 22
Fair Oaks, Battle of, 95
Faraday, Michael, 158, 176
Farm production, 303; increase in, 316
Farman, Henry, 107, 109
Farming devices, mechanical, 304
Farmall tractor, 312, 313, 314
Farmer, Fanny, 293
Farmers Institute, 316
Fatty acids, production of, 300
Federal Communications Commission, 182, 183
Federal Meat Inspection Act, 321
Federal Meat Inspection Service, 321, 322
Federal Power Commission, 224
Federal Smelting and Refining Co., 205
Federal Telegraph Co., 168, 171
Federal Trade Comm., 283, 295
Feldspar, 215, 230
Ferro-alloys, 147
Fertilizer, 51, 138, 143, 229, 305, 320, 322; supply of, 305; uses for, 305
Fessenden, R. A., 167, 169
Fibers, man-made, 146; natural, 146; synthetic, 140
Fibreboard Products, 280
Field artillery, 378
Field, Cyrus W., 49, 150
Field-sequential system of color TV, 183
Fine chemicals, makers of, 132
Finishing agents, use in paper manufacturing, 275
Firearm shops, 14
Firearms, 19
Firestone, Harvey S., 258, 262
Firestone Tire and Rubber Company, 247, 248, 258, 262, 266
Fisk, James, 45, 46
Fisk Tire Company, 248
Five-year Amortization Plan, 386
Flax, 29
Fleming, J. Ambrose, 169, 170, 172
Fleming valve, 169, 170
Fleurus, Battle of, 95
Flight, first power, 104; first trans-Atlantic non-stop, 113; first world-circling, 113; heavier-than-air, 91, 93, 94, 98, 99, 101, 106, 107, 111; lighter-than-air, 9, 93, 113, 114; power, 98; steam-propelled, 99; stratospheric, 181
Florida, increase in manufacturing since World War II, 409
Florida "Marketers," 67
Flotation process, 197
Flotation, selective, 198
Flour, 4, 32
Fluorine, 227, 228
Fluorspar, 211, 215, 228
Flying instruction, 111
Flying exhibitions, 107, 109
Flynn, John T., 373
Foamglass, 322
Food industry, scientific discoveries in, 303
Food, lack of, 388; packaging of, 308,

330-332; refrigerated, transportation of, 306, 319; reinforcement of, 317, 326, 331-332
Ford, Henry, 80, 81, 82, 261
Ford Motor Co., 81, 82, 84, 86, 117, 376, 392, 393
Ford, Edward, 355
Ford, John B., 354, 355
Ford-Stout Air Services, 117
Foreign trade, 75, 76; in 1952, 402
Forest conservation, 298, 299
Forest resources, 276
Forestry industry, 297
Forrestal, James, 385
Fort Meyer, Virginia, trial flights at, 106
Fort Sumter, 27
Fortune, 295
Foulois, Lt. Benjamin D., 107
Fourdrinier, Henry, 272
Fourdrinier, Sealy, 272
Fourdrinier paper machine, 274, 278, 279, 281
Fourteen Points to Germany, 172
France, 8, 9, 32, 106, 365; aid to, in World War II, 396
Franco-Prussian War, 96, 98
Franklin, Benjamin, 93, 176, 285, 289
Free labor system, 36
Freight business, 45, 57
Freight cars, 28
Freight vessels, 38
Fremont, General John C., 354
French Government, 97, 106
French railway system, 378
Frozen food industry, 336
Frye, Sen. William P., 73
Fulton, Robert, 14
Furnaces, gas, 86

G

Gaines' Mill, Battle of, 95
Gair, Robert Co., 280
Gallatin, Albert, 16
Galvani, Luigi, 176
Garfield, Harry A., 371
Garnerin, Andre, 93
Gas, illuminating, 21, 48
Gas for heating, 48
Gas, natural, see Natural gas
Gas, synthetic, use of, 218
Gas turbine, improvement of, 224
Gas turbine locomotive, 66
Gas-oil burners, introduction of, 224
Gasoline, 219, 221; casing-head, 220; high octane, 388
Gaulard, Lucian, 160
Gaissenhainer process, 12
Gelatine, blasting, 128
General Bauxite Co., 207
General Chemical Co., 134, 207
General Electric Corp., 160, 163, 164, 165, 167, 169, 172, 174, 178, 179, 180, 182, 376, 389
General Incorporation Act, 18
General Mills, Inc., 329, 330
General Motors Corp., 84, 86, 121, 148, 376, 385, 389, 391, 392, 405
General Staff, 365
Generating station, first, 159
Generator, development of, 158
German arms production, 385, 386
German attack on Poland, 381
German chemical patents, seizures of, 130
German High Command, attitude of, in 1916, 366
German Navy, 111
Germany, 32, 71, 365, 380
Germicides and du Pont, 133
Giffard, Henri, 96
Ginn & Co., 290
Glass, 351-360; dosimeter, 360; electrapane, 358; fiber, 358; heat-absorbing, 358; heat-resistant, 356; laminated (safety), 355-356; mirropane, 358; optical, 358; photosensitive, 358; plate, 354-355; safety, 83, 147; sewer pipes, 354; telescope, 356; tempered, 357; thermopane, 357; vitrolite, 357; window, 355
Glass chemistry, 353ff., 359
Glenn L. Martin Co., see under Martin
Gliders, 98, 99, 103, 104
Gogebic Range, 38
Gold, 37, 185, 188, 203, 204
Gold Medal flour, 330

INDEX

Gold-bearing ore, 196
Goldenrod, 241, 244
Goodrich, B. F., Co., 248, 252, 258, 263, 264, 266, 306
Goodrich, Dr. B. F., 240, 252, 253
Goodrich, Tew and Company (later B. F. Goodrich Co.,), 253
Goodyear, Charles, 14, 238, 251, 255, 263
Goodyear's Metallic Rubber Shoe Company, 251
Goodyear, Nelson, 238
Goodyear Tire and Rubber Co., 20, 115, 248, 262, 268, 392
Gould, Jay, 45
Government Reserve Rubber, 249
Governmental operation of railroads, 60
Grace, Eugene, 374
Graders, 88
Graf Zeppelin, 113
Graham Act of 1921, 155
Grain, 32; blight-resistant, 307
Grain ships, 72
Gramme, J. T., 158
Grand Coulee Dam, 178
Granite supply, 231
Grant carriage tire patent, 265
Grant, U. S., 151
Graphite, 229
Grasselli Chemical Co., 257
Gravel, production of, 232; uses of, 231
Gray, Elisha, 152, 157
Great Britain, 5, 8, 71, 365; aid to, in World War II, 396
Great Lakes, 68, 70, 71, 72, 78; waterway, 42
Great Northern Railroad, 58, 59, 65
Great Plains, 48
Greeley, Horace, 287
Greely, Gen. A. W., 95
Greenbacks, 28, 37
Griswold, Ely, 209
Guayule, 241
Guggenheim family, 202, 203
Guided missiles, research in, 401; interest in (of aircraft firms), 410
Gulf Steamship Lines, 73
Guncotton, 128
Gunsmiths, 9
Gutenberg, Johannes, 285
Guzmac, Bartholomeo de, 92
Gypsum, 187, 226, 232, 233; uses of, 231
Gypsy moth, 51

H

Haber ammonia process, 229
Hackett, Alice Payne, 292-3
Hadden, Briton, 295
Hagood, Gen. Johnson, 378
Hall, Charles, 206, 207; aluminum process, 208
Hall, John, 14
Hamilton, Alexander, 15
Hammond, G. H., 318
Hancock, Thomas, 237
Hanna, Marcus A., 73
Harcourt Brace & Co., Inc., 293
Harding-Cox election, 174
Hargreave, Lawrence, 99
Hargrove, Marion, 293
Harper & Bros., 290, 293
Harriman, Edward H., 58, 59
Harrow, 19, 310, 312
Hart, E. B., 316
Hartford Rubber Works, 252, 265
Harvard University, 52
Harvester, invention of, 304; sugar beet, 314; wheat, 304, 311
Harvey Machine Co., 405
Haskins, John, 237
Hatch Act, 307, 308
Hay baler, 314
Haynes, Elwood, 80, 82
Hazard, Rowland, 131
Health and comfort, standards of, 302
Hearst, William Randolph, 287
Hearst Consolidated Publications, Inc., 281, 295
Heavy chemicals, 133; makers of, 132-133
Helium, 142
Hell Gate Bridge, 57
Hematite ore, 196
Hemispheric peace, desirability of, 415

Henderson. Leon, 385. 386, 388
Henry, Joseph, 14, 19. 150. 176
Henson, William S., 98
Hepburn Act, 60
Hercules Powder Co., 133
Heroult, Paul, 206-07
Hertz, Rudolph, 166, 167
Hevea Braziliensis. 235 239
Hewitt, Abram S 41
Hide industry, 50
Highways, improvement of, 87
Hill, David E., 263
Hill, James J., 58, 59
Hillman, Sidney, 385, 389
Hill-Morgan lines, 59
Hindenburg, 113, 115
Hinman, John H., 300
Hitler, Adolf, 380
Hoe, Richard, 20, 287
Hoe, Robert, 14
Hog Island, Penna., 75
Hollander paper machine, 272
Hollerith, Herman, 347
Holmes, F. W., 158
Holmes, Oliver Wendell, 61
Holmes-Manley process, 221
Holt, Henry & Co., 293
Holy Tribunal, 93
Homestead Act of 1862, 33
Hood, H. P. & Sons, 327
Hoof and mouth disease, 51
Hoosac railroad tunnel, 191
Hoover, Herbert, 257; Food Administrator, 371
Hormel, George A., 321
Hormel Institute, 320
Hormones, 139
Horse power, farm, 310, 311, 313, 315
Horsford, E. N., 336
Houdaille, Hershey, 392
Houghton Mifflin Co., 290, 293, 296
Household gadgets, 151
Household manufactures, 6
Household production, 10
Housing, defense, 388
Houston, Edwin J., 158, 159
Howard, Roy, 287
Howe, Elias, 14, 19, 20, 29
Hubbard, Gardiner 151, 152, 153 157
Hudson Motor Car Co., 86, 393
Hudson R., 14, 42
Hudson River Railroad, 46
Hudson River Rubber Co., 253
Huenefeld, Baron Gunther von, 112
Huffman's Field, 109, 110
Hulett machines, 72
Hull, Cordell, 76, 385
Humble Oil and Refining Co., 249
Humphrey, G. C., 316
Humphreys, Colonel David, 11
Hunt, Alfred E., 207
Huntington, Collis P., 43, 44
Hussey, Obed, 310
Hyatt, John Wesley, 128
Hydrocarbons, description, 217
Hydro-electric installations, 162
Hydro-electric power, use of, 218
Hydrofluoric acid, 228
Hydrogen, 142; balloons first filled with, 94; isolation of, 132
Hydrogenation, 332; in oil production, 223; of cottonseed oil, 133; process in shortening industry, 332; in soap industry, 338
Hydrolization in soap manufacturing, 339

I

Icarus, 91
Ice cream industry, 324, 327
Ickes, Harold, 394
Iconoscope, 175
Illinois Central Railroad, 33, 58
Illinois Meat Co., 321
Illinois R., 69
Ilmenite Ore, 213-14
Immigrants, 37
Immigration Act of 1864, 33
Immigration of skilled workmen from Europe, 343
Imperial Chemical Industries, Ltd., 133
Incandescent electrical discharges, 159
Induced electric current, discovery of, 158
Industrial capitalism, 36

INDEX

Industrial conversion, 377
Industrial dislocation, 376
Industrial "know-how," 363
Industrial Mobilization Plan of 1931, 384; later revisions, 391
Industrial nations, 363
Industrial preparation, 365
Industrial production, 378; pools in World War II, 393-394
Industrial Rayon Corp., 141, 341
Industrial revolution, 7, 24
Industrial science, 54
Industrial scientists, 363
Industrial system, 15, 22, 25, 363
Industrial unrest in 1917, 372
Inflation, 6, 27
Infra-red lamps, 181
Inorganic products, makers of, 133
Insecticides, 137, 138, 143
Institute of Radio Engineers, 175
Instrumental accuracy, 343
Instruments of precision, 342ff
Insulation materials, manufacture of, 233
Insurance companies, 18
Interborough Subway, 64
Interessen Gemeinschaft Farbenindustrie Aktiengesellschaft, (I. G. Farben), 133, 213, 245, 247
Internal waterways, 68, 69
International Business Machines Corporation (formerly the Computing-Tabulating-Recording Co.), 347, 349
International Harvester Co., 306, 310, 315, 325
International Minerals and Chemical Corp., 231
International News Service, 287
International Niagara Commission, 164
International Paper Co., 280, 283, 288, 298, 299, 300
International Telephone and Telegraph Co., 177
International yacht races, 168
Internationalists, 380
Interstate Commerce Commission, 60, 403
Invention, history of, 91
Inventors, 13; American, 19; mechanical, 25
Investing, 4
Investment banking, 18, 24
Investors, European, 16
Iodine, 227, 228
Iron, 2, 4, 6, 30, 38, 47, 187, 192
Iron district of Alabama, 200; of Lake Superior, 190, 191, 194, 199; of Quebec Province, 200; of the Appalachians, 199; of Venezuela, 200
Iron goods, 18
Iron industry, 12, 17, 194
Iron machinery, 13
Iron masters, 7
Iron ore, 71
Iron, pig, 39
Iron products, 28
Iron, smelting of, 48
Iron works, 14
Irradiated food, 317
Irrigation, 306
Isoprene, 243, 244, 245, 246
Ivory soap, 338

J

Jackson, Dale, 112
Janeway, Eliot, 395
Japan attacks Pacific positions, 382; strength in Pacific, 383
Jarvis, William, 11
Jeffers, William, 249
Jefferson, Thomas, 126
Jefferson's Embargo, 17
Jeffries, Dr. John, 94
Jelatong, 255
Jenks, Joseph, 5
John of Portugal, 93
Johns-Manville Corp., 233, 406
Johnson, Louis, 384
Johnson, S. E., 315
Johnston, Gen. Joseph E., 35
Joint Army and Navy Munitions Board, 383, 395
Joint-stock arrangement, 17

INDEX

Jones Merchant Marine Act of 1920, 75
Jones-White Marine Act, 76
Jordan paper machine, 278; uses for, 278
Journal of Agricultural Chemistry, 309
Julien, Pierre, 96
Juvisy Field, 109

K

KDKA, 174
KYW, 174
Kaiser, Henry J., 393
Kaiser Aluminum and Chemical Co., 208, 405
Kanawha works, 30
Kandrell wheat, 307
Kansas Experiment Station, 307
Kaolin, 209
Keller, Friedrich G., 272
Kelly Air Mail Act of 1925, 114
Kelly, Fred C., 101
Kelly, Lt., 112
Kelly, William, 40, 41
Kelly-Springfield Tire Co., 263, 264, 265
Kelsey, Ben, 114
Kelvin, Lord, 164
Kendall, Amos, 19
Kennecott Copper Corp., 201, 204, 214
Kennecott Wire and Cable Co., (formerly American Electrical Works), 201
Kennedy, Joseph P., 385
Kerosene, 32, 48, 219
Kieselguhr, 127
Kimberly-Clark Corp., 284
Kinescope, 175
Kitimat project for aluminum production, 412
Kitty Hawk, N. C., 103
Knight, Walter H., 162
Knox, Frank, 385
Knudsen, William S., 385, 388
Koehl, Capt. Herman, 112
Koenig, Friedrich, 285, 287
Kok-sagyz, 241
Koppers Company Inc., 249
Korean airlift, 123
Korean peace, 180
Korean War, 401, 403, 406-407
Koroseal, 257
Kraft, James L., 326
Kraft Foods Co., 326
Kraft paper, 278, 280, 282, 284
Kraft wood pulp, 278
Krilium, 143, 144, 305
Kroll titanium process, 213

L

La Condamine, C. M. de, 237
Labor movements, 25
Lac Tio mine, 214
Lacquers and du Pont, 133
LaGuardia Field, 55
Lake Erie, 48, 70, 72
Lake Huron, 70
Lake Michigan, 99
Lake St. Claire-Detroit R. Channel, 70
Lake Shore Railroad, 46
Lake Superior, 30, 38, 70
Lake Victoria, Nyanza, German West Africa, 111
Lambert, Count de, 109
Lamboley, Francis X., 92
Lamp, incandescent, 162
Land grant colleges, 309
Langen, Eugen, 79
Langley plane, 101
Langley, Prof. Samuel P., 100, 101, 102, 106
Langmuir, Dr. Irving, 170, 172
Lard rendering, 318
Latin American Countries, trade with United States, 415
Lavoisier, Antoine Laurent, 126, 127, 132
LeBris, Jean-Marie, 98
Le Chatalier, Henry-Louis, 142
Lead, 187, 195, 203, 204, 205;—mines, Mississippi Valley, 188; mining of, 185; pigments, 204, 227
Leather, 3; artificial, 127; chemicals, 138

Lebaudy brothers, 97
Lederle Laboratories, 138
Lee, Gen. Robert E., 35
Leibwitz, G. W. von, 345
Leland, Henry M., 81
Lend Lease, 383, 385, 390, 394, 396
Lenoir, Etienne, 79
Leopold II, King of Belgium, 239
Levassor-Panhard car, 79, 80
Lever Act of August 1917, 371
Lever Brothers, 338
Lewis, Freeman, 300
Lewis, Sinclair, 292
Libbey, Edward D., 354, 355
Libby-Owens-Ford Glass Co., 233, 355, 357
Liberty Motor, 111
Lief, Alfred, 258
Life, 295
Lighting, 20; arc, 160; incandescent, 160
Lilienthal, Otto, 92, 98, 102
Limestone, 187; supply, 231
Lincoln, Abraham, 27
Linens, 10
Linoleum, manufacture of, 300
Linotype, 286
Lippincott, J. B. Co., 290
Liquid hydrocarbons, sources of, 223
Litchfield, P. W., 264-265
Lithopone, 204
Little Brown & Co., 290
Livingston, Robert R., 10, 11
Lloyd's Register of Shipping, 78
Loaders, mechanical, 193
Loamium, 306
Lockheed Aircraft Corp., 121
Locomotives, 28, 30, 39, 64
Lodge, Sir Oliver, 167
Logging, 276
Logistics, 377; in World War I, 369, 377, 378; in World War II, 382, 383
London Underground Railway, 64
"Long Armistice," 380
Long Island Railroad, 57
Long Island, transmitting center on, 172
Los Angeles, 113
Los Angeles, industrial region, 409
Lotteries, 8
Louis XV, King of France, 237
Louis XVI, King of France, 93
Lowe, T. S. C., 95
Lowell, Francis, 10
Lowery, Grosvenor P., 159
Lubin, Isadore, 389
Lucite, 145
Ludendorff, General Erich von, 367
Luce, Henry, 295
Lumber, 28, 32; mills, 14
Lumbering, 24
Lumbermen, 49
Lunardi, Vincenzo, 94
Lusitania crisis, 366

Mc

McAdoo, William G., 370
MacArthur, Gen. Douglas, 374, 383
McCall Corporation, 296
McClellan, Gen. George B., 95
McCormick, Cyrus, 14, 20, 310, 311
McCormick, Vance C., 371
McCormick-Deering, harvester line, 312
McCormick parlor milker, 325, 326
McCormick-Patterson Publications, 288
McCormick reapers, 33, 311
McGraw-Hill Publishing Co., Inc., 296
MacIntosh, Charles, 237
McNary-Watres Act, 116

M

Machine industry, expansion of, 304;—shops, 14;—tool industry, 389;—tools, 342, 343
Machinery, agricultural, 310; power-driven 22
Macon, 115
Macmillan Co., 290, 293, 296
Macready, Lt. J. A., 112
Magazine publishing, 294-6; statistics, 295
Maginot Line, 381

INDEX

Magnesium, 211-213
Magnesium Development Corp., 213
Magnetite ore, 196
Mail subsidies, 73, 74, 76
Mail transportation, subsidized, 116
Malcolmson, A. W., 82
Manganese, 215
Manitoba, Canada, 71
Manpower, 6, 15, 22, 32, 33, 35, 37, 302, 310, 311, 377-79; immigrant, 20; need for, 302; released, 303; scarcity of, 33; trained, 3, 363
Manton, Henry, 263
Manual telegraphy, 351
Manufacturers, value of, 1810-1860, 21
Manufacturing, 8, 16, 22, 24, 25, 29, 53, 147, 407
Manville, Frank, 109
Maple sugar, 32
Marble supply, 231
Marconi Corporation of America, 172, 173
Marconi, Guglielmo, 166, 167, 168, 171, 176
Marconi Wireless Telegraph Co., 167
Margarine, 332, 333
Marine torpedo, 14
Maritime Commission, 77
Marks, Arthur H., 253, 254, 255
Marquette Range, 38
Marsh Brothers, 311
Martin, Glenn L., 183
Martin, Glenn L., Co., 392
Mass production, 119;—of motor cars, 81
Massachusetts Institute of Technology, 52, 264
Master Car Builders Assoc., 63
Masticator, 237
Match, friction, 21
Materials, conservation of, 180; economy of, 180; requirements by 1975, 417
Maxim, Sir Hiram, 99, 101
Maxwell, James Clerk, 166
Maynard, J. Parker, 128
Means, James, 102
Meat, 3; canning, 320-321
Meat industry, 50, 317-322; center of, 318; evolution of, 319; mechanization in 318, 322; refrigeration introduced, 318; research in, 320; waste, utilization of, 319-320
Meat proteins, nutritional value of, 320
Mechanical methods, 8
Mechanization, 50
Medical research, 139
Melamine, 139
Mellon Institute, 359
Menominee range, 38
Merchant fleet, 77; increased by Liberty Ship program, 1941-1942, 392; production line methods, 393
Merchant marine, 73, 74, 75, 76, 377
Merchant Marine Act of 1936, 77
Mercury, 186
Mergenthaler, Otto, 286, 287
Merino sheep, 11
Merlon, 341
Merrimac R., 14
Mesabi range, 38, 190
Metal industry, reproductive, 12
Metallic minerals, 187
Metallurgical research, 187
Metallurgists, 25
Metals, manufacture of, 186; non-ferrous, 196; precious, 188
Metals Reserve Corps., 388
Meters, 349-350
Methoxychlor, 137
Meusnier, Gen. Jean-Baptiste, 96
Mexican War, 126, 364
Micarta, 178
Michigan School of Mines, 52
Microscopes, 342; dynoptic, 360; electron, 269
Microscopic precision, 35
Miles, Lucius, 263
Military and Naval establishments, American, 380
Military defense, 363, 364, 365, 367, 378, 382-383
Milk, concentrated fluid, 328; evaporated, 324; homogenized, 328; pro-

duction rate of, 323-324; refrigeration of in transportation, 325, 328; —bottle, 327;—container, 327, 337; —coolers, 325;—disperser, refrigerated, 328;—solids, 326
Milking machines, 314, 325-326, 328
Mill, Henry, 345-346
Miller, William, 291
Miller's National Exhibition, 330
Milling industry, 14, 50, 329-332; electrification of, 330; mechanization of, 331; new process in, 329; steel rollers introduced in, 329-330
Mineola, N. Y., 110
Minerals, 185
Miner's lamp, electric portable, 193
Mines, anthracite, 30; development of, 46; schools of, 52
Mining, 2, 15, 24, 25, 33, 185, 187; —industry, 126; non-selective, 190; —operation, 49;—operations of oil ands, 222; selective, 190;—technology, advancement in, 218
Mirroplane, 358
Mississippi R., 27, 69, 70
Mississippi Valley, 31
Missouri Experimental Station, 309
Missouri lead region, 204
Missouri R., 42, 69, 70
Mitchell, Margaret, 293
Mitscherlich, Eilhard, 273
Mixed gas, use of, 224
Mixes (flour), 330, 333
Mobilization, incomplete, 378;—plans, 365
Model T, 82
Molybdenum, 215
Monoplanes, 110
Monsanto Chemical Co., 133, 143, 144, 230, 249, 305, 341, 413
Montgolfier, Joseph & Etienne, 93, 94
Montgomery, John J., 99
Moody, Paul, 10
Morgan, J. P., and Co., 58, 60
Morgenthau, Henry, 384
Morrill Act, 50
Morris & Co., 319, 321
Morse, Samuel F. B., 19, 150, 176
Motor bus, 87, 88
Motor car, 56, 78, 147;—assembly industries, 83;—assembly lines, 84; —body, all-plastic, 148; closed, 93; —industry, 80, 87, 149, 375; need for safety glass, 355-356; cut in production in 1941, 390; geared to defense production, 391-392; affected by Korean War, 404; affected by steel strike of 1952, 405;—manufacturers, 84;—parts manufacturer, 84;—plans, 83;—production, 86
Motor, development of, 158;—manufacturing, 123;—truck, 87, 88, 195
Motorized freight carriers, 56
Movies, talking, 166
Mt. Palomar telescope, 179, 356
Mt. Wilson Observatory, 356
Mulford, Frederick B., 309
Multiplex telegraphic system, 351
Munitions, 388
Munson, Charles S., 407-408
Murrell, Melville M., 92
Mushet, Robert, 40
Mussolini, Benito, 380
Mutual Aircraft Corp., 117
Myers, Charles F., 92
Mylar, 144, 145

N

NC-4, 112
NDAC, 387, 388
Nally, E. J., 173
Napoleonic Wars, 8, 362
Nash Motors, 86
Nashville, Tenn., 42
Nassau Boulevard Airdrome, 110
National Air Transport, 117
National Aniline and Chemical Co., 134, 257
National Bank Act of 1864, 37
National Biscuit Co., 330, 331
National Broadcasting System, 175
National Carbide Corp., 142
National Cash Register Co., 345
National Dairy Products Corp., 327
National Defense Act of 1916, 372
National Defense Advisory Commission, 384-388

INDEX

National Electric Signalling Co., 169
National Industrial Recovery Act, 87
National Lead Co., 205, 213
National Live Stock and Meat Board, 320
National security, 122
National Television Committee, 183
National War Labor Board, 1918, 372
Natural gas, 47, 219, 220, 221, 226
Naugatuck Valley, 11
Naval stores, 4
Nelson, Donald M., 386, 391, 395
Neolite, 266, 340
Neon, 142
Neoprene, 137, 246
Neutrality legislation, 76
Nevins, Allan, 37
New England Glass Co., 354
New England Telephone Co., 153
New Jersey Zinc Company, 202, 214
New products, discovery of, 134
New York Automobile Show (1904), 261
New York Central Railroad, 45, 46, 57, 58, 59, 64, 69, 71
New York *Daily News,* 288
New York *Herald,* 159
New York *Journal and American,* 288
New York *Mirror,* 288
New York *Sunday Mirror,* 288
New York *Sunday* News, 288
New York *Times,* 159, 288
New York *Tribune,* 286
Newport News Shipbuilding and Drydock Co., 78
Newspaper publishing, 279-281, 285-289; circulation, 288
Newsprint, 274, 279, 281-82, 284; manufacture of, 276; production of, 274, 277
Niagara Falls Power Co., 211
Niagara R., 70
Nieuwland, Dr. Julius, 245, 246
Nitrocellulose, 127
Nitrogen, 142, 305;—and its products, 229, 230; supply of, 305
Nitroglycerine, 126, 127, 192
Nitrous oxide, 143
Nobel, Alfred, 17, 128
Non-importation rules, 6
Non-metallic elements, definition, 17; —minerals, manufacture of, 186;—minerals, 187
Norfolk and Western Railroad, 58
North African Campaign, 395
North American Aviation, 121, 392
North Pole flights, 112
Northcliffe, Lord, 109
Northern Pacific Railroad, 43 ,45, 58, 59
Northern Securities Co., 59, 61
Nuclear physics, 90, 184
Nye Committee, 374
Nylon, 137, 146, 341

O

O'Brien, Forrest, 112
Ocean Mail Act, 73
Ocean vessels, 73
Oceanic Steamship Co., 73
Odlum, Floyd B., 393
Oenschlager, George, 254-257
Oersted, H. C., 206
Office of Economic Stabilization, 395
Office of Emergency Management, 1940, 384
Office of Price Administration and Civilian Supply, 388
Office of Production Management, 388
Office of Scientific Research and Development, 398
Office of War Mobilization, 391
Ohio R., 27, 69, 70
Ohio State University, 326
Ohio Valley, 30
Ohio-Mississippi system, 42
Oil, see Petroleum
Oil burners, 86
Olds, R. E., 80, 81, 82
Oleomargarine, see Margarine
Oligopoly, 136; in rubber, 250
Olin Industries, 405
Oliver Iron Mining Co., 200
Oliver, James, 19
Ontario Paper Co., 288
Open-hearth process, 40, 41
Open pit mining, 194

Ophthalmic profession, 359
Optical industry, 353, 358, 359-360
Ordnance, lack of, 388;—plants, 388; —Proving Ground, 165
Oregon, source of pulpwood, 275;—Territory, 21
Ores, low grade, 187
Organic accelerators, 254, 255, 257
Organizational changes in World War II, 391
Organized labor, 61
Orinoco Mining Co., 200
Orlon, 137, 146, 341
Ornithopter 91, 92, 98, 99
Osborne, D. M., Co., 311
Otto, Nicholas A., 79
Overman Act, 369
Ovid, 91
Owen-Illinois Glass Co., 233
Owens, Michael J., 354-355
Owens' bottle making machine, 354
Oxygen, 142, 143; commercial, 142

P

Pacific Gas and Electric Co., 414
Pacifism, 380
Pacinotti, Antonio, 158
Packard Motor Car Co., 86, 111, 391
Paints 133, 140;—and lacquers, 277; manufacture of, 233
Paley, William S., 416
Panama Canal, 69, 74;—Canal Act of 1912, 74
Panhard, 80
Panic of 1873, 354;—of 1893, 163, 263
Paper, 138;—industry, 146, 271, 285, 296, 300;—industry, development of, 273, 274;—industry, place in national economy, 275;—machines, 278
Paperboard, 280, 284; raw materials for, 275
Paper-bound books, popularity of, 300
Papyrus, 272
Para-amino-dimethylaniline, 256
Parchment, 272
Paris Exposition of 1855, 206
Parkes, Alexander, 126
Parkesine, 128
Partial wartime economy, after 1950, 402
Pascal, Blaise, 345
Passaic River, 16
Passenger cars, 39; all-steel, 64; special purpose, 62
Passenger traffic, 123
Pasteur, Louis, 129, 302
Pasteurization, milk, 325, 327
Patent rights, 162
Patent system, 5
Patterson, John H., 345
Patterson, Robert P., 385
Patterson, W. A., 119
Pawtucket Falls, 9
Peal, Samuel, 237
Peanut crops, 308
Pearl Harbor, 120, 80-381
Penaud, Alphonse, 99, 101
Penicillin, production of, in World War II, 398
Peninsular Campaign, 11, 95
Pennsylvania Railroad, 45, 57, 58, 71
Pennsylvania Salt Co., 207
Pennsylvania Steel Co., 40
Perkin Medal, 257
Perkin, W. H., 129
Perkins, Jacob, 14
Pershing, Gen. John J., 368
Pest control, 304, 305; use of the airplane in, 304
Pesticides, 305; uses for, 305
Petroleum, 22, 31, 47, 185, 203, 219, 221, 226; commercial, 308; cracking process, 220-221; extraction of, 221-223;—geologists, 25;—industry, 49, 146, 187, 339;—manufacturing, 48; offshore, 224; pipeline in 1941, 394;—promoters, 49;—refiners, 37; —refining, 48; reserves, 225; role in industry, 218;—shipping, 48; technology, 219, 223
Petroleum-bearing shales, production from, 222
Pettibone, Daniel, 14
Pfeiffer Brothers, 263
Pharmaceuticals, 130, 320

INDEX

Phelan and Collander, 128
Phelps Dodge Corp., 201, 205
Phenol, 226
Philadelphia *Inquirer,* 288
Philippine War, 365
Phillips, Alonso, 21
Phillips, Horatio, 99
Phillips Petroleum Company, 249
Phosphate rock, 186
Phosphates, 187, 229, 230
Photo-electric cell, 166
Photo-engraving, 213
Physicists, 139
Piccard, Auguste, 115
Pickard, G. W., 169
Pilcher, Percy S., 98
Pillsbury, Charles A., 306, 329, 330
Pillsbury, Fred C., 323, 330
Pillsbury, John S., 329, 331
Pillsbury Mills, Inc., 306, 329
Pilots, training of, 109
Pinene, manufacture of, 300
Pipelines, natural gas, 224
Piper Aircraft Co., 121
Pittsburgh Corning Corp., 322
Pittsburgh Plate Glass Co., 233, 354
Pittsburgh Reduction Co., (later Aluminum Co. of America), 207, 209, 211
Pittsburgh, University of, 167
Plant conversion, 377; experts, 25
Plantation system, 33
Plastics, 127, 133, 138, 147, 340; molded, 139; use in clothing, 340
Platinum containers, 359; -- stirrers, 359
Pliofilm, 266
Plow, deep soil, 304; iron, 14, 303; steel-faced, 310
Pneumatic Steel Co., 40
Pocket Books, Inc., 300
Polyuronides, 144
Pooling, 57
Population, 9, 21; farm, 316; increase of, 303, 324
Porkopolis, 318
Port facilities, 378
Postwar problems, 380
Potash Company of America, 231
Potash, supply of, 305
Potassium, 228, 229, 230, 231;—chloride, production of, 231
Potomac River, 100
Poulsen arc transmitter, 168
Powder, smokeless, 127
Power, 4, animal, uses for, 304; electric, 175, 176; hydro-electric, 151, 163; turbo-electric, 211; packaged, 179; sources of, 48
Prefabricated ships, 75
Preparedness legislation, 366
President's Materials Policy Commission, 416
Prevost, Maurice, 110
Priestley, Joseph, 93, 132, 237
Princeton University Press, 294
Printed electrical circuits, 359
Printing, 285-287;—and publishing industry, uses for paper in, 276
Priority lists, 1941, 390, 392
Privateering, 6
Proctor & Gamble Co., 306, 332, 337-339
Profit taxes, 387
Profits in wartime, 373
Promontory Point, Utah, 44
Protective relays in electrical communication, 350
Public Resolution No. 98 (1930), 374
Public Service Co., of Northern Illinois, 414
Publishers' Weekly, 290, 292
Publishing, 285-97; cost statistics, 293-294, 295-296; See also Book publishing, Newspaper publishing, and names of individual publishing houses.
Pulitzer, Joseph, 287
Pullman cars, 46
Pullman, George M., 62
Pulp and paper, production in America, 275
Pulping, chemical, cost of, 277; mechanical, cost of, 277
Pulpwood, source of supply, 275
Punched cards for recording data, 347-348

INDEX

Purchasing power, 25
Putnam's, G. P., Sons, 290

Q

Quebec Iron and Titanium Corp., 214
Quebec North Shore Co., 288
Quimby, Watson F., 92

R

RFC, 388
Radio, 151, 287, 305
Radio and radar devices in World War II, 398
Radio City, 175, 178
Radio Corp. of America, 172, 173, 174, 175, 177, 179, 182
Radio, "ham," 169;—industry, 171;—links, 156;—manufacturing, 174;—pictures, 174;—relay, 182, 183;—relay microwave systems, 404; short-wave, 182;—ultra-high frequency, 156, 182; uses for, 305;—waves, 166
Radiotelegraph circuits, transoceanic, 172
Radiotelephone, 157
Radiotelephony, 171, 173
Railroad, bridges, 30; building, 38; car shifter, mechanical, 195; cars, all steel, 63; construction, 30; demands of modern industry, 66; equipment, 56, 61, 123; extension, 43; freight, 88; promoters, 37; revenue, 88; systems, 47; transportation, 35, 44, 132
Railroads, 18, 22, 35, 42, 56, 69, 71, 73, 87, 88, 317; earnings in 1952, 403; effect on industry, 46; French, 378; modernizing of equipment, 403
Rate wars, 57
Raton Pass, 65
Raymond, Henry J., 287
Rayon, 129, 133, 140, 146, 340, 341; —industry, 139, 297-298, 340-341
Reader's Digest, 295
Reading Railroad, 58
Reading taste, 290-93, 295
Reaper, 14, 20, 303, 310, 311; invention of, 304
Rebates, 57
Reciprocal Trade Agreements Act of 1934, 76
Reclamation, 305
Reconstruction, 150
Reconstruction Finance Corporation, 248, 388, 393, 407
Records, airplane distance, 107; distance, 114; speed, 114
Red D Line, 73
Refrigeration, food, 306, 317, 318, 319, 322, 328, 333, 336
Refrigerator cars, 62, 318-19
Refrigerators, electric, 86
Regenerative circuit, 174
Remington, E., & Sons, 346
Rensselaer Polytechnic Institute, 52
Repeater stations, 156
Reperforator switching in telegraphy, 351
Republic Mining and Manufacturing Co., 207
Research, departments, 53; — laboratory, 136; medical, 139;—personnel, 135
Resloom, 341
Retail sales in 1952, 402
Revere Copper and Brass Co., (formerly Taunton-New Bedford Copper Co.), 201
Revere, Paul, 13
Revolver, 14
Reynolds Metal Co., 208, 405
Richmond, Fredericksburg and Potomac Line, 67
Rittenhouse, William, 273
Ritty, James J., 345
River transportation, 69
Robena mine, 200
Robert, Louis, 272
Robespierre, Maximilien Francois Marie Isidore, de, 126
Rocky Mountains, 48
Rogers, Cal P., 109

Roller process, 330
Rolling mills, 41
Roosevelt, Franklin D., 384, 396
Roosevelt, Theodore, 70, 73, 74
Rosin, production of, 300
Rotary printing press, 14, 20
Rotogravure process, 286
Roxbury India Rubber Co., 237-238
Royal Aeronautical Society, 98
Rubber, 22, 138, 235-270; Amazon, 230;—chemistry, 255; compounds, 238; crepe, 406; crude, 239, 248; fire hose, 253; foam, 252, 266;—industry, 146;—industry in 1952, 406-407; jelatong, 255; jungle, 240; latex, 241; mechanical goods, 252, 266, 407;—plantations, 240, 241, 252, 262, 266;—products, new, 69-70; shoes, 238, 252, 266;—shortage, 243, 244, 388; silicone, 269; synthetic, 147, 39; transmission belting, 253, 407;—vulcanizing, 20;—vulcanizing process, 14; waterproof garments, 38, 252; See also Rubber tires
Rubber Goods Manufacturing Co., 252
Rubber Manufacturers Association, 407
Rubber Reserve Corp., 248, 249, 388
Rubber tires, automobile, 265, 266; bicycle, 239, 253; carriage, 239, 258, 265; clincher, 261; cord, 262; low pressure, 262; pneumatic, 239, 252, 260, 261, 266; rims, 261; safety treads, 261; sales in 1952, 406-407; solid, 252, 260; straight side, 261, 266; truck, 262
Ruberoid Co., 229
Rutgers University Press, 294

S

Sabre engine, 387
Safety glass, 85
St. Clair, Labert, 78
St. Joseph Lead Co., 205
St. Louis Exposition of 1904, 261
St. Louis *Dispatch and Pioneer Press*, 288
St Regis Paper Co., 284
Salt, deposits, 30;—mines, 186
Salvarsan, 130
Sand, production of, 232; uses of, 231
Sanders. Thomas, 151, 152, 153, 157
Santa Fe Railroad, 58, 65
Santos-Dumont, Alberto, 97, 105
Sarnoff, David, 172, 173, 174
Sault St Marie canal, 30, 70
Schafer, Joseph, 322
Scheele, Henry, 132
Scheele, Wilhelm, 93
Schlesinger, Arthur M., 397
Schoellkopf, Jacob, 162
Schoen, C. T., 63
Schoenbein, C. F., 127
Scholfield brothers, 9
Schwab, Charles M., 75, 370
Science, applied, 25; teaching of, 52
Scrapers, 89
Screw propeller, 14
Scribner's, Charles, Sons, 290, 293
Scripps-Howard Newspapers, 281
Seaboard Line Railroad, 67, 72, 403
Sealtest laboratories, 327
Searles Lake region, 230
Sears, Roebuck and Co., 386
Second Bank of the United States, 18
Security, collective, 380
Seed drill, 20
Seeder, invention of, 304
Seeding machines, 304, 312
Seiberling, Charles, 265
Seiberling, Frank, 263-264, 265
Selden, George, 79
Selective extraction, 191
Selective Sequence Electronic Calculator, 348
Selective Service Act, May 18, 1917, 368; Sept. 16, 1940, 383, 388
Selfridge, Lt. Thomas, 106
Semet-Solvay Co., 134
Semolina, 330, 331
Separation of ore, 195, 197
Services of Supply in World War I, 378
Sewing machine, 14, 19, 20
Shakers, 325

INDEX

Shearing machines, 53
Sheep, breeding of, 10; merino, 11
Sheffield Farms Co., 327
Shell Oil Co., 249
Shipbuilding, 6
Shipping, Atlantic coast, 78
Shipping, companies, 73;—lack of in 1914, 74; transoceanic, 78;—under war emergency, 378
Shipyards, 75, 389
Shoe industry, 19;—manufacturing, 29
Sholes, Christopher, 346
Shortages, 378, 388;—in defense areas in 1940, 388
Shortening, 332-3
Shorthand systems, 346
Shovel, caterpillar traction electric, 194
Shriner's closed retort, 335
Shrinkage stopping, 190
Signal Corps, 95, 108
Silliman, Prof. Benjamin, 31
Silver, 37, 185, 188, 195
Silver-bearing ore, 196
Sinclair, Upton, 321
Sinclair Oil Corp., 249
Sisler, Dr. Louis E., 259
Slater, Samuel, 9
Slave system, 35, 36
Slavery, 7
Smith, Betty, 293
Smith, Datus, Jr., 294
Smith-Hughes Act, 316
Smithsonian Institution, 19, 100, 101, 102
Soap industry, 140, 300, 332, 337-339; —manufacturing, 305
Sobrero, A., 127
Society for Establishing Useful Manufactures, 16
Socony-Vacuum Oil Co., 389
Soda, 130-131, 277
Soil, conditioners, 143, 305, 306, 322; —experts, 25;—nutrition, 305
Soil conservation, 309
Solvay, Ernest & Alfred, 130
Solvay Process Co., 131, 134
Sorghum culture, 32
Sound reproduction by wireless, 168
South Carolina, increase in manufacturing, 409
Southern Pacific Railroad, 44, 58
Southern pine, uses for, 276
Southern states, secession of, 27
Soviet Union, 380; eligible for Lend-Lease in 1941; 394
Soy bean, 308
Spaghetti, 330
Specie payments, 28
Spectrograph, 360
Speculation, 6; land, 16
Speculative activity, 28
Spence, David, 256-257
Sperry Corp., 121
Spiegeleisen, 40
Sprague, Frank J., 161
Spreads, 332
Spruce Falls Power and Paper Co., 288
Square-set timbering, 188
Stadelman, G. M., 265
Stainless steel, 41
Stalin, Josef, 380
Stamford Laboratories, 138
Stamping mill, 196
Standard, interchangeable parts, 10
Standard Oil Co., 48, 71, 74;—of Indiana, 221;—of New Jersey, 245, 246, 248, 249;—of Ohio, 219
Standard Oil Development Co., 247
Standardization of parts, 83, 353; in motor car industry, 85
Stanley, William, 160
State experiment stations, 309; Connecticut, 307; Kansas, 307; Missouri, 309; Ohio, 326; Wisconsin, 326; See also Agricultural experiment stations
State Extension Agencies, 323
State fairs, 33
State governments, 18
State highway systems, 87
Steam, buggies, 79;—drill, 191;—engine, 15;—engine, Corliss, 151;—engine, stationary, 14;—locomotive, 64;—locomotives, replacement of, 65;—power, 7, 14, 22, 38, 47, 64;

—shovels, 190;—turbine, 165;—vessel, 71
Steel, 30, 38, 47, 388; age of, 22;—industry, 187;—industry, pattern of, 132; manufacture of, 39—mill operators, 37; shortage in 1918, 375; in 1941, 392; control of, in World War II, 395
Steenback, Harry, 317
Stereoscopic apparatus, 178
Stereotype plate, 286
Stettinius, E. R., Jr., 384, 385
Stevens, Capt. Albert, 115
Stevens, John, 14
Stevenson Act of 1928, 244, 262
Stiegel, William H., 353
Stimson, Henry L., 385
Stock company, informal, 17
Stock farm, improvement in, 308, 323
Stockly, George W., 158
Stone, dimension and crushed, uses for, 232, 234; quarrying of, 186
Stove, cooking, 21; heating, 21
Strategic materials, 386, 404
Stratovision, 183
Stringfellow, John, 98
Strutt, Jedediah, 9
Studebaker Corporation, 86, 392
Sturbridge Village, Mass., 3
Styrene, 247-248, 249
Sublevel caving, 190, 192
Submarines, German, 74; detecting devices, 398; in World War II, 382; menace of, 377
Substitutes developed in World War II, 392
Subway, New York, N. Y., 165
Sugar shortages, 32
Sulfa drugs, 139, 140
Sulphate process, definition of, 277;—in paper manufacturing, 273, 275, 278
Sulphite process, 273-274, 277
Sulphur, 227
Superheterodyne circuit, 174
Superior district, 185
Supply, critical, 180
Supply, Priorities and Allocations Board of, 1941, 390, 392
Supreme Court, 59, 61, 183
Surcouf, 97
Surgical supplies, 143
Sweet's Restaurant, 159
Swift, Gustavus F., 318, 319
Swinehart, James A,, 259
Switchboard, first central, 152

T

TVA (Tennessee Valley Authority), 211, 230, 305
Taconite, 200
Taft, W. H., 70
Taiya project for aluminum production, 411-412
Talloil, manufacture of, 300
Tanks, 387, 388, 389;—for British Army, 387
Tar products, 225, 226
Tariffs, 37, 76; Tariff of 1913, 275
Taylorcraft Corp., 121
Technical and vocational education, 51-53
Technology, 53; American, 378
Telecar mesage delivery service, 352
Telegraph, 14, 19, 22, 24, 49, 150, 176; harmonic, 151
Telegraphic equipment, 19
Telegraphy, wireless, 167, 168
Telephone, 49, 151, 152, 176; dial instrument, 154, 155; wireless, 168
Telescopes, 342
Teletypesetter, 287
Teletypewriter, 287, 351
Television, 151, 182, 183; color, 183; color, field-sequential system, 183; national network, 404;—tubes, 358; uses for, 305
Tesla, Nicola, 164
Tetraethyl lead, 220
Texas cattle fever, 51
Texas Co., 221
Texas Pacific Railroad, 43
Texas Rangers, 20
Textile fabrics, artificial, 340;—industry, 10, 297-298, 340-341

Textiles, 2, 3, 149; rubber-coated, 127
Thermopane, 357
Thiocarbanilide, 256
Thomas, Charles Xavier, 345
Thompson, William Boyce, 109
Thomson, Elihu, 158, 159
Thomson-Houston Co., 163
Thresher, 310, 312
Thumb district, 227
Thurber, Charles, 14
Tillinghast bicycle tire patents, 265
Timber, 3; cutting of, 49
Time, 295
Time, Inc., 295
The Times (London), 285
Timken roller bearings, 66
Timken Roller Bearing Co., 392
Tin, 204; shortage of, 335
Tin cans, 320-321, 333-335
Titanium, 203, 213-215; compounds of, 204, 227; smelting process, 214
Titusville, Penna., 31, 48
Tobacco, 4, 34
Toluol, 226
Total war, characteristics of the Twentieth Century, 362-363; 367, 378
Tractors, 304, 312-314; Diesel-powered, 313; Farmall, 312-314; gasoline powered, 304, 313; touch control on, 314
Trade agreements, 416
Training schools for pilots, 109
Trains, control systems, 67
Transatlantic cable, 49
Transcontinental and Western Airlines, 118
Transcontinental railroad, 31, 43
Transmissions, automatic, 147
Transport, freight, 123
Transportation, 7, 15, 16, 24, 25, 363;—companies, 41;—equipment, 28;—industry, 126;—road, 35;—water, 7
Transworld Airlines, 122
Travel, volume of, 89
Tri-State lead and zinc district, 193
Trolley car, 162;—lines, interurban, 87
Trucking, transcontinental, 87
Trucks, 388; Automotive form, 314; Diesel powered, 314
Truman Committee, 395
Tryptar, 320
Tube manufacturers, 41
Tube mills, 197
Tuf-flex, 357
Tung, 308
Tungsten, 215
Turbine, 53; steam, 165; water, 12
Turnpike companies, 18
Turnpikes, 16, 89
Twain, Mark, 346
Typesetting, 286
Typewriter, 14, 346-347

U

U-boat program, 366
Ultra-high frequency radio, 182
Ultra-violet lamps, 181;—spectrograph, 269
Underwood, William, 334
Union Army, 29
Union Carbide and Carbon Corp., 133, 147, 249
Union Electric Co., of Missouri, 413
Union Pacific Railroad, 43, 44, 58, 65, 66, 249
Union Stock Yards, 317
Union Telegraph Co., 177
United Airlines, 118, 119, 122
United Box Board and Paper Co., 293
United Fruit Co., 74
United Paper Board Co., 283
United Press, 287
The *United States,* 78
United States Air Force (formerly United States Army Air Force), 215, 257, 409-410
United States Armed Forces, 181
United States Army, 7, 106, 108, 115, 173, 366, 371, 375, 376, 383, 385, 386, 390
United States Bonds, 7
United States Bureau of Aeronautics, 212

INDEX

United States Bureau of Agricultural and Industrial Chemistry, 307
United States Bureau of Agricultural Industry, Meat Inspection Div., 333
United States Bureau of Animal Industry, 307
United States Bureau of Dairy Industry, 308, 324, 325, 326
United States Bureau of Human Nutrition and Home Economics, 308
United States Bureau of Mines, 142, 215, 222, 223, 225
United States Census, 16
United States Circuit Court of Appeals, 162
United States Congress, 5, 6, 31, 171, 244, 369, 384, 407
United States Department of Agriculture, 33, 50-51, 244, 307-09, 330; See also names of individual bureaus and divisions
United States Department of Commerce, 77, 415
United States Department of Justice, 213
United States Department of Labor, 389
United States Division of Soils, 308
United States Electric Lighting Co., 162
United States Experiment Stations, 51
United States Food Administration, 1917, 371
United States Fuel Administration, 1917, 371
United States Grain Corporation, 371
United States highways, 87
United States Maritime Commission, 390, 392
United States Military Academy, 52
United States Navy, 112, 113, 167, 173, 215, 360, 364, 366, 375, 376, 382, 383, 386, 390
United States Post Office Dept., 114, 123
United States Potash Co., 230-231
United States Production and Marketing Administration, 308
United States Railroad Administration, 1917, 370
United States Rubber Company, 248, 251-252, 266
United States Shipping Board, 74, 75, 77, 366, 370-371
United States Smelting and Refining Co., 205
United States Steel Corp., 42, 74, 199, 200, 376, 384, 385
United States War Dept., develops heavy bomber program, 387
United States Weather Bureau, 103, 167
United Wireless Service, 168
Universal Atlas Cement Co., 233
University presses, 293-4
Unwin, Stanley, 290
Uranium, 184, 203, 215, 235

V

Vail, Alfred, 19
Van Depoele, C. J., 162
Van Nostrand, D., Co., Inc., 290, 293
Vanadium, 215
Vanderbilt, Cornelius, 45, 46, 58
Varnishes, 140
Vegetables, blight-resistant, 307
Venango County, 31
Vermillion range, 38
Vessels, cargo, 28; merchant, 369; naval, 389; river, 28; steam, 22
Vinyl plastics, 143
Vinyl plastic flooring, 266-267
Virginia City, Nevada, 188
Viscose process, 141
Visking Corp., 336
Vista-Dome, 66
Vitamins, 139, 317, 331
Volta, Alessandro, 176
Vulcalock, 257

W

WEAF, 174, 175
WGY, 174
Wall Street, 18, 28
Waltham, 10

INDEX

Ward Line, 73
Washburn, Cadwallader Co., 329
Washburn, Crosby Co., 329, 330
Washburn, Ichabod, 13
Washington, George, 7, 95
Washington pulpwood production in, 275
Waste products, utilization of, 300, 320, 323, 330, 339
Watch making, 19
Water power, 5, 14, 22, 47;—supply, control of, 305, 306
Waterbury, Connecticut, 19
Water-repellents, 139
Waterways, internal, 42
Watson, Thomas A., 151, 152, 153, 157
Watt and Boulton engine, 14
Weapons, newly-designed American, 378
Weather service aids, 116
Weber, C. O., 255
Welding equipment, 143
Welland Canal, 70
Wenham, Francis, 98
West Indies Steamship Lines, 73
Western Air Express, 117
Western beef, prejudice against, 319
Western Edison Co., 162
Western Electric Co., 154
Western hemlock, uses for, 276
Western Maryland Railroad, 67, 68
Western Pacific Railroad, 65
Western Union Telegraph Co., 150, 152, 153, 168, 351-352
Westinghouse air brake, 46, 62
Westinghouse Electric and Manufacturing Co., 160, 163, 165, 167, 174, 178, 179, 181, 82, 83, 349-350, 376
Westinghouse, George, 60, 161, 164
Whale oil, 21
Wharton School of the University of Pennsylvania, 52
Wheat, harvesting time for, 311; production, 330; spring, 329; See also names of kinds (e.g. Durum wheat)
Whitaker, M. C., 139
Whitman and Barnes, 258-259
Whitney, Asa, 43
Whitney, Eli, 9, 14
Whitney's gin, 7
Whole Books of Psalms, 289
Wickham, Sir Henry, 239-240
Wiley, John, & Sons, Inc., 290, 293
Williamsburg, Virginia, 3
Willys, John N., 81
Willys-Overland Co., 392
Wilmington, North Carolina, 35
Wilson, Thomas E., 320
Wilson & Co., 321
Wilson, Woodrow, 172, 366, 368, 376, 384
Wireless communications, 171
Wire manufacture, 13, 19, 41
Wisconsin, University of, 326
Wistar, Caspar, 353
Wohler, Friedrich, 206
Women in industry, 302-303
Wood, grinding machines in paper manufacturing, 273-274
Wood, James J., 158
Wood, Jethro, 14, 19, 303
Wood pulp, 274-278, 281, 285, 296-300
Wool, 6
Woolen factories, 12, 14, 29
Woolens, 9, 10, 11
Workers, employed, 21
World War I, 109, 130, 133, 194, 201, 229, 244, 304-305, 365-379, 380;—effect of flying, 109
World War II, 76, 87, 175, 201, 212, 214, 242, 363, 380-399, 408
World's Columbian Exposition at Chicago, 80
World-wide telephonic system, 157
Wrapping paper, production of, 277; raw materials for, 275, 278
Wright brothers, 99, 101, 102, 104, 105, 106, 107, 108, 109, 210
Wright Co. of the U. S., 108, 109; develops "ram jet" engine, 411
Wright, Orville, 105, 107
Wright, Wilbur, 107
Writing paper, production of, 277

INDEX

X

Xenon, 142
X-ray, 178;—camera, 181;—image amplifiers, 181
Xylol, 226

Y

Yale University, 52, 167
Young, Owen D., 172

Z

ZR-3, 113
Zeppelin, Count Ferdinand von, 96, 97, 111
Zeppelins, 111, 115
Zinc, 186, 187, 195, 203, 204-205; oxide of, 204, 227
Zworykin, Dr. V. K., 75

TECHNOLOGY AND SOCIETY

An Arno Press Collection

Ardrey, R[obert] L. **American Agricultural Implements.** In two parts. 1894

Arnold, Horace Lucien and Fay Leone Faurote. **Ford Methods and the Ford Shops.** 1915

Baron, Stanley [Wade]. **Brewed in America:** A History of Beer and Ale in the United States. 1962

Bathe, Greville and Dorothy. **Oliver Evans:** A Chronicle of Early American Engineering. 1935

Bendure, Zelma and Gladys Pfeiffer. **America's Fabrics:** Origin and History, Manufacture, Characteristics and Uses. 1946

Bichowsky, F. Russell. **Industrial Research.** 1942

Bigelow, Jacob. **The Useful Arts:** Considered in Connexion with the Applications of Science. 1840. Two volumes in one

Birkmire, William H. **Skeleton Construction in Buildings.** 1894

Boyd, T[homas] A[lvin]. **Professional Amateur:** The Biography of Charles Franklin Kettering. 1957

Bright, Arthur A[aron], Jr. **The Electric-Lamp Industry:** Technological Change and Economic Development from 1800 to 1947. 1949

Bruce, Alfred and Harold Sandbank. **The History of Prefabrication.** 1943

Carr, Charles C[arl]. **Alcoa, An American Enterprise.** 1952

Cooley, Mortimer E. **Scientific Blacksmith.** 1947

Davis, Charles Thomas. **The Manufacture of Paper.** 1886

Deane, Samuel. **The New-England Farmer,** or Georgical Dictionary. 1822

Dyer, Henry. **The Evolution of Industry.** 1895

Epstein, Ralph C. **The Automobile Industry:** Its Economic and Commercial Development. 1928

Ericsson, Henry. **Sixty Years a Builder:** The Autobiography of Henry Ericsson. 1942

Evans, Oliver. **The Young Mill-Wright and Miller's Guide.** 1850

Ewbank, Thomas. **A Descriptive and Historical Account of Hydraulic and Other Machines for Raising Water,** Ancient and Modern. 1842

Field, Henry M. **The Story of the Atlantic Telegraph.** 1893

Fleming, A. P. M. **Industrial Research in the United States of America.** 1917

Van Gelder, Arthur Pine and Hugo Schlatter. **History of the Explosives Industry in America.** 1927

Hall, Courtney Robert. **History of American Industrial Science.** 1954

Hungerford, Edward. **The Story of Public Utilities.** 1928

Hungerford, Edward. **The Story of the Baltimore and Ohio Railroad, 1827-1927.** 1928

Husband, Joseph. **The Story of the Pullman Car.** 1917

Ingels, Margaret. **Willis Haviland Carrier, Father of Air Conditioning.** 1952

Kingsbury, J[ohn] E. **The Telephone and Telephone Exchanges:** Their Invention and Development. 1915

Labatut, Jean and Wheaton J. Lane, eds. **Highways in Our National Life:** A Symposium. 1950

Lathrop, William G[ilbert]. **The Brass Industry in the United States.** 1926

Lesley, Robert W., John B. Lober and George S. Bartlett. **History of the Portland Cement Industry in the United States.** 1924

Marcosson, Isaac F. **Wherever Men Trade:** The Romance of the Cash Register. 1945

Miles, Henry A[dolphus]. **Lowell, As It Was, and As It Is**. 1845

Morison, George S. **The New Epoch:** As Developed by the Manufacture of Power. 1903

Olmsted, Denison **Memoir of Eli Whitney, Esq.** 1846

Passer, Harold C **The Electrical Manufacturers, 1875-1900.** 1953

Prescott, George B[artlett] **Bell's Electric Speaking Telephone.** 1884

Prout, Henry G. **A Life of George Westinghouse.** 1921

Randall, Frank A. **History of the Development of Building Construction in Chicago.** 1949

Riley, John J. **A History of the American Soft Drink Industry:** Bottled Carbonated Beverages, 1807-1957. 1958

Salem, F[rederick] W[illiam] **Beer, Its History and Its Economic Value as a National Beverage.** 1880

Smith, Edgar F. **Chemistry in America.** 1914

Steinman, D[avid] B[arnard] **The Builders of the Bridge:** The Story of John Roebling and His Son. 1950

Taylor, F[rank] Sherwood. **A History of Industrial Chemistry.** 1957

Technological Trends and National Policy, Including the Social Implications of New Inventions. Report of the Subcommittee on Technology to the National Resources Committee. 1937

Thompson, John S. **History of Composing Machines.** 1904

Thompson, Robert Luther. **Wiring a Continent:** The History of the Telegraph Industry in the United States, 1832-1866. 1947

Tilley, Nannie May. **The Bright-Tobacco Industry, 1860-1929.** 1948

Tooker, Elva. **Nathan Trotter:** Philadelphia Merchant, 1787-1853. 1955

Turck, J. A. V. **Origin of Modern Calculating Machines.** 1921

Tyler, David Budlong. **Steam Conquers the Atlantic.** 1939

Wheeler, Gervase. **Homes for the People,** In Suburb and Country. 1855